This version edited and typeset (using LATEX) by Jeff Eldridge, Edmonds Community College.

Report typographical errors to: jeldridg@edcc.edu

Eleventh printing: September 10, 2018 (corrects 7 minor errors)
Tenth printing: March 27, 2017 (corrects 3 minor errors)
Ninth printing: December 29, 2016 (corrects 9 minor errors, adds 12 problems to 2.2)
Eighth printing: August 1, 2016 (corrects 5 minor errors)
Seventh printing: April 1, 2016 (corrects 2 minor errors, adds 6 problems to 1.2)
Sixth printing: January 1, 2016 (corrects 6 minor errors)
Fifth printing: September 14, 2015 (corrects 1 typographical error and 4 images, clarifies one definition)
Fourth printing: March 23, 2015 (corrects 14 errors, adds 12 problems to 3.1 and 18 problems to 3.4)
Third printing: September 7, 2014 (corrects 7 typographical errors)
Second printing: April 1, 2014 (corrects 21 minor errors and improves image quality throughout)
First printing of this version: January 1, 2014

ISBN-13: 978-1494842680
ISBN-10: 1494842688

DALE HOFFMAN

CONTEMPORARY CALCULUS

Contents

0

Welcome to Calculus

Calculus was first developed more than 300 years ago by Sir Isaac Newton and Gottfried Leibniz to help them describe and understand the rules governing the motion of planets and moons. Since then, thousands of other men and women have refined the basic ideas of calculus, developed new techniques to make the calculations easier, and found ways to apply calculus to a wide variety of problems other than planetary motion.

The discovery, development and application of calculus is a great intellectual achievement — and now you have the opportunity to share in that achievement. You should feel exhilarated. You may also be somewhat concerned (a common reaction among students just beginning to study calculus). You need to be concerned enough to work to master calculus, yet confident enough to keep going when you (at first) don't understand something.

Part of the beauty of calculus is that it relies upon a few very simple ideas. Part of the power of calculus is that these simple ideas can help us understand, describe and solve problems in a variety of fields. This book tries to emphasize both the beauty and the power.

In Section 0.1 we will look at three key ideas that will continue throughout the book: finding tangent lines, computing areas and calculating infinite sums. We will also consider a process that underlies each of these problems: the limiting process of approximating a solution and then getting better and better approximations until we finally get an exact solution.

Sections 0.2 (Lines), 0.3 (Functions) and 0.4 (Combinations of Functions) contain review material. These sections emphasize concepts and skills you will need in order to succeed in calculus. You should have worked with most of these concepts in previous courses, but the emphasis and use of the material here may be different than in those earlier classes.

Section 0.5 (Mathematical Language) discusses a few key mathematical phrases. It considers their use and meaning and some of their equivalent forms. It will be difficult to understand the meaning and subtleties of calculus if you don't understand how these phrases are used and what they mean.

0.1 A Preview of Calculus

Calculus can be viewed as an attempt — a historically successful attempt — to solve two fundamental problems. In this section we begin to examine geometric forms of those two problems and some fairly simple attempts to solve them. At first, the problems themselves may not appear very interesting or useful — and the methods for solving them may seem crude — but these simple problems and methods have led to one of the most beautiful, powerful and useful creations in mathematics: Calculus.

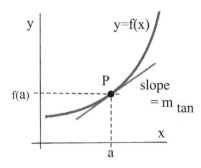

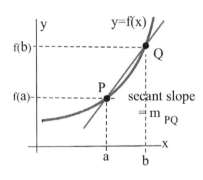

Finding the Slope of a Tangent Line

Suppose we have the graph of a function $y = f(x)$ and we want to find an equation of a line **tangent** to the graph at a particular point P on the graph (see margin). (We will offer a precise definition of "tangent" in Section 1.0; for now, think of the tangent line as a line that touches the curve at P and stays close to the graph of $y = f(x)$ near P.)

We know that P is on the tangent line, so if its x-coordinate is $x = a$, then the y-coordinate of P must be $y = f(a)$: $P = (a, f(a))$. The only other information we need to find an equation of the tangent line is its slope, m_{tan}, but that is where the difficulty arises.

In algebra, we needed two points in order to determine a slope. So far, we only have the point P. Let's simply pick a second point, call it Q, on the graph of $y = f(x)$. If the x-coordinate of Q is b, then the y-coordinate is $f(b)$: $Q = (b, f(b))$. So the slope of the line through P and Q is

$$m_{PQ} = \frac{\text{rise}}{\text{run}} = \frac{f(b) - f(a)}{b - a}$$

If we drew the graph of $y = f(x)$ on a wall, put nails at the points P and Q, and laid a straightedge on the nails, then the straightedge would have slope m_{PQ}. But the slope m_{PQ} can be very different from the value we want (the slope m_{tan} of the tangent line). The key idea is that when the point Q is *close* to the point P, then the slope m_{PQ} should be *close* to the slope we want, m_{tan}. Physically, if we slide the nail at Q along the graph toward the fixed point P, then the slope, $m_{PQ} = \frac{f(b) - f(a)}{b - a}$, of the straightedge gets closer and closer to the slope, m_{tan}, of the tangent line. If the value of b is very close to a, then the point Q is very close to P, and the value of m_{PQ} is very close to the value of m_{tan}.

Rather than defacing walls with graphs and nails, we can instead calculate $m_{PQ} = \frac{f(b) - f(a)}{b - a}$ and examine the values of m_{PQ} as b gets closer and closer to a. We say that m_{tan} is the **limiting value** of m_{PQ} as b gets very close to a, and we write:

$$m_{tan} = \lim_{b \to a} \frac{f(b) - f(a)}{b - a}$$

Eventually we will call the slope m_{tan} of the tangent line the **derivative** of the function $f(x)$ at the point P, and call this part of calculus **differential calculus**. Chapters 2 and 3 begin the study of differential calculus.

The slope of the tangent line to the graph of a function will tell us important information about the function and will allow us to solve problems such as:

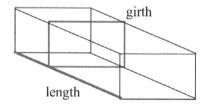

- The U.S. Postal Service requires that the length plus the girth of a package not exceed 84 inches. What is the largest volume that can be mailed in a rectangular box?

- An oil tanker was leaking oil and a 4-inch-thick oil slick had formed. When first measured, the slick had a radius of 200 feet, and the radius was increasing at a rate of 3 feet per hour. At that time, how fast was the oil leaking from the tanker?

Derivatives will even help us solve such "traditional" mathematical problems as finding solutions of equations like $x^2 = 2 + \sin(x)$ and $x^9 + 5x^5 + x^3 + 3 = 0$.

Problems

1. Sketch the lines tangent to the curve shown below at $x = 1, 2$ and 3. Estimate the slope of each of the tangent lines you drew.

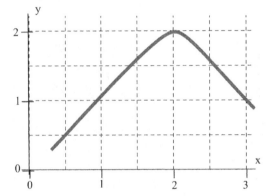

2. A graph of the weight of a "typical" child from age 0 to age 24 months appears below. (Each of your answers should have the units "kg per month.")

   (a) What was the average weight gain from month 0 to month 24?

   (b) What was the average weight gain from month 9 to month 12? From month 12 to month 15?

   (c) Approximately how fast was the child gaining weight at age 12 months? At age 3 months?

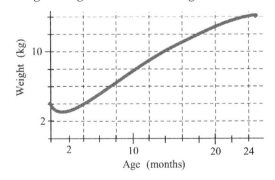

3. The graph below shows the temperature of a cup of coffee during a 10-minute period. (Each of your answers in (a)–(c) should have the units "degrees per minute.")

   (a) What was the average rate of cooling from minute 0 to minute 10?

   (b) What was the average rate of cooling from minute 7 to minute 8? From minute 8 to minute 9?

   (c) What was the rate of cooling at minute 8? At minute 2?

   (d) When was cold milk added to the coffee?

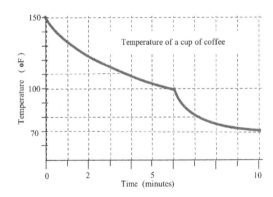

4. Describe a method for determining the slope of a steep hill at a point midway up the hill

   (a) using a ruler, a long piece of string, a glass of water and a loaf of bread.

   (b) using a protractor, a piece of string and a helium-filled balloon.

Finding the Area of a Shape

Suppose we need to find the area of a leaf as part of a study of how much energy a plant gets from sunlight. One method for finding the area would be to trace the shape of the leaf onto a piece of paper and then divide the region into "easy" shapes such as rectangles and triangles (whose areas we can easily calculate). We could add all of these "easy" areas together to approximate the area of the leaf.

A modification of this method would be to trace the shape onto a piece of graph paper and then count the number of squares completely inside the edge of the leaf to get a lower estimate of the area, and count the number of squares that touch the leaf to get an upper estimate of the area. If we repeat this process with smaller and smaller squares, we will have to do more counting and adding, but our estimates should be closer together—and closer to the actual area of the leaf.

(We could also approximate the area of the leaf using a sheet of paper, scissors and an accurate scale. How?)

We can calculate the area A between the graph of a function $y = f(x)$ and the x-axis by using similar methods. We can divide the area into strips of width w and determine the lower and upper values of $y = f(x)$ on each strip. Then we can approximate the area of each rectangle and add all of the little areas together to get A_w, an approximation of the exact area. The key idea is that if w is small, then the rectangles are narrow, and the approximate area A_w should be very close to the actual area A. If we take narrower and narrower rectangles, the approximate areas get closer and closer to the actual area:

$$A = \lim_{w \to 0} A_w$$

The process described above is the basis for a technique called *integration*, and this part of calculus is called *integral calculus*. Integral calculus and integration will begin in Chapter 4.

The process of taking the limit of a sum of "little" quantities will give us important information about a function and will also allow us to solve problems such as:

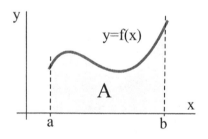

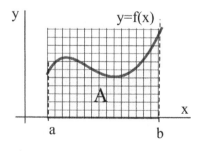

- Find the length of the graph of $y = \sin(x)$ over one period (from $x = 0$ to $x = 2\pi$).

- Find the volume of a torus ("doughnut") of radius 1 inch that has a hole of radius 2 inches.

- A car starts at rest and has an acceleration of $5 + 3\sin(t)$ feet per second per second in the northerly direction at time t seconds. Where will the car be, relative to its starting position, after 100 seconds?

Problems

5. Approximate the area of the leaf on the previous page using

   (a) the grid on the left.

   (b) the grid on the right.

6. A graph showing temperatures during the month of November appears below.

   (a) Approximate the shaded area between the temperature curve and the 65° line from Nov. 15 to Nov. 25.

   (b) The area of the "rectangle" is (base)(height) so what are the units of your answer in part (a)?

   (c) Approximate the shaded area between the temperature curve and the 65° line from Nov. 5 to Nov. 30.

   (d) Who might use or care about these results?

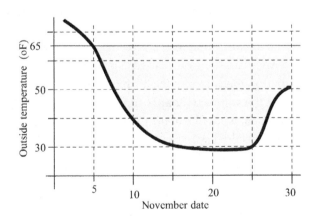

7. Describe a method for determining the volume of a compact fluorescent light bulb using a ruler, a large can, a scale and a jug of water.

A Unifying Process: Limits

We used similar processes to "solve" both the tangent line problem and the area problem. First, we found a way to get an approximate solution, and then we found a way to improve our approximation. Finally, we asked what would happen if we continued improving our approximations "forever": that is, we "took a limit."

For the tangent line problem, we let the point Q get closer and closer and closer to P, the limit as b approached a.

In the area problem, we let the widths of the rectangles get smaller and smaller, the limit as w approached 0. Limiting processes underlie derivatives, integrals and several other fundamental topics in calculus, and we will examine limits and their properties in detail in Chapter 1.

Two Sides of the Same Coin: Differentiation and Integration

Just as the set-up of each of the two basic problems involved a limiting process, the solutions to the two problems are also related. The process of differentiation used to solve the tangent line problem and the process of integration used to solve the area problem turn out to be "opposites" of each other: each process undoes the effect of the other process. The Fundamental Theorem of Calculus in Chapter 4 will show how this "opposite" effect works.

Extensions of the Main Problems

The first five chapters present the two key ideas of calculus, show "easy" ways to calculate derivatives and integrals, and examine some of their applications. And there is more.

Through the ensuing chapters, we will examine new functions and find ways to calculate their derivatives and integrals. We will extend the approximation ideas to use "easy" functions, such as polynomials, to approximate the values of "hard" functions, such as $\sin(x)$ and e^x.

In later chapters, we will extend the notions of "tangent lines" and "areas" to 3-dimensional space as "tangent planes" and "volumes."

> **Success in calculus will require time and effort on your part, but such a beautiful and powerful field is worth that time and effort.**

0.2 Lines in the Plane

The first graphs and functions you encountered in algebra were straight lines and their equations. These lines were easy to graph, and the equations were easy to evaluate and to solve. They described a variety of physical, biological and financial phenomena such as $d = rt$ relating the distance d traveled to the rate r and time t spent traveling, and $C = \frac{5}{9}(F - 32)$ for converting the temperature in degrees Fahrenheit (F) to degrees Celsius (C).

The first part of calculus—differential calculus—will deal with ideas, techniques and applications of tangent lines to the graphs of functions, so it is important that you understand the graphs, properties and equations of straight lines.

The Real Number Line

The real numbers (consisting of all integers, fractions, rational and irrational numbers) can be represented as a line, called the **real number line**. Once we have selected a starting location, called the **origin**, a positive direction (usually up or to the right), and unit of length, then every number can be located as a point on the number line. If we move from a point $x = a$ to a point $x = b$ on the line, then we will have moved an **increment** of $b - a$. We denote this increment with the symbol Δx (read "delta x").

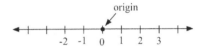

- If b is larger than a, then we will have moved in the positive direction, and $\Delta x = b - a$ will be positive.

- If b is smaller than a, then $\Delta x = b - a$ will be negative and we will have moved in the negative direction.

- Finally, if $\Delta x = b - a = 0$, then $a = b$ and we did not move at all.

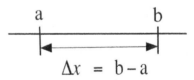

$\Delta x = b - a$

The capital Greek letter delta (Δ) appears often in calculus to represent the "change" in something.

Caution: Δx does not mean Δ *times x*, but rather the *difference* between two x-coordinates.

We can also use the Δ notation and absolute values to express the **distance** that we have moved. On the number line, the distance from $x = a$ to $x = b$ is

$$\text{dist}(a, b) = \begin{cases} b - a & \text{if } b \geq a \\ a - b & \text{if } b < a \end{cases}$$

or:

$$\text{dist}(a, b) = |b - a| = |\Delta x| = \sqrt{(\Delta x)^2}$$

The **midpoint** of the interval from $x = a$ to $x = b$ is the point M such that $\text{dist}(a, M) = \text{dist}(M, b)$, or $|M - a| = |b - M|$. If $a < M < b$,

$$M - a = b - M \Rightarrow 2M = a + b \Rightarrow M = \frac{a + b}{2}$$

It's not difficult to check that this formula also works when $b < M < a$.

Example 1. Find the length and midpoint of the interval from $x = -3$ to $x = 6$.

Solution. $\text{dist}(-3, 6) = |6 - (-3)| = |9| = 9$; $M = \frac{(-3)+6}{2} = \frac{3}{2}$. ◄

Practice 1. Find the length and midpoint of the interval from $x = -7$ to $x = -2$.

Solutions to Practice problems are at the end of each section.

The Cartesian Plane

Two perpendicular number lines, called **coordinate axes**, determine a **real number plane**. The axes intersect at a point called the **origin**. Each point P in the plane can be described by an **ordered pair** (x, y) of numbers that specify how far, and in which directions, we must move from the origin to reach the point P. We can locate the point $P = (x, y)$ in the plane by starting at the origin and moving x units horizontally and then y units vertically. Similarly, we can label each point in the plane with the ordered pair (x, y), which directs us how to reach that point from the origin.

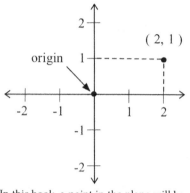

This coordinate system is called the **rectangular coordinate system** or the **Cartesian** coordinate system (after René Descartes), and the resulting plane the **Cartesian plane**.

The coordinate axes divide the plane into four **quadrants**, labeled quadrants I, II, III and IV moving counterclockwise from the upper-right quadrant.

We will often call the horizontal axis the **x-axis** and the vertical axis the **y-axis** and then refer to the plane as the **xy-plane**. This choice of x and y as labels for the axes is a common choice, but we will sometimes prefer to use different labels — and even different units of measurement on the two axes.

In this book, a point in the plane will be labeled either with a name, say P, or with an ordered pair, say (x, y), or with both: $P = (x, y)$.

	quadrant
II	I
III	IV

Increments and Distance Between Points In The Plane

If we move from a point $P = (x_1, y_1)$ in the plane to another point $Q = (x_2, y_2)$, then we will need to consider two **increments** or changes.

- The increment in the x (horizontal) direction is $x_2 - x_1$, denoted by $\Delta x = x_2 - x_1$.

- The increment in the y (vertical) direction is $y_2 - y_1$, denoted by $\Delta y = y_2 - y_1$.

Computing the **distance** between the points $P = (x_1, y_1)$ and $Q = (x_2, y_2)$ involves a simple application of the Pythagorean Theorem:

$$\text{dist}(P, Q) = \sqrt{(\Delta x)^2 + (\Delta y)^2} = \sqrt{(x_2 - x_1)^2 + (y_2 - y_1)^2}$$

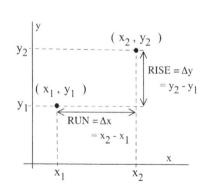

The **midpoint** M of the line segment joining P and Q is:

$$M = \left(\frac{x_1 + x_2}{2}, \frac{y_1 + y_2}{2} \right)$$

where we have just used the one-dimension midpoint formula for each coordinate.

Example 2. Find an equation describing all the points $P = (x, y)$ equidistant from $Q = (2, 3)$ and $R = (5, -1)$.

Solution. The points $P = (x, y)$ must satisfy $\text{dist}(P, Q) = \text{dist}(P, R)$ so:

$$\sqrt{(x - 2)^2 + (y - 3)^2} = \sqrt{(x - 5)^2 + (y - (-1))^2}$$

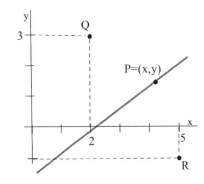

By squaring each side we get:

$$(x - 2)^2 + (y - 3)^2 = (x - 5)^2 + (y + 1)^2$$

Expanding we get:

$$x^2 - 4x + 4 + y^2 - 6y + 9 = x^2 - 10x + 25 + y^2 + 2y + 1$$

and canceling like terms yields:

$$-4x - 6y + 13 = -10x + 2y + 26$$

so $y = 0.75x - 1.625$, the equation of a line. Every point on the line $y = 0.75x - 1.625$ is equally distant from both Q and R.   ◀

Practice 2. Find an equation describing all points $P = (x, y)$ equidistant from $Q = (1, -4)$ and $R = (0, -3)$.

A circle with radius r and center at the point $C = (a, b)$ consists of all points $P = (x, y)$ at a distance of r from the center C: the points P that satisfy $\text{dist}(P, C) = r$.

Example 3. Find an equation of a circle with radius $r = 4$ and center $C = (5, -3)$.

Solution. A circle consists of the set of points $P = (x, y)$ at a fixed distance r from the center point C, so this circle will be the set of points $P = (x, y)$ at a distance of 4 units from the point $C = (5, -3)$; P will be on this circle if $\text{dist}(P, C) = 4$.

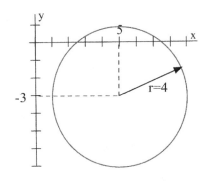

Using the distance formula and rewriting:

$$\sqrt{(x - 5)^2 + (y + 3)^2} = 4 \quad \Rightarrow \quad (x - 5)^2 + (y + 3)^2 = 16$$

which we can also express as $x^2 - 10x + 25 + y^2 + 6y + 9 = 16$.   ◀

Practice 3. Find an equation of a circle with radius $r = 5$ and center $C = (-2, 6)$.

The Slope Between Points in the Plane

In one dimension (on the number line), our only choice was to move in the positive direction (so the x-values were increasing) or in the negative direction. In two dimensions (in the plane), we can move in infinitely many directions, so we need a precise way to describe direction.

The **slope** of the line segment joining $P = (x_1, y_1)$ to $Q = (x_2, y_2)$ is

$$m = \text{slope from } P \text{ to } Q = \frac{\text{rise}}{\text{run}} = \frac{y_2 - y_1}{x_2 - x1} = \frac{\Delta y}{\Delta x}$$

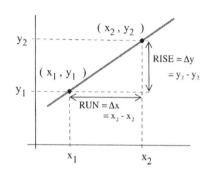

The slope of a line measures how fast we rise or fall as we move from left to right along the line. It measures the rate of change of the y-coordinate with respect to changes in the x-coordinate. Most of our work will occur in two dimensions, and slope will be a very useful concept that will appear often.

If P and Q have the same x-coordinate, then $x_1 = x_2 \Rightarrow x = 0$. The line from P to Q is thus **vertical** and the slope $m = \frac{\Delta y}{\Delta x}$ is **undefined** because $\Delta x = 0$.

If P and Q have the same y-coordinate, then $y_1 = y_2 \Rightarrow \Delta y = 0$, so the line is **horizontal** and the slope is $m = \frac{\Delta y}{\Delta x} = \frac{0}{\Delta x} = 0$ (assuming $\Delta x \neq 0$).

Practice 4. For $P = (-3, 2)$ and $Q = (5, -14)$, find Δx, Δy, and the slope of the line segment from P to Q.

If the coordinates of P or Q contain variables, then the slope m is still given by $m = \frac{\Delta y}{\Delta x}$, but we will need to use algebra to evaluate and simplify m.

Example 4. Find the slope of the line segment from $P = (1, 3)$ to $Q = (1 + h, 3 + 2h)$.

Solution. $y_1 = 3$ and $y_2 = 3 + 2h$, so $\Delta y = (3 + 2h) - (3) = 2h$; $x_1 = 1$ and $x_2 = 1 + h$, so $\Delta x = (1 + h) - (1) = h$. The slope is:

$$m = \frac{\Delta y}{\Delta x} = \frac{2h}{h} = 2$$

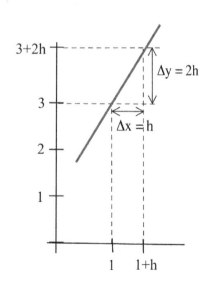

In this example, the value of m is constant (2) and does not depend on the value of h.   ◄

Practice 5. Find the slope and midpoint of the line segment from $P = (2, -3)$ to $Q = (2 + h, -3 + 5h)$.

Example 5. Find the slope between the points $P = (x, x^2 + x)$ and $Q = (a, a^2 + a)$ for $a \neq x$.

Solution. $y_1 = x^2 + x$ and $y_2 = a^2 + a \Rightarrow \Delta y = (a^2 + a) - (x^2 + x)$; $x_1 = x$ and $x_2 = a$, so $\Delta x = a - x$ and the slope is:

$$\begin{aligned} m = \frac{\Delta y}{\Delta x} &= \frac{(a^2 + a) - (x^2 + x)}{a - x} \\ &= \frac{a^2 - x^2 + a - x}{a - x} = \frac{(a - x)(a + x) + (a - x)}{a - x} \\ &= \frac{(a - x)\left((a + x) + 1\right)}{a - x} = (a + x) + 1 \end{aligned}$$

Here the value of m depends on the values of both a and x. ◄

Practice 6. Find the slope between the points $P = (x, 3x^2 + 5x)$ and $Q = (a, 3a^2 + 5a)$ for $a \neq x$.

In application problems, it is important to read the information and the questions very carefully — including the units of measurement of the variables can help you avoid "silly" answers.

Example 6. In 1970, the population of Houston was $1,233,535$ and in 1980 it was $1,595,138$. Find the slope of the line through the points $(1970, 1233535)$ and $(1980, 1595138)$.

Solution. $m = \dfrac{\Delta y}{\Delta x} = \dfrac{1595138 - 1233535}{1980 - 1970} = \dfrac{361603}{10} = 36,160.3$ but $36,160.3$ is just a number that may or may not have any meaning to you. If we include the units of measurement along with the numbers we will get a more meaningful result:

$$\frac{1595138 \text{ people} - 1233535 \text{ people}}{\text{year } 1980 - \text{year } 1970} = \frac{361603 \text{ people}}{10 \text{ years}} = 36,160.3 \frac{\text{people}}{\text{year}}$$

which says that during the decade from 1970 to 1980 the population of Houston grew at an average rate of $36,160$ people per year. ◄

If the x-unit is time (in hours) and the y-unit is distance (in kilometers), then

$$m = \frac{\Delta y \text{ km}}{\Delta x \text{ hours}}$$

so the units for m are $\frac{\text{km}}{\text{hour}}$ ("kilometers per hour"), a measure of velocity, the rate of change of distance with respect to time.

If the x-unit is the number of employees at a bicycle factory and the y-unit is the number of bicycles manufactured, then

$$m = \frac{\Delta y \text{ bicycles}}{\Delta x \text{ employees}}$$

and the units for m are $\frac{\text{bicycles}}{\text{employee}}$ ("bicycles per employee"), a measure of the rate of production per employee.

Equations of Lines

Every (non-vertical) line has the property that the slope of the segment between any two points on the line is the same, and this constant slope property of straight lines leads to ways of finding equations to represent non-vertical lines.

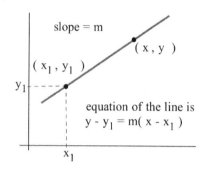

Point-Slope Form

In calculus, we will usually know a point on a line and the slope of that line, so the point-slope form will be the easiest to apply. Other forms of equations for lines can be derived from the point-slope form.

If L is a non-vertical line through a known point $P = (x_1, y_1)$ with a known slope m, then the equation of the line L is:

$$\boxed{\textbf{Point-Slope: } y - y_1 = m(x - x_1)}$$

Example 7. Find an equation of the line through $(2, -3)$ with slope 5.

Solution. We can simply use the point-slope formula: $m = 5$, $y_1 = -3$ and $x_1 = 2$, so $y - (-3) = 5(x - 2)$, which simplifies to $y = 5x - 13$ ◄

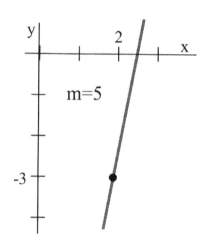

An equation for a **vertical line** through a point $P = (a, b)$ is $x = a$. All points $Q = (x, y)$ on the vertical line through the point P have the same x-coordinate as P.

Two-Point Form

If two points $P = (x_1, y_1)$ and $Q = (x_2, y_2)$ are on the line L, then we can calculate the slope between them and use the first point and the point-slope equation to find an equation for L:

$$\boxed{\textbf{Two-Point: } y - y_1 = m(x - x_1) \text{ where } m = \frac{y_2 - y_1}{x_2 - x_1}}$$

Once we have the slope, m, it does not matter whether we use P or Q as the point. Either choice will result in the same equation for the line once we simplify it.

Slope-Intercept Form

It is common practice to rewrite an equation of a line into the form $y = mx + b$, the **slope-intercept form** of the line. The line $y = mx + b$ has slope m and crosses the y-axis at the point $(0, b)$.

Practice 7. Use the $\frac{\Delta y}{\Delta x}$ definition of slope to calculate the slope of the line $y = mx + b$.

The point-slope and the two-point forms are usually more useful for *finding* an equation of a line, but the slope-intercept form is usually the most useful form for an *answer* because it allows us to easily picture the graph of the line and to quickly calculate y-values given x-values.

Angles Between Lines

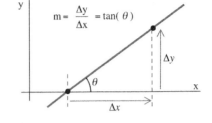

The **angle of inclination** of a line with the x-axis is the smallest angle θ that the line makes with the positive x-axis as measured from the x-axis counterclockwise to the line. Because the slope $m = \frac{\Delta y}{\Delta x}$ and because $\tan(\theta) = \frac{\text{opposite}}{\text{adjacent}}$ in a right triangle, $m = \tan(\theta)$.

The slope of a line is the tangent of its angle of inclination.

Parallel Lines

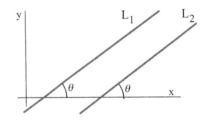

Two parallel lines L_1 and L_2 make equal angles with the x-axis, so their angles of inclination will be equal and hence so will their slopes.

Similarly, if the slopes, m_1 and m_2, of two lines are equal, then the equations of the lines (in slope-intercept form) will always differ by a constant:

$$y_1 - y_2 = (m_1 x + b_1) - (m_2 x + b_2) = (m_1 - m_2)x + (b_1 - b_2) = b_1 - b_2$$

which is a constant, so the lines will be parallel.

The two preceding ideas can be combined into a single statement:

Two non-vertical lines L_1 and L_2 with slopes m_1 and m_2 are parallel if and only if $m_1 = m_2$.

Practice 8. Find an equation of the line that contains the point $(-2, 3)$ and is parallel to the line $3x + 5y = 17$.

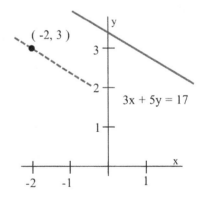

Perpendicular Lines

If two lines are perpendicular, the situation is a bit more complicated.

Assume L_1 and L_2 are two non-vertical lines that intersect at the origin (for simplicity), with $P = (x_1, y_1)$ and $Q = (x_2, y_2)$ points away from the origin on L_1 and L_2, respectively. Then the slopes of L_1 and L_2 will be $m_1 = \frac{y_1}{x_1}$ and $m_2 = \frac{y_2}{x_2}$. The line connecting P and Q forms the third side of triangle OPQ, which will be a right triangle if and only if L_1 and L_2 are perpendicular. In particular, L_1 and L_2 are perpendicular if and only if the triangle OPQ satisfies the Pythagorean Theorem:

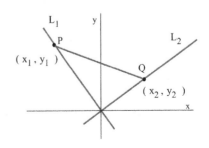

$$(\text{dist}(O, P))^2 + (\text{dist}(O, Q))^2 = (\text{dist}(P, Q))^2$$

or:

$$(x_1 - 0)^2 + (y_1 - 0)^2 + (x_2 - 0)^2 + (y_2 - 0)^2$$
$$= (x_1 - x_2)^2 + (y_1 - y_2)^2$$

Squaring and simplifying, this reduces to $0 = -2x_1 x_2 - 2y_1 y_2$, so:

$$\frac{y_2}{x_2} = -\frac{x_1}{y_1} \Rightarrow m_2 = \frac{y_2}{x_2} = -\frac{x_1}{y_1} = -\frac{1}{\frac{y_1}{x_1}} = -\frac{1}{m_1}$$

We have just proved the following result:

> **Two non-vertical lines L_1 and L_2 with slopes m_1 and m_2 are perpendicular if and only if their slopes are negative reciprocals of each other: $m_2 = -\dfrac{1}{m_1}$.**

Practice 9. Find an equation of the line that goes through the point $(2, -5)$ and is perpendicular to the line $3y - 7x = 2$.

Example 8. Find the distance (that is, the shortest distance) from the point $(1, 8)$ to the line $L : 3y - x = 3$.

Solution. This is a sophisticated problem that requires several steps to solve: First we need a picture of the problem. We will find an equation for the line L^* through the point $(1, 8)$ and perpendicular to L. Then we will find the point P where L and L^* intersect. Finally, we will find the distance from P to $(1, 8)$.

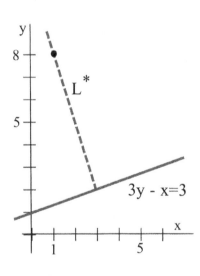

Step 1: L has slope $\frac{1}{3}$ so L^* has slope $m = -\frac{1}{\frac{1}{3}} = -3$, and L^* has equation $y - 8 = -3(x - 1)$, which simplifies to $y = -3x + 11$.

Step 2: We can find the point where L intersects L^* by replacing the y-value in the equation for L with the y-value from our equation for L^*:

$$3(-3x + 11) - x = 3 \Rightarrow x = 3 \Rightarrow y = -3x + 11 = -3(3) + 11 = 2$$

which tells us that L and L^* intersect at $P = (3, 2)$.

Step 3: Finally, the distance from L to $(1, 8)$ is just the distance from the point $(1, 8)$ to the point $P = (3, 2)$, which is

$$\sqrt{(1 - 3)^2 + (8 - 2)^2} = \sqrt{40} \approx 6.325$$

The distance is (exactly) $\sqrt{40}$, or (approximately) 6.325.   ◄

Angle Formed by Intersecting Lines

If two lines that are not perpendicular intersect at a point (and neither line is vertical), then we can use some geometry and trigonometry to determine the angles formed by the intersection of those lines.

Because θ_2 (see figure at right) is an exterior angle of the triangle ABC, θ_2 is equal to the sum of the two opposite interior angles, so $\theta_2 = \theta_1 + \theta \Rightarrow \theta = \theta_2 - \theta_1$. From trigonometry, we then know that:

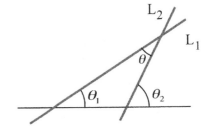

$$\tan(\theta) = \tan(\theta_2 - \theta_1) = \frac{\tan(\theta_2) - \tan(\theta_1)}{1 + \tan(\theta_2)\tan(\theta_1)} = \frac{m_2 - m_1}{1 + m_2 m_1}$$

The range of the arctan function is $\left[-\frac{\pi}{2}, \frac{\pi}{2}\right]$, so $\theta = \arctan\left(\frac{m_2 - m_1}{1 + m_2 m_1}\right)$ always gives the smaller of the angles. The larger angle is $\pi - \theta$ (or $180° - \theta°$ if we measure the angles in degrees).

> **The smaller angle θ formed by two non-perpendicular lines with slopes m_1 and m_2 is:** $\theta = \arctan\left(\dfrac{m_2 - m_1}{1 + m_2 m_1}\right)$

Example 9. Find the point of intersection and the angle between the lines $y = x + 3$ and $y = 2x + 1$.

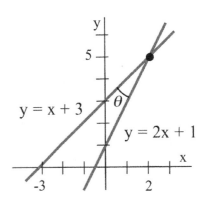

Solution. Solving the first equation for y and then substituting into the second equation:

$$(x + 3) = 2x + 1 \quad \Rightarrow \quad x = 2 \quad \Rightarrow \quad y = 2 + 3 = 5$$

The point of intersection is $(2, 5)$. Because both lines are in slope-intercept form, it is easy to see that $m_1 = 1$ and $m_2 = 2$:

$$\theta = \arctan\left(\frac{m_2 - m_1}{1 + m_2 m_1}\right) = \arctan\left(\frac{2 - 1}{1 + 2 \cdot 1}\right)$$
$$= \arctan\left(\frac{1}{3}\right) \approx 0.322 \text{ radians} = 18.43°$$

The lines intersect at an angle of (approximately) $18.43°$.   ◄

0.2 Problems

1. Estimate the slope of each line shown below.

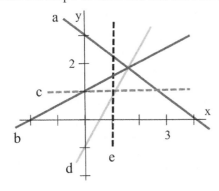

2. Estimate the slope of each line shown below.

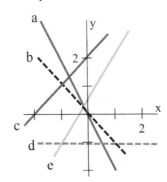

3. Compute the slope of the line that passes through:

   (a) $(2,4)$ and $(5,8)$

   (b) $(-2,4)$ and $(3,-5)$

   (c) $(2,4)$ and (x,x^2)

   (d) $(2,5)$ and $(2+h,1+(2+h)^2)$

   (e) (x,x^2+3) and (a,a^2+3)

4. Compute the slope of the line that passes through:

   (a) $(5,-2)$ and $(3,8)$

   (b) $(-2,-4)$ and $(5,-3)$

   (c) $(x,3x+5)$ and $(a,3a+5)$

   (d) $(4,5)$ and $(4+h,5-3h)$

   (e) $(1,2)$ and $(x,1+x^2)$

   (f) $(2,-3)$ and $(2+h,1-(1+h)^2)$

   (g) (x,x^2) and $(x+h,x^2+2xh+h^2)$

   (h) (x,x^2) and $(x-h,x^2-2xh+h^2)$

5. A small airplane at an altitude of 5,000 feet is flying east at 300 feet per second (a bit over 200 miles per hour), and you are watching it with a small telescope as it passes directly overhead.

   (a) What is the **slope** of the telescope 5, 10 and 20 seconds after the plane passes overhead?

   (b) What is the **slope** of the telescope t seconds after the plane passes overhead?

   (c) After the plane passes overhead, is the slope of the telescope increasing, decreasing or staying the same?

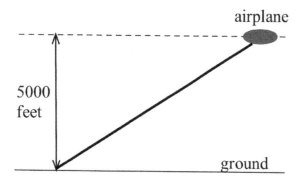

6. You are at the origin, $(0,0)$, and are watching a small bug at the point $(t,1+t^2)$ at time t seconds.

   (a) What is the **slope** of your line of vision when $t=5$, 10 and 15 seconds?

   (b) What is the **slope** of your line of vision at an arbitrary time t?

7. The blocks in a city are all perfect squares. A friend gives you directions to a good restaurant: "Go north 3 blocks, turn east and go 5 blocks, turn south and go 7 blocks, turn west and go 3 blocks." How far away (straight-line distance) is it?

8. At the restaurant (see previous problem), a fellow diner gives you directions to a hotel: "Go north 5 blocks, turn right and go 6 blocks, turn right and go 3 blocks, turn left and go 2 blocks." How far away is the hotel from the restuarant?

9. The bottom of a 20-foot ladder is 4 feet from the base of a wall.

   (a) How far up the wall does the ladder reach?

   (b) What is the slope of the ladder?

   (c) What angle does it make with the ground?

10. Let $P=(1,-2)$ and $Q=(5,4)$. Find:

    (a) the midpoint R of the line segment PQ.

    (b) the point T that is $\frac{1}{3}$ of the way from P to Q:

    $$\text{dist}(P,T)=\frac{1}{3}\,\text{dist}(P,Q)$$

    (c) the point S that is $\frac{2}{5}$ of the way from P to Q.

11. If $P=(2,3)$, $Q=(8,11)$ and $R=(x,y)$, where:

    $$x=2a+8(1-a),\ y=3a+11(1-a),\ 0\le a\le 1$$

    (a) Verify that R is on the line segment PQ.

    (b) Verify that $\text{dist}(P,R)=(1-a)\cdot\text{dist}(P,Q)$.

12. A rectangular box is 24 inches long, 18 inches wide and 12 inches high.

    (a) Find the length of the longest (straight) stick that will fit into the box.

    (b) What angle (in degrees) does that stick make with the base of the box?

13. The lines $y=x$ and $y=4-x$ intersect at $(2,2)$.

    (a) Show that the lines are perpendicular.

    (b) Graph the lines together on your calculator using the "window" $[-10,10]\times[-10,10]$.

    (c) Why do the lines not appear to be perpendicular on the calculator display?

    (d) Find a suitable window so that the lines do appear perpendicular.

14. Two lines both go through the point $(1,2)$, one with slope 3 and one with slope $-\frac{1}{3}$.

   (a) Find equations for the lines.

   (b) Choose a suitable window so that the lines will appear perpendicular, and then graph them together on your calculator..

15. Sketch the line with slope m that goes through the point P, then find an equation for the line.

   (a) $m = 3, P = (2,5)$

   (b) $m = -2, P = (3,2)$

   (c) $m = -\frac{1}{2}, P = (1,4)$

16. Sketch the line with slope m that goes through the point P, then find an equation for the line.

   (a) $m = 5, P = (2,1)$

   (b) $m = -\frac{2}{3}, P = (1,3)$

   (c) $m = \pi, P = (1,-3)$

17. Find an equation for each line.

   (a) L_1 goes through the point $(2,5)$ and is parallel to $3x - 2y = 9$.

   (b) L_2 goes through the point $(-1,2)$ and is perpendicular to $2x = 7 - 3y$.

   (c) L_3 goes through the point $(3,-2)$ and is perpendicular to $y = 1$.

18. Find a value for the constant (A, B or D) so that:

   (a) the line $y = 2x + A$ goes through $(3,10)$.

   (b) the line $y = Bx + 2$ goes through $(3,10)$.

   (c) the line $y = Dx + 7$ crosses the y-axis at the point $(0,4)$.

   (d) the line $Ay = Bx + 1$ goes through the points $(1,3)$ and $(5,13)$.

19. Find the shortest distance between the circles with centers $C_1 = (1,2)$ and $C_2 = (7,10)$ with radii r_1 and r_2 when:

   (a) $r_1 = 2$ and $r_2 = 4$

   (b) $r_1 = 2$ and $r_2 = 7$

   (c) $r_1 = 5$ and $r_2 = 8$

   (d) $r_1 = 3$ and $r_2 = 15$

   (e) $r_1 = 12$ and $r_2 = 1$

20. Find an equation of the circle with center C and radius r when

   (a) $C = (2,7)$ and $r = 4$

   (b) $C = (3,-2)$ and $r = 1$

   (c) $C = (-5,1)$ and $r = 7$

   (d) $C = (-3,-1)$ and $r = 4$

21. Explain how to show, without graphing, whether a point $P = (x,y)$ is inside, on, or outside the circle with center $C = (h,k)$ and radius r.

22. A box with a base of dimensions 2 cm and 8 cm is definitely big enough to hold two semicircular rods with radii of 2 cm (see below).

   (a) Will these same two rods fit in a box 2 cm high and 7.6 cm wide?

   (b) Will they fit in a box 2 cm high and 7.2 cm wide? (Suggestion: Turn one of the rods over.)

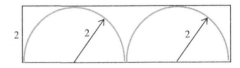

23. Show that an equation of the circle with center $C = (h,k)$ and radius r is $(x - h)^2 + (y - k)^2 = r^2$.

24. Find an equation of the line tangent to the circle $x^2 + y^2 = 25$ at the point P when:

   (a) $P = (3,4)$

   (b) $P = (-4,3)$

   (c) $P = (0,5)$

   (d) $P = (-5,0)$

25. Find an equation of the line tangent to the circle with center $C = (3,1)$ at the point P when:

   (a) $P = (8,13)$

   (b) $P = (-10,1)$

   (c) $P = (-9,6)$

   (d) $P = (3,14)$

26. Find the center $C = (h,k)$ and the radius r of the circle that goes through the three points:

   (a) $(0,1)$, $(1,0)$ and $(0,5)$

   (b) $(1,4)$, $(2,2)$ and $(8,2)$

   (c) $(1,3)$, $(4,12)$ and $(8,4)$

27. How close does

   (a) the line $3x - 2y = 4$ come to the point $(2,5)$?

   (b) the line $y = 5 - 2x$ come to the point $(1,-2)$?

   (c) the circle with radius 3 and center at $(2,3)$ come to the point $(8,3)$?

28. How close does

   (a) the line $2x - 5y = 4$ come to the point $(1,5)$?

   (b) the line $y = 3 - 2x$ come to the point $(5,-2)$?

   (c) the circle with radius 4 and center at $(4,3)$ come to the point $(10,3)$?

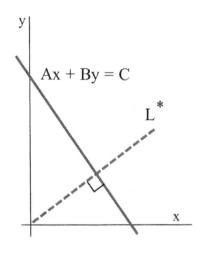

29. Follow the steps below (and refer to the figure) to find a formula for the distance from the origin to the line $Ax + By = C$.

   (a) Show that the line L given by $Ax + By = C$ has slope $m = -\frac{A}{B}$.

   (b) Find the equation of the line L^* that goes through $(0,0)$ and is perpendicular to L.

   (c) Show that L and L^* intersect at the point:
   $$(x,y) = \left(\frac{AC}{A^2 + B^2}, \frac{BC}{A^2 + B^2} \right)$$

   (d) Show that the distance from the origin to the point (x,y) is:
   $$\frac{|C|}{\sqrt{A^2 + B^2}}$$

30. Show that a formula for the distance from the point (p,q) to the line $Ax + By = C$ is:
   $$\frac{|Ap + Bq - C|}{\sqrt{A^2 + B^2}}$$

   (The steps will be similar to those in the previous problem, but the algebra will be more complicated.)

0.2 Practice Answers

1. Length $= \text{dist}(-7,-2) = |(-7)-(-2)| = |-5| = 5$.

   The midpoint is at $\dfrac{(-7)+(-2)}{2} = \dfrac{-9}{2} = -4.5$.

2. $\text{dist}(P,Q) = \text{dist}(P,R) \Rightarrow (x-1)^2 + (y+4)^2 = (x-0)^2 + (y+3)^2$; squaring each side and simplifying eventually yields $y = x - 4$.

3. The point $P = (x,y)$ is on the circle when it is 5 units from the center $C = (-2,6)$, so $\text{dist}(P,C) = 5$. Then $\text{dist}\left((x,y),(-2,6)\right) = 5$, so $\sqrt{(x+2)^2 + (y-6)^2} = 5 \Rightarrow (x+2)^2 + (y-6)^2 = 25$.

4. $\Delta x = 5 - (-3) = 8$ and $\Delta y = -14 - 2 = -16$, so:

$$\text{slope} = \frac{\Delta y}{\Delta x} = \frac{-16}{8} = -2$$

5. $\text{slope} = \dfrac{\Delta y}{\Delta x} = \dfrac{(-3+5h)-(-3)}{(2+h)-2} = \dfrac{5h}{h} = 5$. The midpoint is at
$$\left(\frac{(2)+(2+h)}{2}, \frac{(-3+5h)+(-3)}{2} \right) = \left(2 + \frac{h}{2}, -3 + \frac{5h}{2} \right).$$

6. $\text{slope} = \dfrac{\Delta y}{\Delta x} = \dfrac{(3a^2+5a)-(3x^2+5x)}{a-x} = \dfrac{3(a^2-x^2)+5(a-x)}{a-x} =$
$\dfrac{3(a+x)(a-x)+5(a-x)}{a-x} = 3(a+x)+5$

7. Let $y_1 = mx_1 + b$ and $y_2 = mx_2 + b$. Then:

$$\text{slope} = \frac{\Delta y}{\Delta x} = \frac{(mx_2+b)-(mx_1+b)}{x_2 - x_1} = \frac{m(x_2 - x_1)}{x_2 - x_1} = m$$

8. The line $3x + 5y = 17$ has slope $-\frac{3}{5}$, so the slope of the parallel line is $m = -\frac{3}{5}$. Using the form $y = -\frac{3}{5}x + b$ and the point $(-2,3)$ on the line, we have $3 = -\frac{3}{5}(-2) + b \Rightarrow b = \frac{9}{5} \Rightarrow y = -\frac{3}{5}x + \frac{9}{5}$, or $5y + 3x = 9$.

9. The line $3y - 7x = 2$ has slope $\frac{7}{3}$, so the slope of the perpendicular line is $m = -\frac{3}{7}$. Using the form $y = -\frac{3}{7}x + b$ and the point $(2,-5)$ on the line, we have $-5 = -\frac{3}{7}(2) + b \Rightarrow b = -\frac{29}{7} \Rightarrow y = -\frac{3}{7}x - \frac{29}{7}$, or $7y + 3x = -29$.

0.3   *Functions and Their Graphs*

When you prepared for calculus, you learned to manipulate functions by adding, subtracting, multiplying and dividing them, as well as evaluating functions of functions (composition). In calculus, we will continue to work with functions and their applications. We will create new functions by operating on old ones. We will gather information from the graphs of functions and from derived functions. We will find ways to describe the point-by-point behavior of functions as well as their behavior "close to" certain points and also over entire intervals. We will find tangent lines to graphs of functions and areas between graphs of functions. And, of course, we will see how these ideas can be used in a variety of fields.

This section and the next one review information and procedures you should already know about functions before we begin calculus.

What is a function?

Let's begin with a (very) general definition of a function:

A **function** from a set X to a set Y is a rule for assigning to each element of the set X a single element of the set Y. A function assigns a unique (exactly one) output element from the set Y to each input element from the set X.

The rule that defines a function is often given in the form of an equation, but it could also be given in words or graphically or by a table of values. In practice, functions are given in all of these ways, and we will use all of them in this book.

In the definition of a function, the set X of all inputs is called the **domain** of the function. The set Y of all outputs produced from these inputs is called the **range** of the function. Two different inputs (elements in the domain) can be assigned to the same output (an element in the range) but one input cannot lead to two different outputs.

Most of the time we will work with functions whose domains and ranges are real numbers, but there are other types of functions all around us. Final grades for this course provide an example of a function. For each student, the instructor will assign a final grade based on some rule for evaluating that student's performance. The domain of this function consists of all students registered for the course, and the range consists of grades (either letters like A, B, C, D, F, or numbers like 4.0, 3.7, 2.0, 1.7, 0.0). Two students can receive the same final grade, but only one grade can be assigned to each student.

Function Machines

Functions are abstract structures, but sometimes it is easier to think of them in a more concrete way. One such way is to imagine that a function is a special-purpose computer, a machine that accepts inputs, does something to those inputs according to a defining rule, and produces an output. The output is the value of the function associated with the given input value. If the defining rule for a function f is "multiply the input by itself," $f(\text{input}) = (\text{input})(\text{input})$, then the figure and table in the margin show the results of putting the inputs x, 5, 2.5, a, $c+3$ and $x+h$ into the machine f.

Practice 1. If we have a function machine g whose rule is "divide 3 by the input and add 1," $g(x) = \frac{3}{x} + 1$, what outputs do we get from the inputs x, 5, a, $c+3$ and $x+h$? What happens if the input is 0?

You expect your calculator to behave as a function: each time you press the same sequence of keys (input) you expect to see the same display (output). In fact, if your calculator did not produce the same output each time, you would need a new calculator.

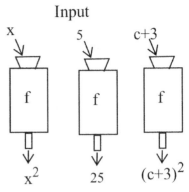

Input

Output

input	output
x	x^2
a	a^2
2.5	6.25
$x+h$	$(x+h)^2$

On many calculators there is a feature that does not produce the same output each time you use it. What is it?

Functions Defined by Equations

If the domain of a function consists of a collection of real numbers (perhaps all real numbers) and the range is also a collection of real numbers, then the function is called a **numerical function**. We can give the rule for a numerical function in several ways, but usually write it as a formula. If the rule for a numerical function, f, is "the output is the input number multiplied by itself," then we could write the rule as

$$f(x) = x \cdot x = x^2$$

The use of an "x" to represent the input is simply a matter of convenience and custom. We could also represent the same function by:

- $f(a) = a^2$

- $f(\#) = \#^2$ or

- $f(\text{input}) = (\text{input})^2$.

For the function f defined by $f(x) = x^2 - x$, we can see that:

- $f(3) = 3^2 - 3 = 6$

- $f(0.5) = (0.5)^2 - (0.5) = -0.25$

- $f(-2) = (-2)^2 - (-2) = 6$

Notice that the two different inputs 3 and -2 both lead to the output of 6. That is allowable for a function.

We can also evaluate f if the input contains variables. If we replace the "x" with something else in the notation "$f(x)$," then we must replace the "x" with the same thing everywhere in the formula:

- $f(c) = c^2 - c$

- $f(a+1) = (a+1)^2 - (a+1) = (a^2 + 2a + 1) - (a+1) = a^2 + a$

- $f(x+h) = (x+h)^2 - (x+h) = (x^2 + 2xh + h^2) - (x+h)$

and, in general: $f(\text{input}) = (\text{input})^2 - (\text{input})$

For more complicated expressions, we can just proceed step-by-step:

$$\frac{f(x+h) - f(x)}{h} = \frac{\left((x+h)^2 - (x+h)\right) - (x^2 - x)}{h}$$
$$= \frac{\left((x^2 + 2xh + h^2) - (x+h)\right) - (x^2 - x)}{h}$$
$$= \frac{2xh + h^2 - h}{h} = \frac{h(2x + h - 1)}{h} = 2x + h - 1$$

Practice 2. For the function g defined by $g(t) = t^2 - 5t$, evaluate $g(1)$, $g(-2)$, $g(w+3)$, $g(x+h)$, $g(x+h) - g(x)$ and $\frac{g(x+h) - g(x)}{h}$.

Functions Defined by Graphs and Tables of Values

The **graph** of a numerical function f consists of a plot of ordered pairs (x, y) where x is in the domain of f and $y = f(x)$, such as the graph of $f(x) = \sin(x)$ for $-4 \le x \le 7$ in the margin. A **table** of values of a numerical function consists of a list of (some of) the ordered pairs (x, y) where $y = f(x)$.

A function can be defined by a graph or by a table of values, and these types of definitions are common in applied fields. The outcome of an experiment will depend on the input, but the experimenter may not know the "rule" for predicting the outcome. In that case, the experimenter usually represents the function of interest as a table of measured outcome values versus input values, or as a graph of the outcomes versus the inputs. The table and graph in the margin show the deflections obtained when weights were loaded at the end of a wooden stick. The next graph shows the temperature of a hot cup of tea as a function of the time as it sits in a 68°F room. In these experiments, the "rule" for the function is that $f(\text{input}) = $ actual outcome of the experiment.

Tables have the advantage of presenting the data explicitly, but it is often difficult to detect patterns simply from lists of numbers.

Graphs usually obscure some of the precision of the data, but more easily allow us to detect patterns visually — we can actually see what is happening with the numbers

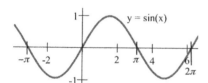

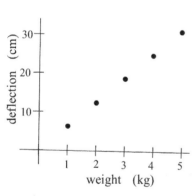

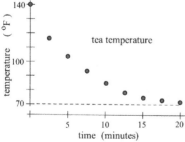

Creating Graphs of Functions

Most people understand and can interpret pictures more quickly than tables of data or equations, so if we have a function defined by a table of values or by an equation, it is often useful (and necessary) to create a picture of the function: a graph.

A Graph from a Table of Values

If we have a table of values for a function, perhaps consisting of measurements obtained from an experiment, then we can plot the ordered pairs in the xy-plane to get a graph consisting of a collection of points.

The table in the margin shows the lengths and weights of trout caught (and released) during several days of fishing. The graph plots those values along with a line that comes "close" to the plotted points. From the graph, you could estimate that a 17-inch trout should weigh slightly more than one pound.

length (in.)	weight (lbs.)
13.5	0.4
14.5	0.9
15.0	0.7
16.0	0.9
18.0	1.2
18.5	1.6
19.5	1.5
20.5	1.7
20.5	2.1

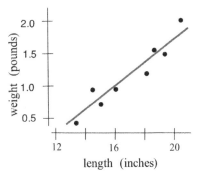

A Graph from an Equation

Creating the graph of a function given by an equation is similar to creating one from a table of values: we need to plot enough points (x, y) where $y = f(x)$ so we can be confident of the shape and location of the graph of the entire function. We can find a point (x, y) that satisfies $y = f(x)$ by picking a value for x and then calculating the value for y by evaluating $f(x)$. Then we can enter the (x, y) value in a table or simply plot the point (x, y).

If you recognize the form of the equation and know something about the shape of graphs of that form, you may not have to plot many points. If you do not recognize the form of the equation, then you will need to plot more points, maybe 10 or 20 or 234: it depends on how complicated the graph appears and on how important it is to you (or your boss) to have an accurate graph. Evaluating $y = f(x)$ at many different values for x and then plotting the points (x, y) is usually not very difficult, but it can be very time-consuming. Fortunately, calculators and computers can often do the evaluation and plotting for you.

Is Every Graph the Graph of a Function?

The definition of "function" requires that each element of the domain (each input value) be sent by the function to exactly one element of the range (to exactly one output value), so for each input x-value there will be exactly one output y-value, $y = f(x)$. The points (x, y_1) and (x, y_2) cannot both be on the graph of f unless $y_1 = y_2$. The graphical interpretation of this result is called the **Vertical Line Test**.

function

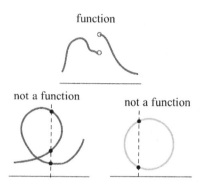

not a function

not a function

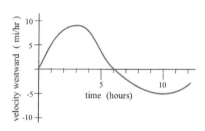

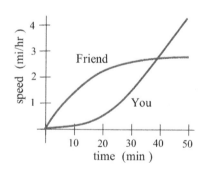

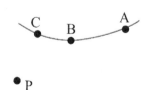

Vertical Line Test for a Function: A graph is the graph of a function if and only if a vertical line drawn through any point in the domain intersects the graph at exactly one point.

The figure in the margin shows the graph of a function, followed by two graphs that are not graphs of functions, along with vertical lines that intersect those graphs at more than one point. Non-functions are not "bad," and are often necessary to describe complicated phenomena.

Reading Graphs Carefully

Calculators and computers can help students, reporters, financial analysts and scientists create graphs quickly and easily. Because of this, graphs are being used more often than ever to present information and justify arguments. This text takes a distinctly graphical approach to the ideas and meaning of calculus. Calculators and computers can help us create graphs, but we need to be able to read them carefully. The next examples illustrate some types of information that can be obtained by carefully reading and understanding graphs.

Example 1. A boat starts from St. Thomas and sails due west with the velocity shown in the margin.

(a) When is the boat traveling the fastest?

(b) What does a negative velocity away from St. Thomas mean?

(c) When is the boat the farthest away from St. Thomas?

Solution. (a) The greatest speed is 10 mph at $t = 3$ hours. (b) It means that the boat is heading back toward St. Thomas. (c) The boat is farthest from St. Thomas at $t = 6$ hours. For $t < 6$ the boat's velocity is positive, and the distance from the boat to St. Thomas is increasing. For $t > 6$ the boat's velocity is negative, and the distance from the boat to St. Thomas is decreasing. ◄

Practice 3. You and a friend start out together and hike along the same trail but walk at different speeds, as shown in the figure.

(a) Who is walking faster at $t = 20$?

(b) Who is ahead at $t = 20$?

(c) When are you and your friend farthest apart?

(d) Who is ahead when $t = 50$?

Example 2. Which has the largest slope: the line through the points A and P, the line through B and P, or the line through C and P?

Solution. The line through C and P: $m_{PC} > m_{PB} > m_{PA}$. ◄

Practice 4. In the figure, the point Q on the curve is fixed, and the point P is moving to the right along the curve toward the point Q. As P moves toward Q, is the indicated value increasing, decreasing, remaining constant, or doing something else?

(a) x-coordinate of P

(b) x-increment from P to Q

(c) slope from P to Q

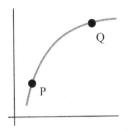

Example 3. The graph of $y = f(x)$ appears in the margin. Let $g(x)$ be the **slope** of the line tangent to the graph of $f(x)$ at the point $(x, f(x))$.

(a) Estimate the values $g(1)$, $g(2)$ and $g(3)$.

(b) For what value(s) of x is $g(x) = 0$?

(c) At what value(s) of x is $g(x)$ largest?

(d) Sketch the graph of $y = g(x)$.

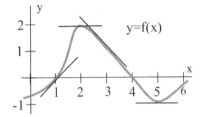

Solution. (a) The figure in the margin shows the graph of $y = f(x)$ with several tangent lines to the graph of f. From this graph, we can estimate that $g(1)$ (the slope of the line tangent to the graph of f at $(1,0)$) is approximately equal to 1. Similarly, $g(2) \approx 0$ and $g(3) \approx -1$.

(b) The slope of the tangent line appears to be horizontal (slope $= 0$) at $x = 2$ and at $x = 5$.

(c) The tangent line to the graph of f appears to have greatest slope (be steepest) near $x = 1.5$.

(d) We can build a table of values of $g(x)$ and then sketch the graph of these values. A graph of $y = g(x)$ appears below. ◄

x	$f(x)$	$g(x)$
0	−1.0	0.5
1	0.0	1.0
2	2.0	0.0
3	1.0	−1.0
4	0.0	−1.0
5	−1.0	0.0
6	−0.5	0.5

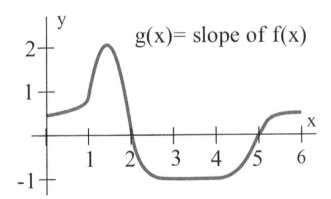

g(x)= slope of f(x)

Practice 5. Water flows into a container (see margin) at a constant rate of 3 gallons per minute. Starting with an empty container, sketch the graph of the height of the water in the container as a function of time.

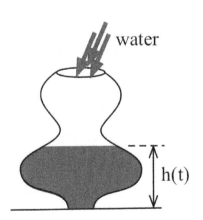

0.3 Problems

In Problems 1–4, match the numerical triples to the graphs. For example, in Problem 1, A: 3, 3, 6 is "over and up" so it matches graph (a).

1. A: 3, 3, 6; B: 12, 6, 6; C: 7, 7, 3 D: 2, 4, 4

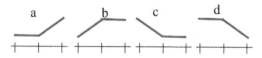

2. A: 7, 10, 7; B: 17, 17, 25; C: 4, 4, 8 D: 12, 8, 16

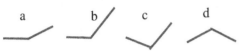

3. A: 7, 14, 10; B: 23, 45, 22; C: 8, 12, 8 D: 6, 9, 3

4. A: 6, 3, 9; B: 8, 1, 1; C: 12, 6, 9 D: 3.7, 1.9, 3.6

5. Water is flowing at a steady rate into each of the bottles shown below. Match each bottle shape with the graph of the height of the water as a function of time

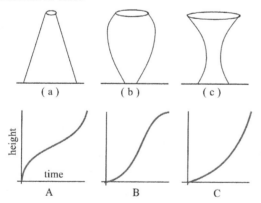

6. Sketch shapes of bottles that will have the water height versus time graphs shown below.

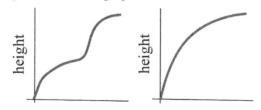

7. If $f(x) = x^2 + 3$, $g(x) = \sqrt{x-5}$ and $H(x) = \frac{x}{x-2}$:
   (a) evaulate $f(1)$, $g(1)$ and $H(1)$.
   (b) graph $f(x)$, $g(x)$ and $H(x)$ for $-5 \leq x \leq 10$.
   (c) evaluate $f(3x)$, $g(3x)$ and $H(3x)$.
   (d) evaluate $f(x+h)$, $g(x+h)$ and $H(x+h)$.

8. Find the slope of the line through the points P and Q when:
   (a) $P = (1,3)$, $Q = (2,7)$
   (b) $P = (x, x^2 + 2)$, $Q = (x+h, (x+h)^2 + 2)$
   (c) $P = (1,3)$, $Q = (x, x^2 + 2)$
   (d) P, Q as in (b) with $x = 2$, $x = 1.1$, $x = 1.002$

9. Find the slope of the line through the points P and Q when:
   (a) $P = (1,5)$, $Q = (2,7)$
   (b) $P = (x, x^2 + 3x - 1)$,
       $Q = (x+h, (x+h)^2 + 3(x+h) - 1)$
   (c) P, Q as in (b) with $x = 1.3$, $x = 1.1$, $x = 1.002$

10. If $f(x) = x^2 + x$ and $g(x) = \frac{3}{x}$, evaluate and simplify $\frac{f(a+h) - f(a)}{h}$ and $\frac{g(a+h) - g(a)}{h}$ when $a = 1$, $a = 2$, $a = -1$, $a = x$.

11. If $f(x) = x^2 - 2x$ and $g(x) = \sqrt{x}$, evaluate and simplify $\frac{f(a+h) - f(a)}{h}$ and $\frac{g(a+h) - g(a)}{h}$ when $a = 1$, $a = 2$, $a = 3$, $a = x$.

12. The temperatures shown below were recorded during a 12-hour period in Chicago.
   (a) At what time was the temperature the highest? Lowest?
   (b) How fast was the temperature rising at 10 a.m.? At 1 p.m.?
   (c) What could have caused the drop in temperature between 1 p.m. and 3 p.m.?

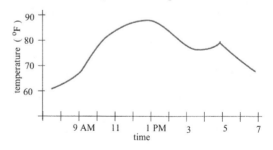

13. The graph below shows the distance of an airplane from an airport during a long flight.

    (a) How far was the airplane from the airport at 1 p.m.? At 2 p.m.?

    (b) How fast was the distance changing at 1 p.m.?

    (c) How could the distance from the plane to the airport remain unchanged from 1:45 p.m. until 2:30 p.m. without the airplane falling?

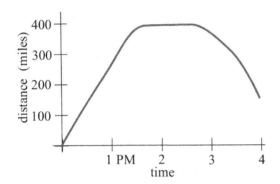

14. The graph below shows the height of a diver above the water level at time t seconds.

    (a) What was the height of the diving board?

    (b) When did the diver hit the water?

    (c) How deep did the diver get?

    (d) When did the diver return to the surface?

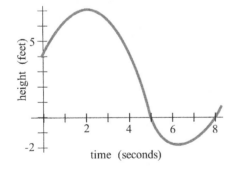

15. Refer to the curve shown below.

    (a) Sketch the lines tangent to the curve at $x = 1$, 2, 3, 4 and 5.

    (b) For what value(s) of x is the value of the function largest? Smallest?

    (c) For what value(s) of x is the slope of the tangent line largest? Smallest?

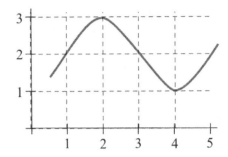

16. The figure below shows the height of the water (above and below mean sea level) at a Maine beach.

    (a) At which time(s) was the most beach exposed? The least?

    (b) At which time(s) was the current the strongest?

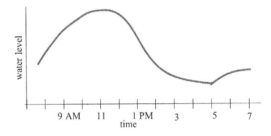

17. Imagine that you are ice skating, **from left to right**, along the path shown below. Sketch the path you will follow if you fall at points A, B and C.

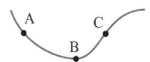

18. Define $s(x)$ to be the **slope** of the line through the points $(0,0)$ and $(x, f(x))$ where $f(x)$ is the function graphed below. For example, $s(3) =$ **slope** of the line through $(0,0)$ and $(3, f(3)) = \frac{4}{3}$.

(a) Evaluate $s(1)$, $s(2)$ and $s(4)$.

(b) For which integer value of x between 1 and 7 is $s(x)$ smallest?

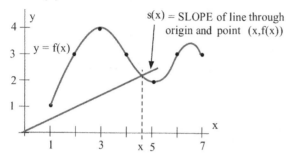

19. Let $f(x) = x + 1$ and define $s(x)$ to be the **slope** of the line through the points $(0,0)$ and $(x, f(x))$, as shown below. For example, $s(2) =$ slope of the line through $(0,0)$ and $(2,3) = \frac{3}{2}$.

(a) Evaluate $s(1)$, $s(3)$ and $s(4)$.

(b) Find a formula for $s(x)$.

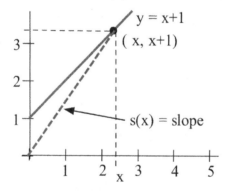

20. Define $A(x)$ to be the area of the rectangle bounded by the coordinate axes, the line $y = 2$ and a vertical line at x, as shown below. For example, $A(3) =$ area of a 2×3 rectangle $= 6$.

(a) Evaluate $A(1)$, $A(2)$ and $A(5)$.

(b) Find a formula for $A(x)$.

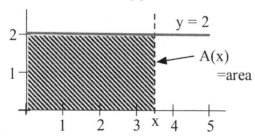

21. Using the graph of $y = f(x)$ below, let $g(x)$ be the slope of the line tangent to the graph of $f(x)$ at the point $(x, f(x))$. Complete the table, estimating values of the slopes as best you can.

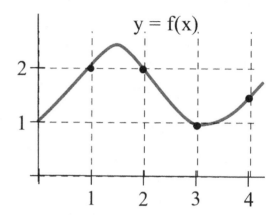

x	$f(x)$	$g(x)$
0	1	1
1		
2		
3		
4		

22. Sketch the graphs of water height versus time for water pouring into a bottle shaped like:

(a) a milk carton

(b) a spherical glass vase

(c) an oil drum (cylinder) lying on its side

(d) a giraffe

(e) you

23. Design bottles whose graphs of (highest) water height versus time will look like those shown below.

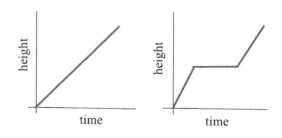

0.3 Practice Answers

1. Create an input-output table using the function rule:

input	output
x	$\frac{3}{x}+1$
5	$\frac{3}{5}+1 = 1.6$
a	$\frac{3}{a}+1$
$c+3$	$\frac{3}{c+3}+1$
$x+h$	$\frac{3}{x+h}+1$

If $x = 0$, then $g(0) = \frac{3}{0}+1$ is not defined (because of division by 0).

2. $g(t) = t^2 - 5t$
    $g(1) = 1^2 - 5(1) = -4$
    $g(-2) = (-2)^2 - 5(-2) = 14$
    $g(w+3) = (w+3)^2 - 5(w+3) = w^2 + 6w + 9 - 5w - 15 = w^2 + w - 6$
    $g(x+h) = (x+h)^2 - 5(x+h) = x^2 + 2xh + h^2 - 5x - 5h$
    $g(x+h) - g(x) = (x^2 + 2xh + h^2 - 5x - 5h) - (x^2 - 5x) = 2xh + h^2 - 5h$
    $\dfrac{g(x+h) - g(x)}{h} = \dfrac{2xh + h^2 - 5h}{h} = 2x + h - 5$

3. (a) Friend (b) Friend (c) At $t = 40$. Before that, your friend is walking faster and increasing the distance between you. Then you start to walk faster than your friend and start to catch up. (d) Friend. You are walking faster than your friend at $t = 50$, but you still have not caught up.

4. (a) The x-coordinate is increasing. (b) The x-increment Δx is decreasing. (c) The slope of the line through P and Q is decreasing.

5. See the figure:

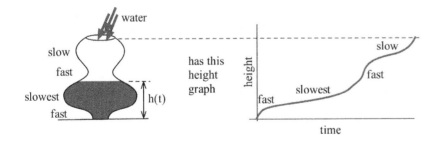

0.4   Combinations of Functions

Sometimes a physical or economic situation behaves differently depending on various circumstances. In these situations, a more complicated formula may be needed to describe the situation.

Multiline Definitions of Functions: Putting Pieces Together

Sales Tax: Some states have different rates of sales tax depending on the type of item purchased. As an example, for many years food purchased at restaurants in Seattle was taxed at a rate of 10%, while most other items were taxed at a rate of 9.5% and food purchased at grocery stores had no tax assessed. We can describe this situation by using a **multiline** function: a function whose defining rule consists of several pieces. Which piece of the rule we need to use will depend on what we buy. In this example, we could define the tax T on an item that costs x to be:

$$T(x) = \begin{cases} 0 & \text{if } x \text{ is the cost of a food at a grocery store} \\ 0.10x & \text{if } x \text{ is the cost of food at a restaurant} \\ 0.095x & \text{if } x \text{ is the cost of any other item} \end{cases}$$

To find the tax on a \$2 can of stew, we would use the first piece of the rule and find that the tax is \$0. To find the tax on a \$30 restaurant bill, we would use the second piece of the rule and find that the tax is \$3.00. The tax on a \$150 textbook requires using the third rule: the tax would be \$14.25.

Wind Chill Index: The rate at which a person's body loses heat depends on the temperature of the surrounding air and on the speed of the air. You lose heat more quickly on a windy day than you do on a day with little or no wind. Scientists have experimentally determined this rate of heat loss as a function of temperature and wind speed, and the resulting function is called the Wind Chill Index, WCI. The WCI is the temperature on a still day (no wind) at which your body would lose heat at the same rate as on the windy day. For example, the WCI value for 30°F air moving at 15 miles per hour is 9°F: your body loses heat as quickly on a 30°F day with a 15 mph wind as it does on a 9°F day with no wind.

If T is the Fahrenheit temperature of the air and v is the speed of the wind in miles per hour, then the WCI can be expressed as a multiline function of the wind speed v (and of the temperature T):

$$\text{WCI} = \begin{cases} T & \text{if } 0 \le v \le 4 \\ 91.4 - \frac{10.45 + 6.69\sqrt{v} - 0.447v}{22}(91.5 - T) & \text{if } 4 < v \le 45 \\ 1.60T - 55 & \text{if } v > 45 \end{cases}$$

The WCI value for a still day ($0 \leq v \leq 4$ mph) is just the air temperature. The WCI for wind speeds above 45 mph are the same as the WCI for a wind speed of 45 mph. The WCI for wind speeds between 4 mph and 45 mph decrease as the wind speeds increase. This WCI function depends on two variables: the temperature and the wind speed; but if the temperature is constant, then the resulting formula for WCI only depends on the wind speed. If the air temperature is 30°F ($T = 30$), then the formula for the Wind Chill Index is:

$$WCI_{30} = \begin{cases} 30° & \text{if } 0 \leq v \leq 4 \text{ mph} \\ 62.19 - 18.70\sqrt{v} + 1.25v & \text{if } 4 \text{ mph} < v \leq 45 \text{ mph} \\ -7° & \text{if } v > 45 \text{ mph} \end{cases}$$

The WCI graphs for temperatures of 40°F, 30°F and 20°F appear below:

From UMAP Module 658, "Windchill," by William Bosch and L.G. Cobb, 1984.

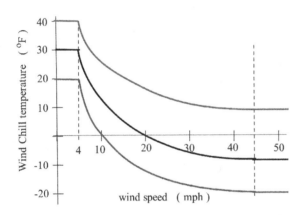

Practice 1. A Hawaiian condo rents for $380 per night during the tourist season (from December 15 through April 30), and for $295 per night otherwise. Define a multiline function that describes these rates.

Example 1. Define $f(x)$ by:

$$f(x) = \begin{cases} 2 & \text{if } x < 0 \\ 2x & \text{if } 0 \leq x < 2 \\ 1 & \text{if } 2 < x \end{cases}$$

Evaluate $f(-3)$, $f(0)$, $f(1)$, $f(4)$ and $f(2)$. Graph $y = f(x)$ on the interval $-1 \leq x \leq 4$.

Solution. To evaluate the function at different values of x, we must first decide which line of the rule applies. If $x = -3 < 0$, then we need to use the first line, so $f(-3) = 2$. When $x = 0$ or $x = 1$, we need the second line of the function definition, so $f(0) = 2(0) = 0$ and $f(1) = 2(1) = 2$. At $x = 4$ we need the third line, so $f(4) = 1$. Finally, at $x = 2$, none of the lines apply: the second line requires $x < 2$ and the third line requires $2 < x$, so $f(2)$ is undefined. The graph of $f(x)$ appears in the margin. Note the "hole" above $x = 2$, which indicates $f(2)$ is not defined by the rule for f. ◀

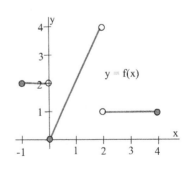

Practice 2. Define $g(x)$ by:

$$g(x) = \begin{cases} x & \text{if } x < -1 \\ 2 & \text{if } -1 \le x < 1 \\ -x & \text{if } 1 < x \le 3 \\ 1 & \text{if } 4 < x \end{cases}$$

Graph $y = g(x)$ for $-3 \le x \le 6$ and evaluate $g(-3)$, $g(-1)$, $g(0)$, $g(\frac{1}{2})$, $g(1)$, $g(\frac{\pi}{3})$, $g(2)$, $g(3)$, $g(4)$ and $g(5)$.

Practice 3. Write a multiline definition for the function whose graph appears in the margin.

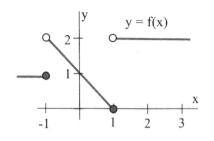

We can think of a multiline function as a machine that first examines the input value to decide which line of the function rule to apply:

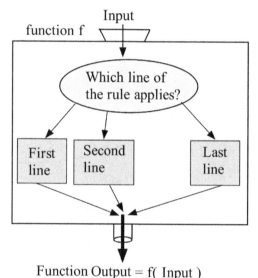

Composition of Functions: Functions of Functions

Basic functions are often combined with each other to describe more complicated situations. Here we will consider the composition of functions — functions of functions.

The **composite** of two functions f and g, written $f \circ g$, is:

$$f \circ g(x) = f(g(x))$$

The domain of the composite function $f \circ g(x) = f(g(x))$ consists of those x-values for which $g(x)$ and $f(g(x))$ are both defined: we can evaluate the composition of two functions at a point x only if each step in the composition is defined.

If we think of our functions as machines, then composition is simply a new machine consisting of an arrangement of the original machines.

The composition $f \circ g$ of the function machines f and g shown in the margin is an arrangement of the machines so that the original input x goes into machine g, the output from machine g becomes the input into machine f, and the output from machine f is our final output. The composition of the function machines $f \circ g(x) = f(g(x))$ is only valid if x is an allowable input into g (that is, x is in the domain of g) and if $g(x)$ is then an allowable input into f.

The composition $g \circ f$ involves arranging the machines so the original input goes into f, and the output from f then becomes the input for g (see right side of margin figure).

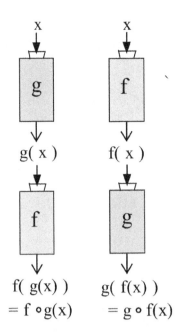

Example 2. For $f(x) = \sqrt{x - 2}$, $g(x) = x^2$ and

$$h(x) = \begin{cases} 3x & \text{if } x < 2 \\ x - 1 & \text{if } 2 \leq x \end{cases}$$

evaluate $f \circ g(3)$, $g \circ f(6)$, $f \circ h(2)$ and $h \circ g(-3)$. Find the formulas and domains of $f \circ g(x)$ and $g \circ f(x)$.

Solution. $f \circ g(3) = f(g(3)) = f(3^2) = f(9) = \sqrt{9 - 2} = \sqrt{7} \approx 2.646$; $g \circ f(6) = g(f(6)) = g(\sqrt{6 - 2}) = g(\sqrt{4}) = g(2) = 2^2 = 4$; $f \circ h(2) = f(h(2)) = f(2 - 1) = f(1) = \sqrt{1 - 2} = \sqrt{-1}$, which is undefined; $h \circ g(-3) = h(g(-3)) = h(9) = 9 - 1 = 8$; $f \circ g(x) = f(g(x)) = f(x^2) = \sqrt{x^2 - 2}$, and the domain of $f \circ g$ consists of those x-values for which $x^2 - 2 \geq 0$, so the domain of $f \circ g$ is all x such that $x \geq \sqrt{2}$ or $x \leq -\sqrt{2}$; $g \circ f(x) = g(f(x)) = g(\sqrt{x - 2}) = (\sqrt{x - 2})^2 = x - 2$, but this last equality is true only when $x - 2 \geq 0 \Rightarrow x \geq 2$, so the domain of $g \circ f$ is all $x \geq 2$. ◄

Practice 4. For $f(x) = \frac{x}{x-3}$, $g(x) = \sqrt{1 + x}$ and

$$h(x) = \begin{cases} 2x & \text{if } x \leq 1 \\ 5 - x & \text{if } 1 < x \end{cases}$$

evaluate $f \circ g(3)$, $f \circ g(8)$, $g \circ f(4)$, $f \circ h(1)$, $f \circ h(3)$, $f \circ h(2)$ and $h \circ g(-1)$. Find formulas for $f \circ g(x)$ and $g \circ f(x)$.

Shifting and Stretching Graphs

Some common compositions are fairly straightforward; you should recognize the effect of these compositions on graphs of the functions.

Example 3. The margin figure shows the graph of $y = f(x)$. Graph

(a) $2 + f(x)$

(b) $3 \cdot f(x)$

(c) $f(x - 1)$

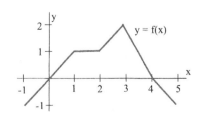

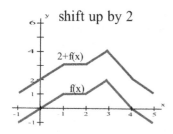

shift up by 2

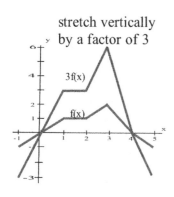

stretch vertically by a factor of 3

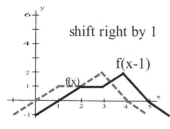

shift right by 1

Solution. All of the new graphs appear in the margin.

(a) Adding 2 to all of the values of $f(x)$ rigidly shifts the graph of $f(x)$ upward 2 units.

(b) Multiplying all of the values of $f(x)$ by 3 leaves all of the roots (zeros) of f fixed: if x is a root of f then $f(x) = 0 \Rightarrow 3 \cdot f(x) = 3(0) = 0$ so x is also a root of $3 \cdot f(x)$. If x is not a root of f, then the graph of $3f(x)$ looks like the graph of $f(x)$ stretched vertically by a factor of 3.

(c) The graph of $f(x-1)$ is the graph of $f(x)$ rigidly shifted 1 unit to the right. We could also get these results by examining the graph of $y = f(x)$, creating a table of values for $f(x)$ and the new functions:

x	$f(x)$	$2 + f(x)$	$3f(x)$	$x - 1$	$f(x - 1)$
-1	-1	1	-3	-2	$f(-2)$ not defined
0	0	2	0	-1	$f(0 - 1) = -1$
1	1	3	3	0	$f(1 - 1) = 0$
2	1	3	3	1	$f(2 - 1) = 1$
3	2	4	6	2	$f(3 - 1) = 1$
4	0	2	0	3	$f(4 - 1) = 2$
5	-1	1	-3	4	$f(5 - 1) = 0$

and then graphing those new functions.   ◀

If k is a positive constant, then

• the graph of $y = k + f(x)$ will be the graph of $y = f(x)$ rigidly shifted up by k units

• the graph of $y = kf(x)$ will have the same roots as the graph of $f(x)$ and will be the graph of $y = f(x)$ vertically stretched by a factor of k

• the graph of $y = f(x - k)$ will be the graph of $y = f(x)$ rigidly shifted right by k units

• the graph of $y = f(x + k)$ will be the graph of $y = f(x)$ rigidly shifted left by k units

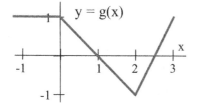

Practice 5. The figure in the margin shows the graph of $g(x)$. Graph:

(a) $1 + g(x)$

(b) $2g(x)$

(c) $g(x - 1)$

(d) $-3g(x)$

Iteration of Functions

Certain applications feed the output from a function machine back into the same machine as the new input. Each time through the machine is called an **iteration** of the function.

Example 4. Suppose $f(x) = \dfrac{\frac{5}{x} + x}{2}$ and we start with the input $x = 4$ and repeatedly feed the output from f back into f. What happens?

Solution. Creating a table:

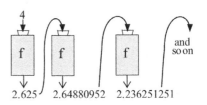

iteration	input	output
1	4	$f(4) = \dfrac{\frac{5}{4}+4}{2} = 2.625000000$
2	2.625000000	$f(f(4)) = \dfrac{\frac{5}{2.625}+2.625}{2} = 2.264880952$
3	2.264880952	$f(f(f(4))) = 2.236251251$
4	2.236251251	2.236067985
5	2.236067985	2.236067977
6	2.236067977	2.236067977

Once we have obtained the output 2.236067977, we will just keep getting the same output (to 9 decimal places). You might recognize this output value as an approximation of $\sqrt{5}$.

    This algorithm always finds $\pm\sqrt{5}$. If we start with any positive input, the values will eventually get as close to $\sqrt{5}$ as we want. Starting with any negative value for the input will eventually get us close to $-\sqrt{5}$. We cannot start with $x = 0$, as $\frac{5}{0}$ is undefined.   ◀

Practice 6. What happens if we start with the input value $x = 1$ and iterate the function $f(x) = \dfrac{\frac{9}{x} + x}{2}$ several times? Do you recognize the resulting number? What do you think will happen to the iterates of $g(x) = \dfrac{\frac{A}{x} + x}{2}$? (Try several positive values of A.)

Two Useful Functions: Absolute Value and Greatest Integer

Two functions (one of which should be familiar to you, the other perhaps not) possess useful properties that let us describe situations in which an object abruptly changes direction or jumps from one value to another value. Their graphs will have corners and breaks, respectively.

    The **absolute value function** evaluated at a number x, $y = f(x) = |x|$, is the distance between the number x and 0.

    If x is greater than or equal to 0, then $|x|$ is simply $x - 0 = x$. If x is negative, then $|x|$ is $0 - x = -x = -1 \cdot x$, which is positive because:

$$-1 \cdot (\text{negative number}) = \text{a positive number}$$

Some calculators and computer programming languages represent the absolute value function by $\text{abs}(x)$ or $\text{ABS}(x)$.

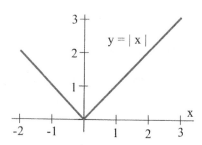

Definition of $|x|$:

$$|x| = \begin{cases} x & \text{if } x \geq 0 \\ -x & \text{if } x < 0 \end{cases}$$

We can also write: $|x| = \sqrt{x^2}$.

The domain of $y = f(x) = |x|$ consists of all real numbers. The range of $f(x) = |x|$ consists of all numbers larger than or equal to zero (all non-negative numbers). The graph of $y = f(x) = |x|$ (see margin) has no holes or breaks, but it does have a sharp corner at $x = 0$.

The absolute value will be useful for describing phenomena such as reflected light and bouncing balls that change direction abruptly or whose graphs have corners. The absolute value function has a number of properties we will use later.

This last property is widely known as the **triangle inequality**.

Properties of $|x|$: For all real numbers a and b,

- $|a| = 0 \cdot |a| = 0$ if and only if $a = 0$

- $|ab| = |a| \cdot |b|$

- $|a + b| \leq |a| + |b|$

Taking the absolute value of a function has an interesting effect on the graph of the function: for any function $f(x)$, we have

$$|f(x)| = \begin{cases} f(x) & \text{if } f(x) \geq 0 \\ -f(x) & \text{if } f(x) < 0 \end{cases}$$

In other words, if $f(x) \geq 0$, then $|f(x)| = f(x)$, so the graph of $|f(x)|$ is the same as the graph of $f(x)$. If $f(x) < 0$, then $|f(x)| = -f(x)$, so the graph of $|f(x)|$ is just the graph of $f(x)$ "flipped" about the x-axis, and it lies above the x-axis. The graph of $|f(x)|$ will always be on or above the x-axis.

Example 5. The figure in the margin shows the graph of $f(x)$. Graph:

(a) $|f(x)|$

(b) $|1 + f(x)|$

(c) $1 + |f(x)|$

Solution. The graphs appear below:

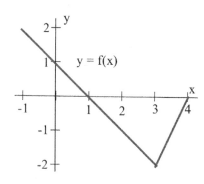

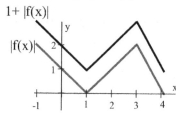

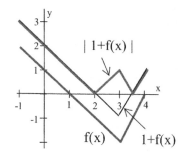

In (b), we rigidly shift the graph of f up 1 unit before taking the absolute value. In (c), we take the absolute value before rigidly shifting the graph up 1 unit.   ◀

Practice 7. The figure in the margin shows the graph of $g(x)$. Graph:

(a) $|g(x)|$

(b) $|g(x-1)|$

(c) $g(|x|)$

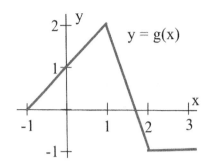

The **greatest integer function** evaluated at a number x, $y = f(x) = \lfloor x \rfloor$, is the largest integer less than or equal to x.

The value of $\lfloor x \rfloor$ is always an integer and $\lfloor x \rfloor$ is always less than or equal to x. For example, $\lfloor 3.2 \rfloor = 3$, $\lfloor 3.9 \rfloor = 3$ and $\lfloor 3 \rfloor = 3$. If x is positive, then $\lfloor x \rfloor$ **truncates** x (drops the fractional part of x). If x is negative, the situation is different: $\lfloor -4.2 \rfloor \neq -4$ because -4 is not less than or equal to -4.2: $\lfloor -4.2 \rfloor = -5$, $\lfloor -4.7 \rfloor = -5$ and $\lfloor -4 \rfloor = -4$.

Historically, many textbooks have used the square brackets [ ] to represent the greatest integer function, while calculators and many programming languages use $\text{INT}(x)$.

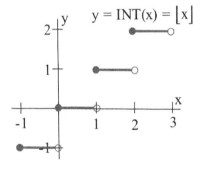

Definition of $\lfloor x \rfloor$:

$$\lfloor x \rfloor = \begin{cases} x & \text{if } x \text{ is an integer} \\ \text{largest integer strictly less than } x & \text{if } x \text{ is not an integer} \end{cases}$$

The domain of $f(x) = \lfloor x \rfloor$ is all real numbers. The range of $f(x) = \lfloor x \rfloor$ is only the integers. The graph of $y = f(x) = \lfloor x \rfloor$ appears in the margin. It has a jump break—a "step"—at each integer value of x, so $f(x) = \lfloor x \rfloor$ is called a **step function**. Between any two consecutive integers, the graph is horizontal with no breaks or holes.

The greatest integer function is useful for describing phenomena that change values abruptly, such as postage rates as a function of weight. As of January 26, 2014, the cost to mail a first-class retail "flat" (such as a manila envelope) was $0.98 for the first ounce and another $0.21 for each additional ounce.

The $\lfloor x \rfloor$ function can also be used for functions whose graphs are "square waves," such as the on and off of a flashing light.

Example 6. Graph $y = \lfloor 1 + 0.5 \sin(x) \rfloor$.

Solution. One way to create this graph is to first graph $y = 1 + 0.5 \sin(x)$, the thin curve in margin figure, and then apply the greatest integer function to y to get the thicker "square wave" pattern.   ◀

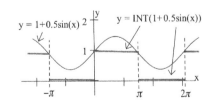

Practice 8. Sketch the graph of $y = \lfloor x^2 \rfloor$ for $-2 \leq x \leq 2$.

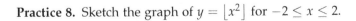

A Really "Holey" Function

The graph of $\lfloor x \rfloor$ has a break or jump at each integer value, but how many breaks can a function have? The next function illustrates just how broken or "holey" the graph of a function can be.

Define a function $h(x)$ as:

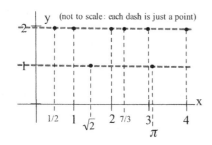

$$h(x) = \begin{cases} 2 & \text{if } x \text{ is a rational number} \\ 1 & \text{if } x \text{ is an irrational number} \end{cases}$$

Then $h(3) = 2$, $h\left(\frac{5}{3}\right) = 2$ and $h\left(-\frac{2}{5}\right) = 2$, because 3, $\frac{5}{3}$ and $-\frac{2}{5}$ are all rational numbers. Meanwhile, $h(\pi) = 1$, $h(\sqrt{7}) = 1$ and $h(\sqrt{2}) = 1$, because π, $\sqrt{7}$ and $\sqrt{2}$ are all irrational numbers. These and some other points are plotted in the margin figure.

In order to analyze the behavior of $h(x)$ the following fact about rational and irrational numbers is useful.

Fact: Every interval contains both rational and irrational numbers.

Equivalently: If a and b are real numbers and $a < b$, then there is

- a rational number R between a and b ($a < R < b$)

- an irrational number I between a and b ($a < I < b$).

The above fact tells us that between any two places where $y = h(x) = 1$ (because x is rational) there is a place where $y = h(x)$ is 2, because there is an irrational number between any two distinct rational numbers. Similarly, between any two places where $y = h(x) = 2$ (because x is irrational) there is a place where $y = h(x) = 1$, because there is a rational number between any two distinct irrational numbers.

The graph of $y = h(x)$ is impossible to actually draw, because every two points on the graph are separated by a hole. This is also an example of a function that your computer or calculator cannot graph, because in general it can not determine whether an input value of x is irrational.

Example 7. Sketch the graph of

$$g(x) = \begin{cases} 2 & \text{if } x \text{ is a rational number} \\ x & \text{if } x \text{ is an irrational number} \end{cases}$$

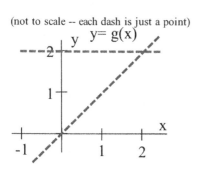

Solution. A sketch of the graph of $y = g(x)$ appears in the margin. When x is rational, the graph of $y = g(x)$ looks like the "holey" horizontal line $y = 2$. When x is irrational, the graph of $y = g(x)$ looks like the "holey" line $y = x$. ◀

Practice 9. Sketch the graph of

$$h(x) = \begin{cases} \sin(x) & \text{if } x \text{ is a rational number} \\ x & \text{if } x \text{ is an irrational number} \end{cases}$$

0.4 Problems

1. If T is the Celsius temperature of the air and v is the speed of the wind in kilometers per hour, then

$$\text{WCI} = \begin{cases} T & \text{if } 0 \le v \le 6.5 \\ 33 - \frac{10.45 + 5.29\sqrt{v} - 0.279v}{22}(33 - T) & \text{if } 6.5 < v \le 72 \\ 1.6T - 19.8 & \text{if } v > 72 \end{cases}$$

(a) Determine the Wind Chill Index for

   i. a temperature of 0°C and a wind speed of 49 km/hr

   ii. a temperature of 11°C and a wind speed of 80 km/hr.

(b) Write a multiline function definition for the WCI if the temperature is 11°C.

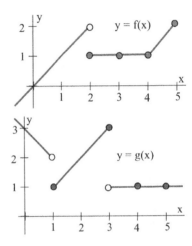

2. Use the graph of $y = f(x)$ in the margin to evaluate $f(0)$, $f(1)$, $f(2)$, $f(3)$, $f(4)$ and $f(5)$. Write a multiline function definition for f.

3. Use the graph of $y = g(x)$ in the margin to evaluate $g(0)$, $g(1)$, $g(2)$, $g(3)$, $g(4)$ and $g(5)$. Write a multiline function definition for g.

4. Use the values given in the table below, along with $h(x) = 2x + 1$, to determine the missing values of $f \circ g$, $g \circ f$ and $h \circ g$.

x	$f(x)$	$g(x)$	$f \circ g(x)$	$g \circ f(x)$	$h \circ g(x)$
-1	2	0			
0	1	2			
1	-1	1			
2	0	2			

5. Use the graphs shown below and the function $h(x) = x - 2$ to determine the values of

(a) $f(f(1))$, $f(g(2))$, $f(g(0))$, $f(g(1))$

(b) $g(f(2))$, $g(f(3))$, $g(g(0))$, $g(f(0))$

(c) $f(h(3))$, $f(h(4))$, $h(g(0))$, $h(g(1))$

6. Use the graphs shown below and the function $h(x) = 5 - 2x$ to determine the values of

(a) $h(f(0))$, $f(h(1))$, $f(g(2))$, $f(f(3))$

(b) $g(f(0))$, $g(f(1))$, $g(h(2))$, $h(f(3))$

(c) $f(g(0))$, $f(g(1))$, $f(h(2))$, $h(g(3))$

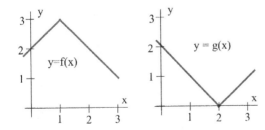

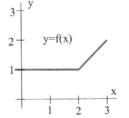

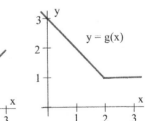

7. Defining $h(x) = x - 2$, $f(x)$ as:

$$f(x) = \begin{cases} 3 & \text{if } x < 1 \\ x - 2 & \text{if } 1 \le x < 3 \\ 1 & \text{if } 3 \le x \end{cases}$$

and $g(x)$ as:

$$g(x) = \begin{cases} x^2 - 3 & \text{if } x < 0 \\ \lfloor x \rfloor & \text{if } 0 \le x \end{cases}$$

(a) evaluate $f(x)$, $g(x)$ and $h(x)$ for $x = -1, 0, 1$, 2, 3 and 4.

(b) evaluate $f(g(1))$, $f(h(1))$, $h(f(1))$, $f(f(2))$, $g(g(3.5))$.

(c) graph $f(x)$, $g(x)$ and $h(x)$ for $-5 \le x \le 5$.

8. Defining $h(x) = 3$, $f(x)$ as:

$$f(x) = \begin{cases} x + 1 & \text{if } x < 1 \\ 1 & \text{if } 1 \le x < 3 \\ 2 - x & \text{if } 3 \le x \end{cases}$$

and $g(x)$ as:

$$g(x) = \begin{cases} |x + 1| & \text{if } x < 0 \\ 2x & \text{if } 0 \le x \end{cases}$$

(a) evaluate $f(x)$, $g(x)$ and $h(x)$ for $x = -1, 0, 1$, 2, 3 and 4.

(b) evaluate $f(g(1))$, $f(h(1))$, $h(f(1))$, $f(f(2))$, $g(g(3.5))$.

(c) graph $f(x)$, $g(x)$ and $h(x)$ for $-5 \le x \le 5$.

9. You are planning to take a one-week vacation in Europe, and the tour brochure says that Monday and Tuesday will be spent in England, Wednesday in France, Thursday and Friday in Germany, and Saturday and Sunday in Italy. Let $L(d)$ be the location of the tour group on day d and write a multiline function definition for $L(d)$.

10. A state has just adopted the following state income tax system: no tax on the first $10,000 earned, 1% of the next $10,000 earned, 2% of the next $20,000 earned, and 3% of all additional earnings. Write a multiline function definition for $T(x)$, the state income tax due on earnings of x dollars.

11. Write a multiline function definition for the curve $y = f(x)$ shown below.

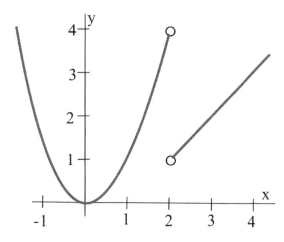

12. Define $B(x)$ to be the **area** of the rectangle whose lower left corner is at the origin and whose upper right corner is at the point $(x, f(x))$ for the function f shown below. For example, $B(3) = 6$. Evaluate $B(1)$, $B(2)$, $B(4)$ and $B(5)$.

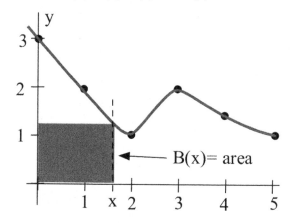

13. Define $B(x)$ to be the **area** of the rectangle whose lower left corner is at the origin and whose upper right corner is at the point $\left(x, \frac{1}{x}\right)$.

(a) Evaluate $B(1)$, $B(2)$ and $B(3)$.

(b) Show that $B(x) = 1$ for all $x > 0$.

14. For $f(x) = |9 - x|$ and $g(x) = \sqrt{x - 1}$:

(a) evaluate $f \circ g(1)$, $f \circ g(3)$, $f \circ g(5)$, $f \circ g(7)$, $f \circ g(0)$.

(b) evaluate $f \circ f(2)$, $f \circ f(5)$, $f \circ f(-2)$.

(c) Does $f \circ f(x) = |x|$ for all values of x?

15. The function $g(x)$ is graphed below. Graph

(a) $g(x) - 1$

(b) $g(x - 1)$

(c) $|g(x)|$

(d) $\lfloor g(x) \rfloor$

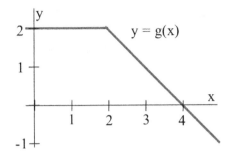

16. The function $f(x)$ is graphed below. Graph

(a) $f(x) - 2$

(b) $f(x - 2)$

(c) $|f(x)|$

(d) $\lfloor f(x) \rfloor$

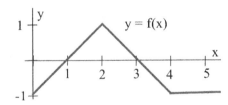

17. Find A and B so that $f(g(x)) = g(f(x))$ when:

(a) $f(x) = 3x + 2$ and $g(x) = 2x + A$

(b) $f(x) = 3x + 2$ and $g(x) = Bx - 1$

18. Find C and D so that $f(g(x)) = g(f(x))$ when:

(a) $f(x) = Cx + 3$ and $g(x) = Cx - 1$

(b) $f(x) = 2x + D$ and $g(x) = 3x + D$

19. Graph $y = f(x) = x - \lfloor x \rfloor$ for $-1 \leq x \leq 3$. This function is called the "fractional part of x" and its graph an example of a "sawtooth" graph.

20. The function $f(x) = \lfloor x + 0.5 \rfloor$ rounds off x to the **nearest** integer, while $g(x) = \dfrac{\lfloor 10x + 0.5 \rfloor}{10}$ rounds

off x to the nearest tenth (the first decimal place). What function will round off x to:

(a) the nearest hundredth (two decimal places)?

(b) the nearest thousandth (three decimal places)?

21. Modify the function in Example 6 to produce a "square wave" graph with a "long on, short off, long on, short off" pattern.

22. Many computer languages contain a "signum" or "sign" function defined by

$$sgn(x) = \begin{cases} 1 & \text{if } x > 0 \\ 0 & \text{if } x = 0 \\ -1 & \text{if } x < 0 \end{cases}$$

(a) Graph $sgn(x)$.

(b) Graph $sgn(x - 2)$.

(c) Graph $sgn(x - 4)$.

(d) Graph $sgn(x - 2) \cdot sgn(x - 4)$.

(e) Graph $1 - sgn(x - 2) \cdot sgn(x - 4)$.

(f) For real numbers a and b with $a < b$, describe the graph of $1 - sgn(x - a) \cdot sgn(x - b)$.

23. Define $g(x)$ to be the **slope** of the line tangent to the graph of $y = f(x)$ (shown below) at (x, y).

(a) Estimate $g(1)$, $g(2)$, $g(3)$ and $g(4)$.

(b) Graph $y = g(x)$ for $0 \leq x \leq 4$.

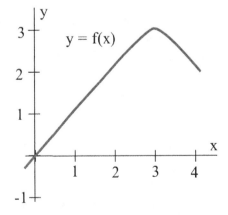

24. Define $h(x)$ to be the **slope** of the line tangent to the graph of $y = f(x)$ (see figure below) at (x, y).

    (a) Estimate $h(1)$, $h(2)$, $h(3)$ and $h(4)$.

    (b) Graph $y = h(x)$ for $0 \le x \le 4$.

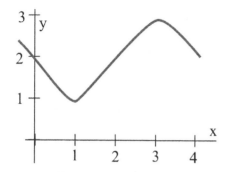

25. Using the **cos** (cosine) button on your calculator several times produces iterates of $f(x) = \cos(x)$. What number will the iterates approach if you use the **cos** button 20 or 30 times starting with

    (a) $x = 1$?

    (b) $x = 2$?

    (c) $x = 10$?

    (Be sure your calculator is in radian mode.)

26. Let $f(x) = 1 + \sin(x)$.

    (a) What happens if you start with $x = 1$ and repeatedly feed the output from f back into f?

    (b) What happens if you start with $x = 2$ and examine the iterates of f?

    (Be sure your calculator is in radian mode.)

27. Starting with $x = 1$, do the iterates of $f(x) = \frac{x^2 + 1}{2x}$ approach a number? What happens if you start with $x = 0.5$ or $x = 4$?

28. Let $f(x) = \frac{x}{2} + 3$.

    (a) What are the iterates of f if you start with $x = 2$? $x = 4$? $x = 6$?

    (b) Find a number c so that $f(c) = c$. This value of c is called a **fixed point** of f.

    (c) Find a fixed point of $g(x) = \frac{x}{2} + A$.

29. Let $f(x) = \frac{x}{3} + 4$.

    (a) What are the iterates of f if you start with $x = 2$? $x = 4$? $x = 6$?

    (b) Find a number c so that $f(c) = c$.

    (c) Find a fixed point of $g(x) = \frac{x}{3} + A$.

30. Some iterative procedures are geometric rather than numerical. Start with an equilateral triangle with sides of length 1, as shown at left in the figure below.

    • Remove the middle third of each line segment.

    • Replace the removed portion with two segments with the same length as the removed segment.

    The first two iterations of this procedure are shown at center and right in the figure below. Repeat these steps several more times, each time removing the middle third of each line segment and replacing it with two new segments. What happens to the length of the shape with each iteration? (The result of iterating over and over with this procedure is called Koch's Snowflake, named for Helga von Koch.)

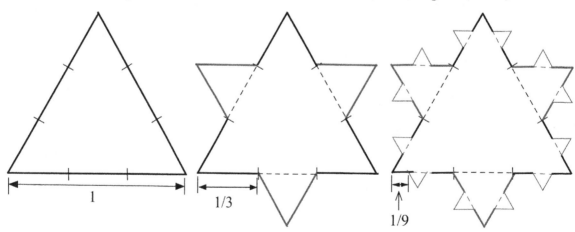

31. Sketch the graph of

$$p(x) = \begin{cases} 3-x & \text{if } x \text{ is a rational number} \\ 1 & \text{if } x \text{ is an irrational number} \end{cases}$$

32. Sketch the graph of

$$q(x) = \begin{cases} x^2 & \text{if } x \text{ is a rational number} \\ x+11 & \text{if } x \text{ is an irrational number} \end{cases}$$

0.4 Practice Answers

1. $C(x)$ is the cost for one night on date x:

$$C(x) = \begin{cases} \$380 & \text{if } x \text{ is between December 15 and April 30} \\ \$295 & \text{if } x \text{ is any other date} \end{cases}$$

2. For the graph, see the figure in the margin.

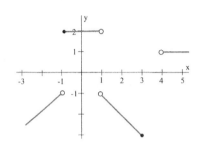

x	$g(x)$	x	$g(x)$
-3	-3	$\frac{\pi}{3}$	$-\frac{\pi}{3}$
-1	2	2	-2
0	2	3	-3
$\frac{1}{2}$	2	4	undefined
1	undefined	5	1

3. Define $f(x)$ as:

$$f(x) = \begin{cases} 1 & \text{if } x \le -1 \\ 1-x & \text{if } -1 < x \le 1 \\ 2 & \text{if } 1 < x \end{cases}$$

4. $f \circ g(3) = f(2) = \frac{2}{-1} = -2; f \circ g(8) = f(3)$ is undefined; $g \circ f(4) = g(4) = 5; f \circ h(1) = f(2) = \frac{2}{-1} = -2; f \circ h(3) = f(2) = -2; f \circ h(2) = f(3)$ is undefined; $h \circ g(-1) = h(0) = 0; f \circ g(x) = f(\sqrt{1+x}) = \frac{1+x}{\sqrt{1+x}-3}, g \circ f(x) = g\left(\frac{x}{x-3}\right) = \sqrt{1 + \frac{x}{x-3}}$

5. See the figure below:

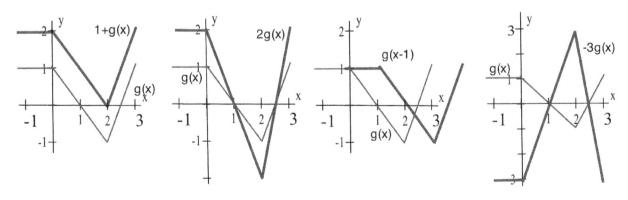

6. Using $f(x) = \dfrac{\frac{9}{x} + x}{2}$, $f(1) = \dfrac{\frac{9}{1}+1}{2} = 5$, $f(5) = \dfrac{\frac{9}{5}+5}{2} = 3.4$, $f(3.4) \approx$ 3.023529412 and $f(3.023529412) \approx 3.000091554$. The next iteration gives $f(3.000091554) \approx 3.000000001$: these values are approaching 3, the square root of 9.

With $A = 6$, $f(x) = \dfrac{\frac{6}{x} + x}{2}$, so $f(1) = \dfrac{\frac{6}{1}+1}{2} = 3.5$, $f(3.5) = \dfrac{\frac{6}{3.5}+3.5}{2} = 2.607142857$, and the next iteration gives $f(2.607142857) \approx$ 2.45425636. Then $f(2.45425636) \approx 2.449494372$, $f(2.449494372) \approx$ 2.449489743 and $f(2.449489743) \approx 2.449489743$ (the output is the same as the input to 9 decimal places): these values are approaching 2.449489743, an approximation of $\sqrt{6}$.

For any positive value A, the iterates of $f(x) = \dfrac{\frac{A}{x} + x}{2}$ (starting with any positive x) will approach $\sqrt{A}$.

7. The figure below shows some intermediate steps and final graphs:

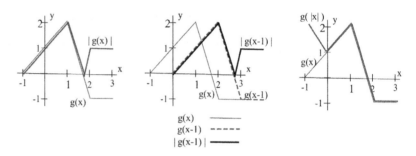

8. The figure in the margin shows the graph of $y = x^2$ and the (thicker) graph of $y = \lfloor x^2 \rfloor$.

9. The figure below shows the "holey" graph of $y = x$ with a hole at each rational value of x and the "'holey" graph of $y = \sin(x)$ with a hole at each irrational value of x. Together they form the graph of $r(x)$.

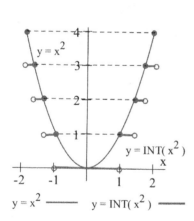

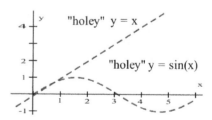

(This is a very crude image, since we can't really see the individual holes, which have zero width.)

0.5 Mathematical Language

The calculus concepts we will explore in this book are simple and powerful, but sometimes subtle. To succeed in calculus you will need to master some techniques, but (more importantly) you will need to understand ideas and be able to work with the ideas in words and pictures — very clear words and pictures.

You also need to understand some of the common linguistic constructions used in mathematics. In this section, we will discuss a few of the most common mathematical phrases, the meanings of these phrases, and some of their equivalent forms.

Your calculus instructor is going to use these types of statements, and it is very important that you understand exactly what your instructor means. You have reached a level in mathematics where the precise use of language is important.

Equivalent Statements

Two statements are **equivalent** if they always have the same logical value (a logical value is either "true" or "false"): that is, they are both true or are both false.

The statements "$x = 3$" and "$x + 2 = 5$" are equivalent statements, because if one of them is true then so is the other — and if one of them is false then so is the other.

The statements "$x = 3$" and "$x^2 - 4x + 3 = 0$" are not equivalent, because $x = 1$ makes the second statement true but the first one false.

AND and OR

In everyday language, we use the words "and" and "or" all the time, but in mathematics we must use them very carefully.

> The compound statement "A **and** B are true" is equivalent to "both of A and B are true."

If A or if B or if both are false, then the statement "A and B are true" is false. The statement "$x^2 = 4$ and $x > 0$" is true when $x = 2$ and is false for every other value of x.

> The compound statement "A **or** B is true" is equivalent to "at least one of A or B is true."

If both A and B are false, then the statement "A or B is true" is false.

The statement "$x^2 = 4$ or $x > 0$" is true if $x = -2$ or x is any positive number. The statement is false when $x = -3$ (and for lots of other values of x).

Practice 1. Which values of x make each statement true?

(a) $x < 5$

(b) $x + 2 = 6$

(c) $x^2 - 10x + 24 = 0$

(d) "(a) and (b)"

(e) "(a) or (c)"

Negation of a Statement

For some simple statements, we can construct the **negation** just by adding the word "not."

statement	negation of the statement
x is equal to 3 ($x = 3$)	x is **not** equal to 3 ($x \neq 3$)
x is less than 5 ($x < 5$)	x is **not** less than 5 ($x \not< 5$)
	x is greater than or equal to 5 ($x \geq 5$)

When the statement contains words such as "all," "no" or "some," then the negation becomes more complicated.

statement	negation of the statement
All x satisfy A.	At least one x does not satisfy A.
Every x satisfies A.	There is an x that does not satisfy A.
	Some x does not satisfy A.
No x satisfies A.	At least one x satisfies A.
Every x does not satisfy A.	Some x satisfies A.
There is an x that satisfies A.	No x satisfies A.
At least one x satisfies A.	Every x does not satisfy A.
Some x satisfies A.	

We can also negate compound statements containing "and" and "or."

statement	negation of the statement
A and B are both true.	At least one of A or B is not true.
A and B and C are all true.	At least one of A or B or C is not true.
A or B is true.	Both A and B are not true.

Practice 2. Write the negation of each statement.

(a) $x + 5 = 3$

(b) All prime numbers are odd.

(c) $x^2 < 4$

(d) x divides 2 and x divides 3.

(e) No mathematician can sing well.

If... Then...: A Very Common Structure in Mathematics

The most common and basic structure used in mathematical language is the "If {some hypothesis} then {some conclusion}" sentence. Almost every result in mathematics can be stated using one or more "If...then..." sentences.

> **"If** A **then** B" means that when the hypothesis A is true, the conclusion B must also be true.

When the hypothesis is false, the "If...then..." sentence makes no claim about the truth or falsity of the conclusion — the conclusion may be either true or false.

Even in everyday life you have probably encountered "If...then..." statements for a long time. A parent might try to encourage a child with a statement like "If you clean your room then I will buy you an ice cream cone."

To show that an "If...then..." statement is not valid (not true), all we need to do is find a **single** example where the hypothesis is true and the conclusion is false. Such an example with a true hypothesis and false conclusion is called a **counterexample** for the "If...then..." statement. A valid "If...then..." statement has no counterexample.

> A **counterexample** to the statement "If A then B" is an example in which A is true and B is false.

The only way for the statement "If you clean your room then I will buy you an ice cream cone" to be false is if the child cleaned the room and the parent did not buy the ice cream cone. If the child did not clean the room but the parent bought the ice cream cone anyway, we would say that the statement was true.

The statement "If n is a positive integer, then $n^2 + 5n + 5$ is a prime number" has hypothesis "n is a positive integer" and conclusion "$n^2 + 5n + 5$ is a prime number." This "If...then..." statement is false, because replacing n with the number 5 will make the hypothesis true and the conclusion false. The number 5 is a counterexample for the statement.

Every invalid "If... then..." statement has at least one counterexample, and the most convincing way to show that a statement is not valid is to find a counterexample to the statement.

Several other language structures are equivalent to the "If... then..." form. The statements below all mean the same as "If {A} then {B}":

- "All A are B."
- "Every A is B."
- "Each A is B."
- "Whenever A, then B."
- "B whenever A."
- "A only if B."
- "A implies B."
- "A $\Rightarrow$ B" (the symbol "$\Rightarrow$" means "implies")

Practice 3. Restate "If {a shape is a square} then {the shape is a rectangle}" as many ways as you can.

"If... then..." statements occur hundreds of times in every mathematics book, including this one. It is important that you are able to recognize the various forms of "If... then..." statements and that you are able to distinguish the hypotheses from the conclusions.

Contrapositive Form of an "If... Then..." Statement

The statement "If A then B" means that if the hypothesis A is true, then the conclusion B is guaranteed to be true.

Suppose we know that in a certain town the statement "If {a building is a church} then {the building is green}" is a true statement. What can we validly conclude about a red building? Based on the information we have, we can validly conclude that the red building is "not a church." because every church is green. We can also conclude that a blue building is not a church. In fact, we can conclude that every "not green" building is "not a church." That is, if the conclusion of a valid "If... then..." statement is false, then the hypothesis must also be false.

> The **contrapositive** form of "If A then B" is
> "If {negation of B} then {negation of A}"
> or "If {B is false} then {A is false}."

> The statement "If A then B" and its contrapositive
> "If {not B} then {not A}" are equivalent.

What about a green building in the aforementioned town? The green building may or may not be a church—perhaps every post office is also painted green. Or perhaps every building in town is green, in which case the statement "If {a building is a church} then {the building is green}" is certainly true.

Practice 4. Write the contrapositive form of each statement.

(a) If a function is differentiable then it is continuous.

(b) All men are mortal.

(c) If $x = 3$ then $x^2 - 5x + 6 = 0$

(d) If {2 divides x and 3 divides x} then {6 divides x}.

Converse of an "If...then..." Statement

If we switch the hypothesis and the conclusion of an "If A then B" statement, we get the **converse** "If B then A." For example, the converse of "If {a building is a church} then {the building is green}" is "If {a building is green} then {the building is a church}."

The converse of an "If...then..." statement is **not equivalent** to the original "If...then..." statement. For example, the statement "If $x = 2$ then $x^2 = 4$" is true, but the converse statement "If $x^2 = 4$ then $x = 2$" is not true because $x = -2$ makes the hypothesis of the converse true and the conclusion false

> The **converse** of "If A then B" is "If B then A."

> The statement "If A then B" and its converse "If B then A."
> are **not equivalent**.

Wrap-up

The precise use of language by mathematicians (and mathematics books) is an attempt to clearly communicate ideas from one person to another, but this requires that both people understand the use and rules of the language. If you don't understand this usage, the communication of the ideas will almost certainly fail.

0.5 Problems

In Problems 1–2, define the sets A, B and C as $A = \{1,2,3,4,5\}$, $B = \{0,2,4,6\}$ and $C = \{-2,-1,0,1,2,3\}$. List all values of x that satisfy each statement.

1. (a) x is in A **and** x is in B
   (b) x is in A **or** x is in C
   (c) x is not in B **and** x is in C

2. (a) x is not in B or C
   (b) x is in B and C but not in A
   (c) x is not in A but is in B or C

In Problems 3–5, list or describe all the values of x that make each statement true.

3. (a) $x^2 + 3 > 1$
   (b) $x^3 + 3 > 1$
   (c) $\lfloor x \rfloor \le |x|$

4. (a) $\frac{x^2+3x}{x} = x + 3$
   (b) $x > 4$ and $x < 9$
   (c) $|x| = 3$ and $x < 0$

5. (a) $x + 5 = 3$ or $x^2 = 9$
   (b) $x + 5 = 3$ and $x^2 = 9$
   (c) $|x + 3| = |x| + 3$

In Problems 6–8, write the **contrapositive** of each statement. If false, give a **counterexample**.

6. (a) If $x > 3$ then $x^2 > 9$.
   (b) Every solution of $x^2 - 6x + 8 = 0$ is even.

7. (a) If $x^2 + x - 6 = 0$ then $x = 2$ or $x = -3$.
   (b) All triangles have 3 sides.

8. (a) Every polynomial has at least one zero.
   (b) If I exercise and eat right then I will be healthy.

In Problems 9–11, write the **contrapositive** of each statement.

9. (a) If your car is properly tuned, it will get at least 24 miles per gallon.
   (b) You can have pie if you eat your vegetables.

10. (a) A well-prepared student will miss less than 15 points on the final exam.
    (b) I feel good when I jog.

11. (a) If you love your country, you will vote for me.
    (b) If guns are outlawed then only outlaws will have guns.

In 12–15, write the **negation** of each statement.

12. (a) It is raining.
    (b) Some equations have solutions.
    (c) $f(x)$ and $g(x)$ are polynomials.

13. (a) $f(x)$ or $g(x)$ is positive.
    (b) x is positive.
    (c) 8 is a prime number.

14. (a) Some months have six Mondays.
    (b) All quadratic equations have solutions.
    (c) The absolute value of a number is positive.

15. (a) For all numbers a and b, $|a + b| = |a| + |b|$.
    (b) All snakes are poisonous.
    (c) No dog can climb trees.

16. Write an "If...then..." statement that is true but whose converse is false.

17. Write an "If...then..." statement that is true and whose converse is true.

18. Write an "If...then..." statement that is false and whose converse is false.

In 19–22, determine whether each statement is true or false. If false, give a counterexample.

19. (a) If a and b are real numbers then:
$$(a + b)^2 = a^2 + b^2$$
    (b) If $a > b$ then $a^2 > b^2$.
    (c) If $a > b$ then $a^3 > b^3$.

20. (a) For all real numbers a and b, $|a + b| = |a| + |b|$
    (b) For all real numbers a and b, $\lfloor a \rfloor + \lfloor b \rfloor = \lfloor a + b \rfloor$.
    (c) If $f(x)$ and $g(x)$ are linear functions, then $f(g(x))$ is a linear function.

21. (a) If $f(x)$ and $g(x)$ are linear functions then $f(x) + g(x)$ is a linear function.
    (b) If $f(x)$ and $g(x)$ are linear functions then $f(x) \cdot g(x)$ is a linear function.
    (c) If x divides 6 then x divides 30.

22. (a) If x divides 50 then x divides 10.

    (b) If x divides yz then x divides y or z.

    (c) If x divides a^2 then x divides a.

In 23–26, rewrite each statement in the form of an "If... then..." statement and determine whether it is true or false. If the statement is false, give a counterexample.

23. (a) The sum of two prime numbers is a prime.

    (b) The sum of two prime numbers is never a prime number.

    (c) Every prime number is odd.

    (d) Every prime number is even.

24. (a) Every square has 4 sides.

    (b) All 4-sided polygons are squares.

    (c) Every triangle has 2 equal sides.

    (d) Every 4-sided polygon with equal sides is a square.

25. (a) Every solution of $x + 5 = 9$ is odd.

    (b) Every 3-sided polygon with equal sides is a triangle.

    (c) Every calculus student studies hard.

    (d) All (real) solutions of $x^2 - 5x + 6 = 0$ are even.

26. (a) Every line in the plane intersects the x-axis.

    (b) Every (real) solution of $x^2 + 3 = 0$ is even.

    (c) All birds can fly.

    (d) No mammal can fly.

0.5 Practice Answers

1. (a) All values of x less than 5. (b) $x = 4$ (c) Both $x = 4$ and $x = 6$. (d) $x = 4$ (e) $x = 6$ and all x less than 5.

2. (a) $x + 5 < 3$ (b) At least one prime number is even. (There is an even prime number.) (c) $x^2 = 4$ (d) x does not divide 2 or x does not divide 3. (e) At least one mathematician can sing well. (There is a mathematician who can sing well.)

3. Here are several ways:

- All squares are rectangles.
- Every square is a rectangle.
- Each square is a rectangle.
- Whenever a shape is a square, then it is a rectangle.
- A shape is a rectangle whenever it is a square.
- A shape is a square only if it is a rectangle.
- A shape is a square implies that it is a rectangle.
- Being a square implies being a rectangle.

4. (a) If a function is not continuous then it is not differentiable.

    (b) All immortals are not men.

    (c) $x^2 - 5x + 6 \neq 0 \Rightarrow x \neq 3$

    (d) If {6 does not divide x} then {2 does not divide x or 3 does not divide x}.

1

Limits and Continuity

1.0 Tangent Lines, Velocities, Growth

In Section 0.2, we estimated the slope of a line tangent to the graph of a function *at a point*. At the end of Section 0.3, we constructed a new function that gave the slope of the line tangent to the graph of a given function *at each point*. In both cases, before we could calculate a slope, we had to **estimate** the tangent line from the graph of the given function, a method that required an accurate graph and good estimating. In this section we will begin to look at a more precise method of finding the slope of a tangent line that does not require a graph or any estimation by us. We will start with a non-applied problem and then look at two applications of the same idea.

The Slope of a Line Tangent to a Function at a Point

Our goal is to find a way of exactly determining the slope of the line that is tangent to a function (to the graph of the function) at a point in a way that does not require us to actually have the graph of the function.

Let's start with the problem of finding the slope of the line L (see margin figure), which is tangent to $f(x) = x^2$ at the point $(2, 4)$. We could estimate the slope of L from the graph, but we won't. Instead, we can see that the line through $(2, 4)$ and $(3, 9)$ on the graph of f is an approximation of the slope of the tangent line, and we can calculate that slope exactly:

$$m = \frac{\Delta y}{\Delta x} = \frac{9 - 4}{3 - 2} = 5$$

But $m = 5$ is only an *estimate* of the slope of the tangent line — and not a very good estimate. It's too big. We can get a better estimate by picking a second point on the graph of f closer to $(2, 4)$ — the point $(2, 4)$ is fixed and it must be one of the two points we use. From the figure in the margin, we can see that the slope of the line through the points $(2, 4)$ and $(2.5, 6.25)$ is a better approximation of the slope of the

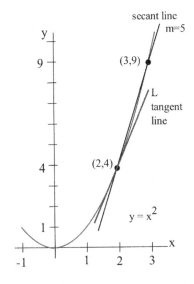

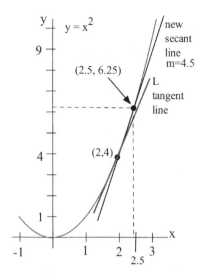

tangent line at $(2, 4)$:

$$m = \frac{\Delta y}{\Delta x} = \frac{6.25 - 4}{2.5 - 2} = \frac{2.25}{0.5} = 4.5$$

This is a better estimate, but still an approximation.

We can continue picking points closer and closer to $(2, 4)$ on the graph of f, and then calculating the slopes of the lines through each of these points (x, y) and the point $(2, 4)$:

points to the left of $(2,4)$			points to the right of $(2,4)$		
x	$y = x^2$	slope	x	$y = x^2$	slope
1.5	2.25	3.5	3	9	5
1.9	3.61	3.9	2.5	6.25	4.5
1.99	3.9601	3.99	2.01	4.0401	4.01

The only thing special about the x-values we picked is that they are numbers close — and very close — to $x = 2$. Someone else might have picked other nearby values for x. As the points we pick get closer and closer to the point $(2, 4)$ on the graph of $y = x^2$, the slopes of the lines through the points and $(2, 4)$ are better approximations of the slope of the tangent line, and these slopes are getting closer and closer to 4.

Practice 1. What is the slope of the line through $(2, 4)$ and (x, y) for $y = x^2$ and $x = 1.994$? For $x = 2.0003$?

We can bypass much of the calculating by not picking the points one at a time: let's look at a general point near $(2, 4)$. Define $x = 2 + h$ so h is the increment from 2 to x (see margin figure). If h is small, then $x = 2 + h$ is close to 2 and the point $(2 + h, f(2 + h)) = (2 + h, (2 + h)^2)$ is close to $(2, 4)$. The slope m of the line through the points $(2, 4)$ and $(2 + h, (2 + h)^2)$ is a good approximation of the slope of the tangent line at the point $(2, 4)$:

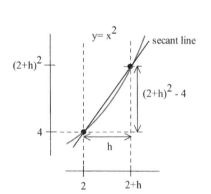

$$m = \frac{\Delta y}{\Delta x} = \frac{(2 + h)^2 - 4}{(2 + h) - 2} = \frac{(4 + 4h + h^2) - 4}{h}$$
$$= \frac{4h + h^2}{h} = \frac{h(4 + h)}{h} = 4 + h$$

If h is very small, then $m = 4 + h$ is a very good approximation to the slope of the tangent line, and $m = 4 + h$ also happens to be very close to the value 4. The value $m = 4 + h$ is called the slope of the **secant line** through the two points $(2, 4)$ and $(2 + h, (2 + h)^2)$. The limiting value 4 of $m = 4 + h$ as h gets smaller and smaller is called the slope of the **tangent line** to the graph of f at $(2, 4)$.

Example 1. Find the slope of the line tangent to $f(x) = x^2$ at the point $(1, 1)$ by evaluating the slope of the secant line through $(1, 1)$ and $(1 + h, f(1 + h))$ and then determining what happens as h gets very small (see margin figure).

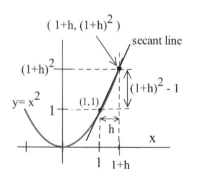

Solution. The slope of the secant line through the points $(1,1)$ and $(1+h, f(1+h))$ is:

$$m = \frac{f(1+h) - 1}{(1+h) - 1} = \frac{(1+h)^2 - 1}{h} = \frac{(1 + 2h + h^2) - 1}{h}$$
$$= \frac{2h + h^2}{h} = \frac{h(2+h)}{h} = 2 + h$$

As h gets very small, the value of m approaches the value 2, the slope of tangent line at the point $(1,1)$. ◄

Practice 2. Find the slope of the line tangent to the graph of $y = f(x) = x^2$ at the point $(-1, 1)$ by finding the slope of the secant line, m_{sec}, through the points $(-1,1)$ and $(-1 + h, f(-1+h))$ and then determining what happens to m_{sec} as h gets very small.

Falling Tomato

Suppose we drop a tomato from the top of a 100-foot building (see margin figure) and record its position at various times during its fall:

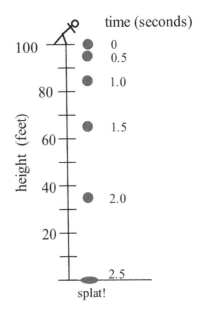

time (sec)	height (ft)
0.0	100
0.5	96
1.0	84
1.5	64
2.0	36
2.5	0

Some questions are easy to answer directly from the table:

(a) How long did it take for the tomato to drop 100 feet?
   (2.5 seconds)

(b) How far did the tomato fall during the first second?
   ($100 - 84 = 16$ feet)

(c) How far did the tomato fall during the last second?
   ($64 - 0 = 64$ feet)

(d) How far did the tomato fall between $t = 0.5$ and $t = 1$?
   ($96 - 84 = 12$ feet)

Other questions require a little calculation:

(e) What was the average velocity of the tomato during its fall?

$$\text{average velocity} = \frac{\text{distance fallen}}{\text{total time}} = \frac{\Delta\text{position}}{\Delta\text{time}} = \frac{-100 \text{ ft}}{2.5 \text{ s}} = -40 \,\frac{\text{ft}}{\text{sec}}$$

(f) What was the average velocity between $t = 1$ and $t = 2$ seconds?

$$\text{average velocity} = \frac{\Delta\text{position}}{\Delta\text{time}} = \frac{36 \text{ ft} - 84 \text{ ft}}{2 \text{ s} - 1 \text{ s}} = \frac{-48 \text{ ft}}{1 \text{ s}} = -48 \,\frac{\text{ft}}{\text{sec}}$$

Some questions are more difficult.

(g) How fast was the tomato falling 1 second after it was dropped?

This question is significantly different from the previous two questions about average velocity. Here we want the **instantaneous velocity**, the velocity at an instant in time. Unfortunately, the tomato is not equipped with a speedometer, so we will have to give an approximate answer.

One crude approximation of the instantaneous velocity after 1 second is simply the average velocity during the entire fall, $-40\frac{ft}{sec}$. But the tomato fell slowly at the beginning and rapidly near the end, so this estimate may or may not be a good answer.

We can get a better approximation of the instantaneous velocity at $t = 1$ by calculating the average velocities over a short time interval near $t = 1$. The average velocity between $t = 0.5$ and $t = 1$ is:

$$\frac{-12 \text{ feet}}{0.5 \text{ sec}} = -24 \frac{ft}{sec}$$

and the average velocity between $t = 1$ and $t = 1.5$ is

$$\frac{-20 \text{ feet}}{0.5 \text{ sec}} = -40 \frac{ft}{sec}$$

so we can be reasonably sure that the instantaneous velocity is between $-24 \frac{ft}{sec}$ and $-40 \frac{ft}{sec}$.

In general, the shorter the time interval over which we calculate the average velocity, the better the average velocity will approximate the instantaneous velocity. The average velocity over a time interval is:

$$\frac{\Delta \text{position}}{\Delta \text{time}}$$

which is the slope of the secant line through two points on the graph of height versus time (see margin figure).

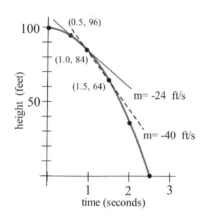

$$\text{average velocity} = \frac{\Delta \text{position}}{\Delta \text{time}}$$
$$= \text{slope of the secant line through two points}$$

The instantaneous velocity at a particular time and height is the slope of the tangent line to the graph at the point given by that time and height.

instantaneous velocity = slope of the line tangent to the graph

Practice 3. Estimate the instantaneous velocity of the tomato 2 seconds after it was dropped.

Growing Bacteria

Suppose we set up a machine to count the number of bacteria growing on a Petri plate (see margin figure). At first there are few bacteria, so the population grows slowly. Then there are more bacteria to divide, so the population grows more quickly. Later, there are more bacteria and less room and nutrients available for the expanding population, so the population grows slowly again. Finally, the bacteria have used up most of the nutrients and the population declines as bacteria die.

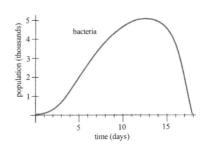

The population graph can be used to answer a number of questions:

(a) What is the bacteria population at time $t = 3$ days?
(about 500 bacteria)

(b) What is the population increment from $t = 3$ to $t = 10$ days?
(about 4,000 bacteria)

(c) What is the **rate** of population growth from $t = 3$ to $t = 10$ days?

To answer this last question, we compute the average change in population during that time:

$$\text{average change in population} = \frac{\text{change in population}}{\text{change in time}}$$
$$= \frac{\Delta \text{population}}{\Delta \text{time}} = \frac{4000 \text{ bacteria}}{7 \text{ days}} \approx 570 \frac{\text{bacteria}}{\text{day}}$$

This is the slope of the secant line through $(3, 500)$ and $(10, 4500)$.

> $$\text{average population growth rate} = \frac{\Delta \text{population}}{\Delta \text{time}}$$
> $$= \text{slope of the secant line through two points}$$

Now for a more difficult question:

(d) What is the rate of population growth on the third day, at $t = 3$?

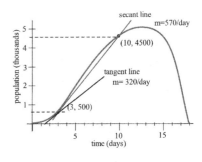

This question asks for the instantaneous rate of population change, the slope of the line tangent to the population curve at $(3, 500)$. If we sketch a line approximately tangent to the curve at $(3, 500)$ and pick two points near the ends of the tangent line segment (see margin figure), we can estimate that the instantaneous rate of population growth is approximately $320 \frac{\text{bacteria}}{\text{day}}$.

> $$\text{instantaneous population growth rate} =$$
> $$\text{slope of the line tangent to the graph}$$

Practice 4. Find approximate values for:

(a) the average change in population between $t = 9$ and $t = 13$.

(b) the rate of population growth at $t = 9$ days.

The tangent line problem, the instantaneous velocity problem and the instantaneous growth rate problem are all similar. In each problem we wanted to know how rapidly something was **changing at an instant in time**, and each problem turned out to involve finding the **slope of a tangent line**. The approach in each problem was also the same: find an approximate solution and then examine what happens to the approximate solution over shorter and shorter intervals. We will often use this approach of finding a limiting value, but before we can use it effectively we need to describe the concept of a limit with more precision.

1.0 Problems

1. (a) What is the slope of the line through $(3, 9)$ and (x, y) for $y = x^2$ when:

   i. $x = 2.97$?

   ii. $x = 3.001$?

   iii. $x = 3 + h$?

   (b) What happens to this last slope when h is very small (close to 0)?

   (c) Sketch the graph of $y = x^2$ for x near 3.

2. (a) What is the slope of the line through $(-2, 4)$ and (x, y) for $y = x^2$ when:

   i. $x = -1.98$?

   ii. $x = -2.03$?

   iii. $x = -2 + h$?

   (b) What happens to this last slope when h is very small (close to 0)?

   (c) Sketch the graph of $y = x^2$ for x near -2.

3. (a) What is the slope of the line through $(2, 4)$ and (x, y) for $y = x^2 + x - 2$ when:

   i. $x = 1.99$?

   ii. $x = 2.004$?

   iii. $x = 2 + h$?

   (b) What happens to this last slope when h is very small (close to 0)?

   (c) Sketch the graph of $y = x^2 + x - 2$ for x near 2.

4. (a) What is the slope of the line through $(-1, -2)$ and (x, y) for $y = x^2 + x - 2$ when:

   i. $x = -0.98$?

   ii. $x = -1.03$?

   iii. $x = -1 + h$?

   (b) What happens to this last slope when h is very small (close to 0)?

   (c) Sketch the graph of $y = x^2 + x - 2$ for x near -1.

5. The figure below shows the temperature during a day in Ames.

   (a) What was the average change in temperature from 9 a.m. to 1 p.m.?

   (b) Estimate how fast the temperature was rising at **10 a.m.** and **at 7 p.m.**

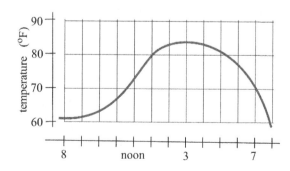

6. The figure below shows the distance of a car from a measuring position located on the edge of a straight road.

   (a) What was the average velocity of the car from $t = 0$ to $t = 30$ seconds?

   (b) What was the average velocity from $t = 10$ to $t = 30$ seconds?

   (c) About how fast was the car traveling **at** $t = 10$ seconds? **At** $t = 20$? **At** $t = 30$?

   (d) What does the horizontal part of the graph between $t = 15$ and $t = 20$ seconds tell you?

   (e) What does the negative velocity at $t = 25$ represent?

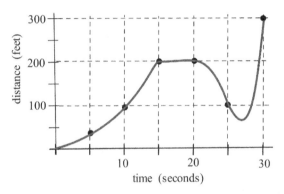

7. The figure below shows the distance of a car from a measuring position located on the edge of a straight road.

   (a) What was the average velocity of the car from $t = 0$ to $t = 20$ seconds?

   (b) What was the average velocity from $t = 10$ to $t = 30$ seconds?

   (c) About how fast was the car traveling **at** $t = 10$ seconds? **At** $t = 20$? **At** $t = 30$?

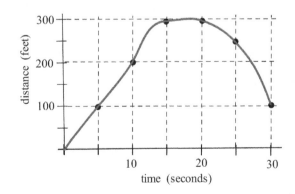

8. The figure below shows the composite developmental skill level of chessmasters at different ages as determined by their performance against other chessmasters. (From "Rating Systems for Human Abilities," by W.H. Batchelder and R.S. Simpson, 1988. UMAP Module 698.)

   (a) At what age is the "typical" chessmaster playing the best chess?

   (b) At approximately what age is the chessmaster's skill level increasing most rapidly?

   (c) Describe the development of the "typical" chessmaster's skill in words.

   (d) Sketch graphs that you **think** would reasonably describe the performance levels versus age for an athlete, a classical pianist, a rock singer, a mathematician and a professional in your major field.

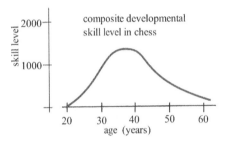

9. Define $A(x)$ to be the **area** bounded by the t- (horizontal) and y-axes, the horizontal line $y = 3$, and the vertical line at x (see figure below). For example, $A(4) = 12$ is the area of the 4×3 rectangle.

   (a) Evaluate $A(0)$, $A(1)$, $A(2)$, $A(2.5)$ and $A(3)$.

   (b) What area would $A(4) - A(1)$ represent?

   (c) Graph $y = A(x)$ for $0 \leq x \leq 4$.

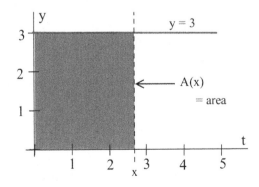

10. Define $A(x)$ to be the **area** bounded by the t-(horizontal) and y-axes, the line $y = t + 1$, and the vertical line at x (see figure). For example, $A(4) = 12$.

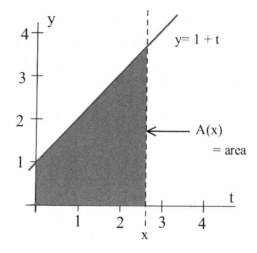

(a) Evaluate $A(0)$, $A(1)$, $A(2)$, $A(2.5)$ and $A(3)$.

(b) What area would $A(3) - A(1)$ represent in the figure?

(c) Graph $y = A(x)$ for $0 \le x \le 4$.

1.0 Practice Answers

1. If $x = 1.994$, then $y = 3.976036$, so the slope between $(2,4)$ and (x,y) is:

$$\frac{4-y}{2-x} = \frac{4 - 3.976036}{2 - 1.994} = \frac{0.023964}{0.006} \approx 3.994$$

If $x = 2.0003$, then $y \approx 4.0012$, so the slope between $(2,4)$ and (x,y) is:

$$\frac{4-y}{2-x} = \frac{4 - 4.0012}{2 - 2.0003} = \frac{-0.0012}{-0.0003} \approx 4.0003$$

2. Computing m_{sec}:

$$\frac{f(-1+h) - (1)}{(-1+h) - (-1)} = \frac{(-1+h)^2 - 1}{h} = \frac{1 - 2h + h^2 - 1}{h} = \frac{h(-2+h)}{h} = -2 + h$$

As $h \to 0$, $m_{sec} = -2 + h \to -2$.

3. The average velocity between $t = 1.5$ and $t = 2.0$ is:

$$\frac{36 - 64 \text{ feet}}{2.0 - 1.5 \text{ sec}} = -56 \frac{\text{feet}}{\text{sec}}$$

The average velocity between $t = 2.0$ and $t = 2.5$ is:

$$\frac{0 - 36 \text{ feet}}{2.5 - 2.0 \text{ sec}} = -72 \frac{\text{feet}}{\text{sec}}$$

The velocity at $t = 2.0$ is somewhere between $-56 \frac{\text{feet}}{\text{sec}}$ and $-72 \frac{\text{feet}}{\text{sec}}$, probably around the middle of this interval:

$$\frac{(-56) + (-72)}{2} = -64 \frac{\text{feet}}{\text{sec}}$$

4. (a) When $t = 9$ days, the population is approximately $P = 4,200$ bacteria. When $t = 13$, $P \approx 5,000$. The average change in population is approximately:

$$\frac{5000 - 4200 \text{ bacteria}}{13 - 9 \text{ days}} = \frac{800 \text{ bacteria}}{4 \text{ days}} = 200 \frac{\text{bacteria}}{\text{day}}$$

   (b) To find the rate of population growth at $t = 9$ days, sketch the line tangent to the population curve at the point $(9, 4200)$ and then use $(9, 4200)$ and another point on the tangent line to calculate the slope of the line. Using the approximate values $(5, 2800)$ and $(9, 4200)$, the slope of the tangent line at the point $(9, 4200)$ is approximately:

$$\frac{4200 - 2800 \text{ bacteria}}{9 - 5 \text{ days}} = \frac{1400 \text{ bacteria}}{4 \text{ days}} \approx 350 \frac{\text{bacteria}}{\text{day}}$$

1.1 The Limit of a Function

Calculus has been called the study of continuous change, and the **limit** is the basic concept that allows us to describe and analyze such change. An understanding of limits is necessary to understand derivatives, integrals and other fundamental topics of calculus.

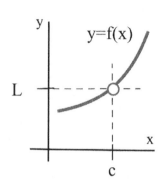

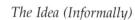

L

c

The symbol → means "approaches" or "gets very close to."

The Idea (Informally)

The limit of a function at a point describes the behavior of the function when the input variable is near—**but does not equal**—a specified number (see margin figure). If the values of $f(x)$ get closer and closer— as close as we want—to one number L as we take values of x very close to (but not equal to) a number c, then

> we say: "the limit of $f(x)$, as x approaches c, is L"
>
> and we write: $\lim\limits_{x \to c} f(x) = L$

It is very important to note that:

> $f(c)$ is a single number that describes the behavior (value) of f **at** the point $x = c$

while:

> $\lim\limits_{x \to c} f(x)$ is a single number that describes the behavior of f **near, but not at** the point $x = c$

If we have a graph of the function $f(x)$ near $x = c$, then it is usually easy to determine $\lim\limits_{x \to c} f(x)$.

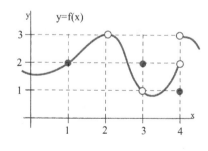

Example 1. Use the graph of $y = f(x)$ given in the margin to determine the following limits:

(a) $\lim\limits_{x \to 1} f(x)$ (b) $\lim\limits_{x \to 2} f(x)$ (c) $\lim\limits_{x \to 3} f(x)$ (d) $\lim\limits_{x \to 4} f(x)$

Solution. Each of these limits involves a different issue, as you may be able to tell from the graph.

(a) $\lim\limits_{x \to 1} f(x) = 2$: When x is very close to 1, the values of $f(x)$ are very close to $y = 2$. In this example, it happens that $f(1) = 2$, but that is irrelevant for the limit. The only thing that matters is what happens for x close to 1 but with $x \neq 1$.

(b) $f(2)$ is undefined, but we only care about the behavior of $f(x)$ for x *close to* 2 and not equal to 2. When x is close to 2, the values of $f(x)$ are close to 3. If we restrict x close enough to 2, the values of y will be as close to 3 as we want, so $\lim\limits_{x \to 2} f(x) = 3$.

(c) When x is close to 3, the values of $f(x)$ are close to 1, so $\lim\limits_{x \to 3} f(x) = 1$. For this limit it is completely irrelevant that $f(3) = 2$: we only care about what happens to $f(x)$ for x close to and not equal to 3.

(d) This one is harder and we need to be careful. When x is close to 4 and slightly **less than** 4 (x is just to the left of 4 on the x-axis) then the values of $f(x)$ are close to 2. But if x is close to 4 and slightly **larger than** 4 then the values of $f(x)$ are close to 3.

If we know only that x is very close to 4, then we cannot say whether $y = f(x)$ will be close to 2 or close to 3 — it depends on whether x is on the right or the left side of 4. In this situation, the $f(x)$ values are not close to a single number so we say $\lim\limits_{x \to 4} f(x)$ **does not exist**.

In (d), it is irrelevant that $f(4) = 1$. The limit, as x approaches 4, would still be undefined if $f(4)$ was 3 or 2 or anything else. ◄

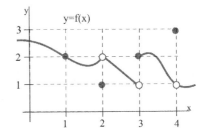

Practice 1. Use the graph of $y = f(x)$ in the margin to determine the following limits:

(a) $\lim\limits_{x \to 1} f(x)$ (b) $\lim\limits_{t \to 2} f(t)$ (c) $\lim\limits_{x \to 3} f(x)$ (d) $\lim\limits_{w \to 4} f(w)$

Example 2. Determine the value of $\lim\limits_{x \to 3} \dfrac{2x^2 - x - 1}{x - 1}$.

Solution. We need to investigate the values of $f(x) = \frac{2x^2 - x - 1}{x - 1}$ when x is close to 3. If the $f(x)$ values get arbitrarily close to — or even equal to — some number L, then L will be the limit.

One way to keep track of both the x and the $f(x)$ values is to set up a table and to pick several x values that get closer and closer (but not equal) to 3.

We can pick some values of x that approach 3 from the left, say $x = 2.91, 2.9997, 2.999993$ and 2.9999999, and some values of x that approach 3 from the right, say $x = 3.1, 3.004, 3.0001$ and 3.000002. The only thing important about these particular values for x is that they get closer and closer to 3 without actually equaling 3. You should try some other values "close to 3" to see what happens. Our table of values is:

x	$f(x)$	x	$f(x)$
2.9	6.82	3.1	7.2
2.9997	6.9994	3.004	7.008
2.999993	6.999986	3.0001	7.0002
2.9999999	6.9999998	3.000002	7.000004
↓	↓	↓	↓
3	7	3	7

As the x values get closer and closer to 3, the $f(x)$ values are getting closer and closer to 7. In fact, we can get $f(x)$ as close to 7 as we want ("arbitrarily close") by taking the values of x very close ("sufficiently close") to 3. We write:

$$\lim_{x \to 3} \frac{2x^2 - x - 1}{x - 1} = 7$$

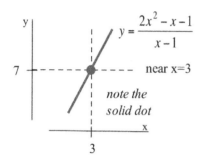

Instead of using a table of values, we could have graphed $y = f(x)$ for x close to 3 (see margin) and used the graph to answer the limit question. This graphical approach is easier, particularly if you have a calculator or computer do the graphing work for you, but it is really very similar to the "table of values" method: in each case you need to evaluate $y = f(x)$ at many values of x near 3. ◀

In the previous example, you might have noticed that if we just evaluate $f(3)$, then we get the correct answer, 7. That works for this particular problem, but it often fails. The next example (identical to the previous one, except $x \to 1$) illustrates one such difficulty.

Example 3. Find $\lim_{x \to 1} \dfrac{2x^2 - x - 1}{x - 1}$.

Solution. You might try to evaluate $f(x) = \frac{2x^2-x-1}{x-1}$ at $x = 1$, but $f(1) = \frac{0}{0}$, so f is not defined at $x = 1$.

It is tempting — **but wrong** — to conclude that this function does not have a limit as x approaches 1.

Table Method: Trying some "test" values for x that get closer and closer to 1 from both the left and the right, we get:

x	$f(x)$	x	$f(x)$
0.9	2.82	1.1	3.2
0.9998	2.9996	1.003	3.006
0.999994	2.999988	1.0001	3.0002
0.9999999	2.9999998	1.000007	3.000014
↓	↓	↓	↓
1	3	1	3

The function f is not defined at $x = 1$, but when x gets close to 1, the values of $f(x)$ get very close to 3. We can get $f(x)$ as close to 3 as we want by taking x very close to 1, so:

$$\lim_{x \to 1} \frac{2x^2 - x - 1}{x - 1} = 3$$

Graph Method: We can graph $y = f(x) = \frac{2x^2 - x - 1}{x - 1}$ for x close to 1 (see margin) and notice that whenever x is close to 1, the values of $y = f(x)$ are close to 3; f is not defined at $x = 1$, so the graph has a hole above $x = 1$, but we only care about what $f(x)$ is doing for x *close to* but **not equal to** 1.

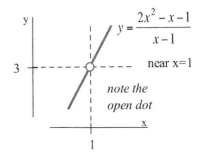

Algebra Method: We could have found the same result by noting:

$$f(x) = \frac{2x^2 - x - 1}{x - 1} = \frac{(2x + 1)(x - 1)}{x - 1} = 2x + 1$$

as long as $x \neq 1$. The "$x \to 1$" part of the limit means that x is *close to* 1 but **not equal to** 1, so our division step is valid and:

$$\lim_{x \to 1} \frac{2x^2 - x - 1}{x - 1} = \lim_{x \to 1} [2x + 1] = 3$$

which is the same answer we obtained using the first two methods. ◀

Three Methods for Evaluating Limits

The previous example utilized three different methods, each of which led us to the same answer for the limit.

The Algebra Method

The algebra method involves algebraically simplifying the function before trying to evaluate its limit. Often, this simplification just means factoring and dividing, but sometimes more complicated algebraic or even trigonometric steps are needed.

The Table Method

To evaluate a limit of a function $f(x)$ as x approaches c, the table method involves calculating the values of $f(x)$ for "enough" values of x very close to c so that we can "confidently" determine which value $f(x)$ is approaching. If $f(x)$ is well behaved, we may not need to use very many values for x. However, this method is usually used with complicated functions, and then we need to evaluate $f(x)$ for lots of values of x.

A computer or calculator can often make the function evaluations easier, but their calculations are subject to "round off" errors. The result of any computer calculation that involves both large and small numbers should be viewed with some suspicion. For example, the function

$$f(x) = \frac{((0.1)^x + 1) - 1}{(0.1)^x} = \frac{(0.1)^x}{(0.1)^x} = 1$$

for every value of x, and my calculator gives the correct answer for some values of x: $f(3) = 1$, and $f(8)$ and $f(9)$ both equal 1.

But my calculator says $((0.1)^{10} + 1) - 1 = 0$, so it evaluates $f(10)$ to be 0, definitely an incorrect value.

Your calculator may evaluate $f(10)$ correctly, but try $f(35)$ or $f(107)$.

> **Calculators are too handy to be ignored, but they are too prone to these types of errors to be believed uncritically. Be careful.**

The Graph Method

The graph method is closely related to the table method, but we create a graph of the function instead of a table of values, and then we use the graph to determine which value $f(x)$ is approaching.

Which Method Should You Use?

In general, the algebraic method is preferred because it is precise and does not depend on which values of x we chose or the accuracy of our graph or precision of our calculator. **If you can evaluate a limit algebraically, you should do so.** Sometimes, however, it will be very difficult to evaluate a limit algebraically, and the table or graph methods offer worthwhile alternatives. Even when you can algebraically evaluate the limit of a function, it is still a good idea to graph the function or evaluate it at a few points just to verify your algebraic answer.

The table and graph methods have the same advantages and disadvantages. Both can be used on complicated functions that are difficult to handle algebraically or whose algebraic properties you don't know.

Often both methods can be easily programmed on a calculator or computer. However, these two methods are very time-consuming by hand and are prone to round-off errors on computers. You need to know how to use these methods when you can't figure out how to use the algebraic method, but you need to use these two methods warily.

Example 4. Evaluate each limit.

(a) $\displaystyle\lim_{x \to 0} \frac{x^2 + 5x + 6}{x^2 + 3x + 2}$ 　　　　(b) $\displaystyle\lim_{x \to -2} \frac{x^2 + 5x + 6}{x^2 + 3x + 2}$

Solution. The function in each limit is the same but x is approaching a different number in each of them.

(a) Because $x \to 0$, we know that x is getting closer and closer to 0, so the values of the x^2, $5x$ and $3x$ terms get as close to 0 as we want. The numerator approaches 6 and the denominator approaches 2, so the values of the whole function get arbitrarily close to $\frac{6}{2} = 3$, the limit.

(b) As x approaches -2, the numerator and denominator approach 0, and a small number divided by a small number can be almost anything — the ratio depends on the size of the top compared to the size of the bottom. More investigation is needed.

Table Method: If we pick some values of x close to (but not equal to) -2, we get the table:

x	$x^2 + 5x + 6$	$x^2 + 3x + 2$	$\frac{x^2+5x+6}{x^2+3x+2}$
-1.97	0.0309	-0.0291	-1.061856
-2.005	-0.004975	0.005025	-0.990050
-1.9998	0.00020004	-0.00019996	-1.00040008
-2.00003	-0.00002999	0.0000300009	-0.9996666
$\downarrow$	$\downarrow$	$\downarrow$	$\downarrow$
-2	0	0	-1

Even though the numerator and denominator are each getting closer and closer to 0, their ratio is getting arbitrarily close to -1, which is the limit.

Graph Method: The graph of $y = f(x) = \frac{x^2+5x+6}{x^2+3x+2}$ in the margin shows that the values of $f(x)$ are very close to -1 when the x-values are close to -2.

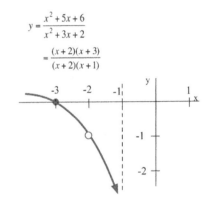

$$y = \frac{x^2 + 5x + 6}{x^2 + 3x + 2}$$
$$= \frac{(x+2)(x+3)}{(x+2)(x+1)}$$

Algebra Method: Factoring the numerator and denominator:

$$f(x) = \frac{x^2 + 5x + 6}{x^2 + 3x + 2} = \frac{(x+2)(x+3)}{(x+2)(x+1)}$$

We know $x \to -2$ so $x \neq -2$ and we can divide the top and bottom by $(x+2)$. Then

$$f(x) = \frac{(x+3)}{(x+1)} \to \frac{1}{-1} = -1$$

as $x \to -2$. ◀

You should remember the technique used in the previous example:

> If $\displaystyle\lim_{x \to c} \frac{\text{polynomial}}{\text{another polynomial}} = \frac{0}{0}$,
> try dividing the top and bottom by $x - c$.

Practice 2. Evaluate each limit.

(a) $\lim\limits_{x\to 2} \dfrac{x^2 - x - 2}{x - 2}$

(b) $\lim\limits_{t\to 0} \dfrac{t \cdot \sin(t)}{t^2 + 3t}$

(c) $\lim\limits_{w\to 2} \dfrac{w - 2}{\ln\left(\frac{w}{2}\right)}$

One-Sided Limits

Sometimes, what happens to us at a place depends on the direction we use to approach that place. If we approach Niagara Falls from the upstream side, then we will be 182 feet higher and have different worries than if we approach from the downstream side. Similarly, the values of a function near a point may depend on the direction we use to approach that point.

If we let x approach 3 from the left (x is close to 3 and $x < 3$) then the values of $\lfloor x \rfloor = \text{INT}(x)$ equal 2 (see margin).

If we let x approach 3 from the right (x is close to 3 and $x > 3$) then the values of $\lfloor x \rfloor = \text{INT}(x)$ equal 3.

On the number line we can approach a point from the left or the right, and that leads to **one-sided limits**.

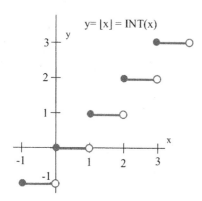

$y = \lfloor x \rfloor = \text{INT}(x)$

Definition of Left and Right Limits:

The **left limit** as x approaches c of $f(x)$ is L if the values of $f(x)$ get as close to L as we want when x is very close to but left of c ($x < c$):

$$\lim\limits_{x\to c^-} f(x) = L$$

The **right limit**, $\lim\limits_{x\to c^+} f(x)$, requires that x lie to the right of c ($x > c$).

Example 5. Evaluate $\lim\limits_{x\to 2^-} x - \lfloor x \rfloor$ and $\lim\limits_{x\to 2^+} x - \lfloor x \rfloor$.

Solution. The left-limit notation $x \to 2^-$ requires that x be close to 2 and that x be to the left of 2, so $x < 2$. If $1 < x < 2$, then $\lfloor x \rfloor = 1$ and:

$$\lim\limits_{x\to 2^-} x - \lfloor x \rfloor = \lim\limits_{x\to 2^-} x - 1 = 2 - 1 = 1$$

If x is close to 2 and is to the right of 2, then $2 < x < 3$, so $\lfloor x \rfloor = 2$ and:

$$\lim\limits_{x\to 2^+} x - \lfloor x \rfloor = \lim\limits_{x\to 2^+} x - 2 = 2 - 2 = 0$$

A graph of $f(x) = x - \lfloor x \rfloor$ appears in the margin.   ◀

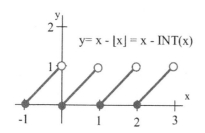

$y = x - \lfloor x \rfloor = x - \text{INT}(x)$

If the left and right limits of $f(x)$ have the same value at $x = c$:

$$\lim_{x \to c^-} f(x) = \lim_{x \to c^+} f(x) = L$$

then the value of $f(x)$ is close to L whenever x is close to c, and it does not matter whether x is left or right of c, so

$$\lim_{x \to c} f(x) = L$$

Similarly, if:

$$\lim_{x \to c} f(x) = L$$

then $f(x)$ is close to L whenever x is close to c and less than c, and whenever x is close to c and greater than c, so:

$$\lim_{x \to c^-} f(x) = \lim_{x \to c^+} f(x) = L$$

We can combine these two statements into a single theorem.

One-Sided Limit Theorem:

$$\lim_{x \to c} f(x) = L \text{ if and only if } \lim_{x \to c^-} f(x) = \lim_{x \to c^+} f(x) = L$$

This theorem has an important corollary.

Corollary:

If $\lim_{x \to c^-} f(x) \neq \lim_{x \to c^+} f(x)$, then $\lim_{x \to c} f(x)$ does not exist.

One-sided limits are particularly useful for describing the behavior of functions that have steps or jumps.

To determine the limit of a function involving the greatest integer or absolute value or a multiline definition, definitely consider both the left and right limits.

Practice 3. Use the graph in the margin to evaluate the one- and two-sided limits of f at $x = 0, 1, 2$ and 3.

Practice 4. Defining $f(x)$ as:

$$f(x) = \begin{cases} 1 & \text{if } x < 1 \\ x & \text{if } 1 < x < 3 \\ 2 & \text{if } 3 < x \end{cases}$$

find the one- and two-sided limits of f at 1 and 3.

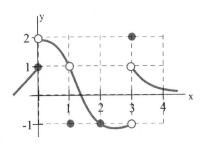

1.1 Problems

1. Use the graph below to determine the limits.

   (a) $\lim\limits_{x\to 1} f(x)$ (b) $\lim\limits_{x\to 2} f(x)$

   (c) $\lim\limits_{x\to 3} f(x)$ (d) $\lim\limits_{x\to 4} f(x)$

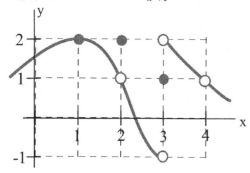

2. Use the graph below to determine the limits.

   (a) $\lim\limits_{x\to 1} f(x)$ (b) $\lim\limits_{x\to 2} f(x)$

   (c) $\lim\limits_{x\to 3} f(x)$ (d) $\lim\limits_{x\to 4} f(x)$

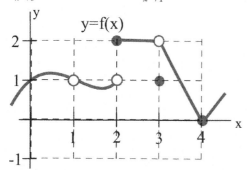

3. Use the graph below to determine the limits.

   (a) $\lim\limits_{x\to 1} f(2x)$ (b) $\lim\limits_{x\to 2} f(x-1)$

   (c) $\lim\limits_{x\to 3} f(2x-5)$ (d) $\lim\limits_{x\to 0} f(4+x)$

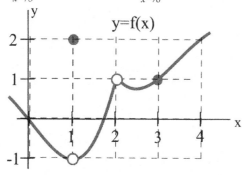

4. Use the graph below to determine the limits.

   (a) $\lim\limits_{x\to 1} f(3x)$ (b) $\lim\limits_{x\to 2} f(x+1)$

   (c) $\lim\limits_{x\to 3} f(2x-4)$ (d) $\lim\limits_{x\to 0} |f(4+x)|$

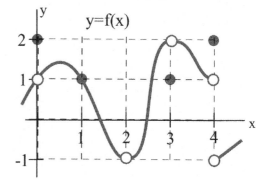

In Problems 5–11, evaluate each limit.

5. (a) $\lim\limits_{x\to 1} \dfrac{x^2+3x+3}{x-2}$ (b) $\lim\limits_{x\to 2} \dfrac{x^2+3x+3}{x-2}$

6. (a) $\lim\limits_{x\to 0} \dfrac{x+7}{x^2+9x+14}$ (b) $\lim\limits_{x\to 3} \dfrac{x+7}{x^2+9x+14}$

   (c) $\lim\limits_{x\to -4} \dfrac{x+7}{x^2+9x+14}$ (d) $\lim\limits_{x\to -7} \dfrac{x+7}{x^2+9x+14}$

7. (a) $\lim\limits_{x\to 1} \dfrac{\cos(x)}{x}$ (b) $\lim\limits_{x\to \pi} \dfrac{\cos(x)}{x}$

   (c) $\lim\limits_{x\to -1} \dfrac{\cos(x)}{x}$

8. (a) $\lim\limits_{x\to 7} \sqrt{x-3}$ (b) $\lim\limits_{x\to 9} \sqrt{x}-3$

   (c) $\lim\limits_{x\to 9} \dfrac{\sqrt{x}-3}{x-9}$

9. (a) $\lim\limits_{x\to 0^-} |x|$ (b) $\lim\limits_{x\to 0^+} |x|$

   (c) $\lim\limits_{x\to 0} |x|$

10. (a) $\lim\limits_{x\to 0^-} \dfrac{|x|}{x}$ (b) $\lim\limits_{x\to 0^+} \dfrac{|x|}{x}$

    (c) $\lim\limits_{x\to 0} \dfrac{|x|}{x}$

11. (a) $\lim\limits_{x \to 5} |x - 5|$ (b) $\lim\limits_{x \to 3} \dfrac{|x - 5|}{x - 5}$

   (c) $\lim\limits_{x \to 5} \dfrac{|x - 5|}{x - 5}$

12. Find the one- and two-sided limits of:

$$f(x) = \begin{cases} x & \text{if } x < 0 \\ \sin(x) & \text{if } 0 < x \le 2 \\ 1 & \text{if } 2 < x \end{cases}$$

   as $x \to 0, 1$ and 2.

13. Find the one- and two-sided limits of:

$$g(x) = \begin{cases} 1 & \text{if } x \le 2 \\ \frac{8}{x} & \text{if } 2 < x < 4 \\ 6 - x & \text{if } 4 < x \end{cases}$$

   as $x \to 1, 2, 4$ and 5.

In 14–17, use a calculator or computer to get approximate answers accurate to 2 decimal places.

14. (a) $\lim\limits_{x \to 0} \dfrac{2^x - 1}{x}$ (b) $\lim\limits_{x \to 1} \dfrac{\log_{10}(x)}{x - 1}$

15. (a) $\lim\limits_{x \to 0} \dfrac{3^x - 1}{x}$ (b) $\lim\limits_{x \to 1} \dfrac{\ln(x)}{x - 1}$

16. (a) $\lim\limits_{x \to 5} \dfrac{\sqrt{x - 1} - 2}{x - 5}$ (b) $\lim\limits_{x \to 0} \dfrac{\sin(3x)}{5x}$

17. (a) $\lim\limits_{x \to 16} \dfrac{\sqrt{x} - 4}{x - 16}$ (b) $\lim\limits_{x \to 0} \dfrac{\sin(7x)}{2x}$

18. Define $A(x)$ to be the **area** bounded by the t- and y-axes, the "bent line" in the figure below, and the vertical line $t = x$. For example, $A(4) = 10$.

   (a) Evaluate $A(0)$, $A(1)$, $A(2)$ and $A(3)$.
   (b) Graph $y = A(x)$ for $0 \le x \le 4$.
   (c) What area does $A(3) - A(1)$ represent?

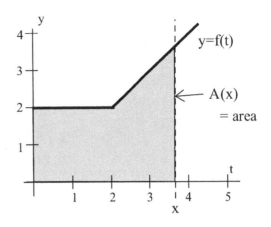

19. Define $A(x)$ to be the **area** bounded by the t- and y-axes, the line $y = \frac{1}{2}t + 2$ and the vertical line $t = x$ (See figure below). For example, $A(4) = 12$.

   (a) Evaluate $A(0)$, $A(1)$, $A(2)$ and $A(3)$.
   (b) Graph $y = A(x)$ for $0 \le x \le 4$.
   (c) What area does $A(3) - A(1)$ represent?

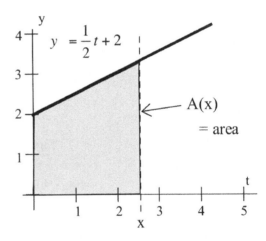

20. Sketch the graph of $f(t) = \sqrt{4t - t^2}$ for $0 \le t \le 4$ (you should get a semicircle). Define $A(x)$ to be the area bounded below by the t-axis, above by the graph $y = f(t)$ and on the right by the vertical line at $t = x$.

   (a) Evaluate $A(0)$, $A(2)$ and $A(4)$.
   (b) Sketch a graph $y = A(x)$ for $0 \le x \le 4$.
   (c) What area does $A(3) - A(1)$ represent?

1.1 Practice Answers

1. (a) 2

   (b) 2

   (c) does not exist (no limit)

   (d) 1

2. (a) $\lim\limits_{x \to 2} \dfrac{(x+1)(x-2)}{x-2} = \lim\limits_{x \to 2} (x+1) = 3$

   (b) $\lim\limits_{t \to 0} \dfrac{t \sin(t)}{t(t+3)} = \lim\limits_{t \to 0} \dfrac{\sin(t)}{t+3} = \dfrac{0}{3} = 0$

   (c) $\lim\limits_{w \to 2} \dfrac{w-2}{\ln\left(\frac{w}{2}\right)} = 2$ To see this, make a graph or a table:

w	$\frac{w-2}{\ln\left(\frac{w}{2}\right)}$	w	$\frac{w-2}{\ln\left(\frac{w}{2}\right)}$
2.2	2.098411737	1.9	1.949572575
2.01	2.004995844	1.99	1.994995823
2.003	2.001499625	1.9992	1.999599973
2.0001	2.00005	1.9999	1.99995
$\downarrow$	$\downarrow$	$\downarrow$	$\downarrow$
2	2	2	2

3. $\lim\limits_{x \to 0^-} f(x) = 1$ $\lim\limits_{x \to 0^+} f(x) = 2$ $\lim\limits_{x \to 0} f(x)$ DNE

   $\lim\limits_{x \to 1^-} f(x) = 1$ $\lim\limits_{x \to 1^+} f(x) = 1$ $\lim\limits_{x \to 1} f(x) = 1$

   $\lim\limits_{x \to 2^-} f(x) = -1$ $\lim\limits_{x \to 2^+} f(x) = -1$ $\lim\limits_{x \to 2} f(x) = -1$

   $\lim\limits_{x \to 3^-} f(x) = -1$ $\lim\limits_{x \to 3^+} f(x) = 1$ $\lim\limits_{x \to 3} f(x)$ DNE

4. $\lim\limits_{x \to 1^-} f(x) = 1$ $\lim\limits_{x \to 1^+} f(x) = 1$ $\lim\limits_{x \to 1} f(x) = 1$

   $\lim\limits_{x \to 3^-} f(x) = 3$ $\lim\limits_{x \to 3^+} f(x) = 2$ $\lim\limits_{x \to 3} f(x)$ DNE

1.2 Properties of Limits

This section presents results that make it easier to calculate limits of combinations of functions or to show that a limit does not exist. The main result says we can determine the limit of "elementary combinations" of functions by calculating the limit of each function separately and recombining these results to get our final answer.

Main Limit Theorem:

If $\quad\lim\limits_{x \to a} f(x) = L \;$ and $\; \lim\limits_{x \to a} g(x) = M$

then

(a) $\lim\limits_{x \to a} [f(x) + g(x)] = L + M$

(b) $\lim\limits_{x \to a} [f(x) - g(x)] = L - M$

(c) $\lim\limits_{x \to a} k \cdot f(x) = k \cdot L$

(d) $\lim\limits_{x \to a} f(x) \cdot g(x) = L \cdot M$

(e) $\lim\limits_{x \to a} \dfrac{f(x)}{g(x)} = \dfrac{L}{M} \quad$ (if $M \neq 0$)

(f) $\lim\limits_{x \to a} [f(x)]^n = L^n$

(g) $\lim\limits_{x \to a} \sqrt[n]{f(x)} = \sqrt[n]{L}$

When n is an even integer in part (g) of the Main Limit Theorem, we need $L \geq 0$ and $f(x) \geq 0$ for x near a.

The Main Limit Theorem says we get the same result if we first perform the algebra and then take the limit or if we take the limits first and then perform the algebra: for example, (a) says that the limit of the sum equals the sum of the limits.

A proof of the Main Limit Theorem is not inherently difficult, but it requires a more precise definition of the limit concept than we have at the moment, and it then involves a number of technical difficulties.

Practice 1. For $f(x) = x^2 - x - 6$ and $g(x) = x^2 - 2x - 3$, evaluate:

(a) $\lim\limits_{x \to 1} [f(x) + g(x)]$

(b) $\lim\limits_{x \to 1} f(x) \cdot g(x)$

(c) $\lim\limits_{x \to 1} \dfrac{f(x)}{g(x)}$

(d) $\lim\limits_{x \to 3} [f(x) + g(x)]$

(e) $\lim\limits_{x \to 3} f(x) \cdot g(x)$

(f) $\lim\limits_{x \to 3} \dfrac{f(x)}{g(x)}$

(g) $\lim\limits_{x \to 2} [f(x)]^3$

(h) $\lim\limits_{x \to 2} \sqrt{1 - g(x)}$

Limits of Some Very Nice Functions: Substitution

As you may have noticed in the previous example, for some functions $f(x)$ it is possible to calculate the limit as x approaches a simply by substituting $x = a$ into the function and then evaluating $f(a)$, but sometimes this method does not work. The following results help to (partially) answer the question about when such a substitution is valid.

Two Easy Limits:

$$\lim_{x \to a} k = k \quad \text{and} \quad \lim_{x \to a} x = a$$

We can use the preceding Two Easy Limits and the Main Limit Theorem to prove the following Substitution Theorem.

Substitution Theorem For Polynomial and Rational Functions:

If $P(x)$ and $Q(x)$ are polynomials and a is any number

then $\lim\limits_{x \to a} P(x) = P(a)$ and $\lim\limits_{x \to a} \dfrac{P(x)}{Q(x)} = \dfrac{P(a)}{Q(a)}$

as long as $Q(a) \neq 0$.

The Substitution Theorem says that we can calculate the limits of polynomials and rational functions by substituting (as long as the substitution does not result in a division by 0).

Practice 2. Evaluate each limit.

(a) $\lim\limits_{x \to 2} \left[5x^3 - x^2 + 3 \right]$

(b) $\lim\limits_{x \to 2} \dfrac{x^3 - 7x}{x^2 + 3x}$

(c) $\lim\limits_{x \to 2} \dfrac{x^2 - 2x}{x^2 - x - 2}$

Limits of Other Combinations of Functions

So far we have concentrated on limits of single functions and elementary combinations of functions. If we are working with limits of other combinations or compositions of functions, the situation becomes slightly more difficult, but sometimes these more complicated limits have useful geometric interpretations.

Example 1. Use the graph in the margin to evaluate each limit.

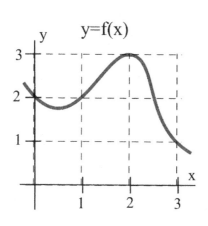

(a) $\lim\limits_{x \to 1} [3 + f(x)]$

(b) $\lim\limits_{x \to 1} f(2 + x)$

(c) $\lim\limits_{x \to 0} f(3 - x)$

(d) $\lim\limits_{x \to 2} [f(x + 1) - f(x)]$

Solution. (a) $\lim\limits_{x \to 1} [3 + f(x)]$ requires a straightforward application of part (a) of the Main Limit Theorem:

$$\lim\limits_{x \to 1} [3 + f(x)] = \lim\limits_{x \to 1} 3 + \lim\limits_{x \to 1} f(x) = 3 + 2 = 5$$

(b) We first need to examine what happens to the quantity $2 + x$ as $x \to 1$ before we can consider the limit of $f(2 + x)$. When x is very close to 1, the value of $2 + x$ is very close to 3, so the limit of $f(2 + x)$ as $x \to 1$ is equivalent to the limit of $f(w)$ as $w \to 3$ (where $w = 2 + x$) and it is clear from the graph that $\lim\limits_{w \to 3} f(w) = 1$, so:

$$\lim\limits_{x \to 1} f(2 + x) = \lim\limits_{w \to 3} f(w) = 1$$

In most situations it is not necessary to formally substitute a new variable w for the quantity $2 + x$, but it is still necessary to think about what happens to the quantity $2 + x$ as $x \to 1$.

(c) As $x \to 0$ the quantity $3 - x$ will approach 3, so we want to know what happens to the values of f when the input variable is approaching 3:

$$\lim\limits_{x \to 0} f(3 - x) = 1$$

(d) Using part (b) of the Main Limit Theorem:

$$\lim\limits_{x \to 2} [f(x + 1) - f(x)] = \lim\limits_{x \to 2} f(x + 1) - \lim\limits_{x \to 2} f(x)$$
$$= \lim\limits_{w \to 3} f(w) - \lim\limits_{x \to 2} f(x) = 1 - 3 = -2$$

Notice the use of the substitution $w = x + 1$ above. ◀

Practice 3. Use the graph in the margin to evaluate each limit.

(a) $\lim\limits_{x \to 1} f(2x)$ (c) $\lim\limits_{x \to 0} 3 \cdot f(4 + x)$

(b) $\lim\limits_{x \to 2} f(x - 1)$ (d) $\lim\limits_{x \to 2} f(3x - 2)$

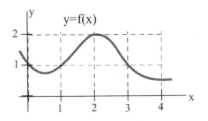

Example 2. Use the graph in the margin to evaluate each limit.

(a) $\lim\limits_{h \to 0} f(3 + h)$ (c) $\lim\limits_{h \to 0} [f(3 + h) - f(3)]$

(b) $\lim\limits_{h \to 0} f(3)$ (d) $\lim\limits_{h \to 0} \dfrac{f(3 + h) - f(3)}{h}$

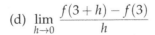
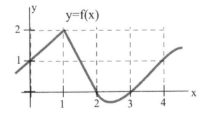

Solution. The last limit is a special type of limit we will encounter often in this book, while the first three parts are the steps we need to evaluate it.

(a) As $h \to 0$, the quantity $w = 3 + h$ will approach 3, so

$$\lim_{h \to 0} f(3 + h) = \lim_{w \to 3} f(w) = 1$$

(b) $f(3)$ is a constant (equal to 1) and does not depend on h in any way, so:

$$\lim_{h \to 0} f(3) = f(3) = 1$$

(c) This limit is just an algebraic combination of the first two limits:

$$\lim_{h \to 0} \left[f(3 + h) - f(3) \right] = \lim_{h \to 0} f(3 + h) - \lim_{h \to 0} f(3) = 1 - 1 = 0$$

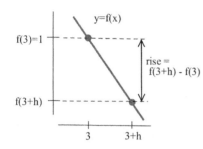

The quantity $f(3 + h) - f(3)$ also has a geometric interpretation: it is the change in the y-coordinates, the Δy, between the points $(3, f(3))$ and $(3 + h, f(3 + h))$ (see margin figure).

(d) As $h \to 0$, the numerator and denominator of $\dfrac{f(3 + h) - f(3)}{h}$ both approach 0, so we cannot immediately determine the value of the limit. But if we recognize that $f(3 + h) - f(3) = \Delta y$ for the two points $(3, f(3))$ and $(3 + h, f(3 + h))$ and that $h = \Delta x$ for the same two points, then we can interpret $\dfrac{f(3 + h) - f(3)}{h}$ as $\frac{\Delta y}{\Delta x}$, which is the slope of the secant line through the two points:

$$\lim_{h \to 0} \frac{f(3 + h) - f(3)}{h} = \lim_{\Delta x \to 0} \left[\text{slope of the secant line}\right]$$

$$= \text{slope of the tangent line at } (3, f(3))$$

$$\approx -1$$

This last limit represents the slope of line tangent to the graph of f at the point $(3, f(3))$.

It is a pattern we will encounter often. ◄

Tangent Lines as Limits

If we have two points on the graph of the function $y = f(x)$:

$$(x, f(x)) \text{ and } (x + h, f(x + h))$$

then $\Delta y = f(x + h) - f(x)$ and $\Delta x = (x + h) - (x) = h$, so the slope of the secant line through those points is:

$$m_{\text{sec}} = \frac{\Delta y}{\Delta x}$$

and the slope of the line tangent to the graph of f at the point $(x, f(x))$ is, by definition,

$$m_{\text{tan}} = \lim_{\Delta x \to 0} \left[\text{slope of the secant line}\right] = \lim_{h \to 0} \frac{f(x + h) - f(x)}{h}$$

Example 3. Give a geometric interpretation for the following limits and **estimate** their values for the function whose graph appears in the margin.

(a) $\displaystyle\lim_{h\to0} \frac{f(1+h)-f(1)}{h}$ (b) $\displaystyle\lim_{h\to0} \frac{f(2+h)-f(2)}{h}$

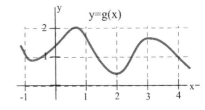

Solution. (a) The limit represents the slope of the line tangent to the graph of $f(x)$ at the point $(1, f(1))$, so $\displaystyle\lim_{h\to0} \frac{f(1+h)-f(1)}{h} \approx 1$. (b) The limit represents the slope of the line tangent to the graph of $f(x)$ at the point $(2, f(2))$, so $\displaystyle\lim_{h\to0} \frac{f(2+h)-f(2)}{h} \approx -1$. ◀

Practice 4. Give a geometric interpretation for the following limits and estimate their values for the function whose graph appears in the margin.

(a) $\displaystyle\lim_{h\to0} \frac{g(1+h)-g(1)}{h}$ (c) $\displaystyle\lim_{h\to0} \frac{g(h)-g(0)}{h}$

(b) $\displaystyle\lim_{h\to0} \frac{g(3+h)-g(3)}{h}$

Comparing the Limits of Functions

Sometimes it is difficult to work directly with a function. However, if we can compare our difficult function with easier ones, then we can use information about the easier functions to draw conclusions about the difficult one. If the complicated function is always between two functions whose limits are equal, then we know the limit of the complicated function.

Squeezing Theorem:

If $g(x) \le f(x) \le h(x)$ for all x near (but not equal to) c

and $\displaystyle\lim_{x\to c} g(x) = \lim_{x\to c} h(x) = L$

then $\displaystyle\lim_{x\to c} f(x) = L.$

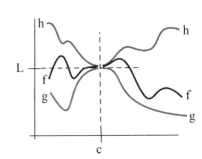

The margin figure shows the idea behind the proof of this theorem: the function $f(x)$ gets "squeezed" between the smaller function $g(x)$ and the bigger function $h(x)$. Because $g(x)$ and $h(x)$ converge to the same limit, L, so must $f(x)$.

We can use the Squeezing Theorem to evaluate some "hard" limits by squeezing a "difficult" function in between two "nicer" functions with "easier" limits.

Example 4. Use the inequality $-|x| \leq \sin(x) \leq |x|$ to determine:

(a) $\lim\limits_{x \to 0} \sin(x)$ 　　　　　　(b) $\lim\limits_{x \to 0} \cos(x)$

Solution. (a) $\lim\limits_{x \to 0} |x| = 0$ and $\lim\limits_{x \to 0} -|x| = 0$ so, by the Squeezing Theorem, $\lim\limits_{x \to 0} \sin(x) = 0$. (b) If $-\frac{\pi}{2} < x < \frac{\pi}{2}$, then $\cos(x) = \sqrt{1 - \sin^2(x)}$, so $\lim\limits_{x \to 0} \cos(x) = \lim\limits_{x \to 0} \sqrt{1 - \sin^2(x)} = \sqrt{1 - 0^2} = 1$. ◀

Example 5. Evaluate $\lim\limits_{x \to 0} x \cdot \sin\left(\dfrac{1}{x}\right)$.

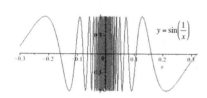

Solution. In the graph of $\sin\left(\frac{1}{x}\right)$ (see margin), the y-values change very rapidly for values of x near 0, but they all lie between -1 and 1:

$$-1 \leq \sin\left(\frac{1}{x}\right) \leq 1$$

so, if $x > 0$, multiplying this inequality by x we get:

$$-x \leq x \cdot \sin\left(\frac{1}{x}\right) \leq x$$

which we can rewrite as:

$$-|x| \leq x \cdot \sin\left(\frac{1}{x}\right) \leq |x|$$

because $|x| = x$ when $x > 0$.

If $x < 0$, when we multiply the original inequality by x we get:

$$-x \geq x \cdot \sin\left(\frac{1}{x}\right) \geq x \quad \Rightarrow \quad |x| \geq x \cdot \sin\left(\frac{1}{x}\right) \geq -|x|$$

because $|x| = -x$ when $x < 0$. Either way we have:

$$-|x| \leq x \cdot \sin\left(\frac{1}{x}\right) \leq |x|$$

for all $x \neq 0$, and in particular for x near 0.

Both "easy" functions ($-|x|$ and $|x|$) approach 0 as $x \to 0$, so

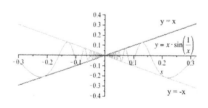

$$\lim\limits_{x \to 0} x \cdot \sin\left(\frac{1}{x}\right) = 0$$

by the Squeezing Theorem. ◀

Practice 5. If $f(x)$ is always between $x^2 + 2$ and $2x + 1$, what can you say about $\lim\limits_{x \to 1} f(x)$?

Problem 27 guides you through the steps to prove this relation.

Practice 6. Use the relation $\cos(x) \leq \dfrac{\sin(x)}{x} \leq 1$ to show that:

$$\lim\limits_{x \to 0} \frac{\sin(x)}{x} = 1$$

List Method for Showing that a Limit Does Not Exist

If the limit of $f(x)$, as x approaches c, exists and equals L, then we can guarantee that the values of $f(x)$ are as close to L as we want by restricting the values of x to be very, very close to c. To show that a limit, as x approaches c, does **not** exist, we need to show that no matter how closely we restrict the values of x to c, the values of $f(x)$ are not **all** close to a single, finite value L.

One way to demonstrate that $\lim\limits_{x \to c} f(x)$ does not exist is to show that the left and right limits exist but are not equal.

Another method of showing that $\lim\limits_{x \to c} f(x)$ does not exist uses two (infinite) lists of numbers, $\{a_1, a_2, a_3, a_4, \ldots\}$ and $\{b_1, b_2, b_3, b_4, \ldots\}$, that become arbitrarily close to the value c as the subscripts get larger, but with the lists of function values, $\{f(a_1), f(a_2), f(a_3), f(a_4), \ldots\}$ and $\{f(b_1), f(b_2), f(b_3), f(b_4), \ldots\}$ approaching two different numbers as the subscripts get larger.

Example 6. For $f(x)$ defined as:

$$f(x) = \begin{cases} 1 & \text{if } x < 1 \\ x & \text{if } 1 < x < 3 \\ 2 & \text{if } 3 < x \end{cases}$$

show that $\lim\limits_{x \to 3} f(x)$ does not exist.

Solution. We could use one-sided limits to show that this limit does not exist, but instead we will use the list method.

One way to define values of $\{a_1, a_2, a_3, a_4, \ldots\}$ that approach 3 from the right is to define $a_1 = 3 + 1$, $a_2 = 3 + \frac{1}{2}$, $a_3 = 3 + \frac{1}{3}$, $a_4 = 3 + \frac{1}{4}$ and, in general, $a_n = 3 + \frac{1}{n}$. Then $a_n > 3$ so $f(a_n) = 2$ for all subscripts n, and the values in the list $\{f(a_1), f(a_2), f(a_3), f(a_4), \ldots\}$ are approaching 2—in fact, all of the $f(a_n)$ values equal 2.

We can define values of $\{b_1, b_2, b_3, b_4, \ldots\}$ that approach 3 from the left by $b_1 = 3 - 1$, $b_2 = 3 - \frac{1}{2}$, $b_3 = 3 - \frac{1}{3}$, $b_4 = 3 - \frac{1}{4}$, and, in general, $b_n = 3 - \frac{1}{n}$. Then $b_n < 3$ so $f(b_n) = b_n = 3 - \frac{1}{n}$ for each subscript n, and the values in the list $\{f(b_1), f(b_2), f(b_3), f(b_4), \ldots\} = \left\{2, 2.5, 2\frac{2}{3}, 2\frac{3}{4}, 2\frac{4}{5}, \ldots, 3 - \frac{1}{n}, \ldots\right\}$ approach 3.

Because the values in the lists $\{f(a_1), f(a_2), f(a_3), f(a_4), \ldots\}$ and $\{f(b_1), f(b_2), f(b_3), f(b_4), \ldots\}$ approach two different numbers, we can conclude that $\lim\limits_{x \to 3} f(x)$ does not exist. ◀

Example 7. Define $h(x)$ as:

$$h(x) = \begin{cases} 2 & \text{if } x \text{ is a rational number} \\ 1 & \text{if } x \text{ is an irrational number} \end{cases}$$

(the "holey" function introduced in Section 0.4). Use the list method to show that $\lim\limits_{x \to 3} h(x)$ does not exist.

Solution. Let $\{a_1, a_2, a_3, a_4, \ldots\}$ be a list of rational numbers that approach 3: for example, $a_1 = 3 + 1$, $a_2 = 3 + \frac{1}{2}$, $a_3 = 3 + \frac{1}{3}, \ldots, a_n = 3 + \frac{1}{n}$. Then $f(a_n) = 2$ for all n, so:

$$\{f(a_1), f(a_2), f(a_3), f(a_4), \ldots\} = \{2, 2, 2, 2, \ldots\}$$

and the $f(a_n)$ values "approach" 2.

If $\{b_1, b_2, b_3, b_4, \ldots\}$ is a list of irrational numbers that approach 3 (for example, $b_1 = 3 + \pi$, $b_2 = 3 + \frac{\pi}{2}, \ldots, b_n = 3 + \frac{\pi}{n}$) then:

$$\{f(b_1), f(b_2), f(b_3), f(b_4), \ldots\} = \{1, 1, 1, 1, \ldots\}$$

and the $f(b_n)$ values "approach" 1.

Because the $f(a_n)$ and $f(b_n)$ values approach different numbers, the limit as $x \to 3$ does not exist. A similar argument will work as x approaches any number c, so for every c we can show that $\lim_{x \to c} (x)$ does not exist. The "holey" function does not have a limit as x approaches *any* value c.   ◀

1.2 Problems

1. Use the functions f and g defined by the graphs below to determine the following limits.

   (a) $\lim_{x \to 1} [f(x) + g(x)]$ (b) $\lim_{x \to 1} f(x) \cdot g(x)$

   (c) $\lim_{x \to 1} \dfrac{f(x)}{g(x)}$ (d) $\lim_{x \to 1} f(g(x))$

3. Use the function h defined by the graph below to determine the following limits.

   (a) $\lim_{x \to 2} h(2x - 2)$ (b) $\lim_{x \to 2} [x + h(x)]$

   (c) $\lim_{x \to 2} h(1 + x)$ (d) $\lim_{x \to 3} h\left(\dfrac{x}{2}\right)$

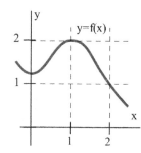
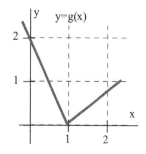
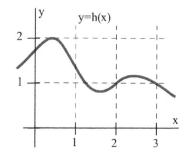

2. Use the functions f and g defined by the graphs above to determine the following limits.

   (a) $\lim_{x \to 2} [f(x) + g(x)]$ (b) $\lim_{x \to 2} f(x) \cdot g(x)$

   (c) $\lim_{x \to 2} \dfrac{f(x)}{g(x)}$ (d) $\lim_{x \to 2} f(g(x))$

4. Use the function h defined by the graph above to determine the following limits.

   (a) $\lim_{x \to 2} h(5 - x)$ (b) $\lim_{x \to 0} [h(3 + x) - h(3)]$

   (c) $\lim_{x \to 2} x \cdot h(x - 1)$ (d) $\lim_{x \to 0} \dfrac{h(3 + x) - h(3)}{x}$

5. Label the parts of the graph of f (below) that are described by

(a) $2 + h$

(b) $f(2)$

(c) $f(2 + h)$

(d) $f(2 + h) - f(2)$

(e) $\dfrac{f(2 + h) - f(2)}{(2 + h) - 2}$

(f) $\dfrac{f(2 - h) - f(2)}{(2 - h) - 2}$

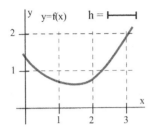

6. Label the parts of the graph of g (below) that are described by

(a) $a + h$

(b) $g(a)$

(c) $g(a + h)$

(d) $g(a + h) - g(a)$

(e) $\dfrac{g(a + h) - g(a)}{(a + h) - a}$

(f) $\dfrac{g(a - h) - g(2)}{(a - h) - a}$

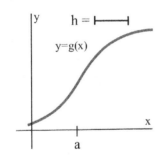

7. Use the graph below determine:

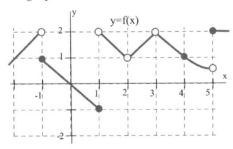

(a) $\lim\limits_{x \to 1^+} f(x)$ (b) $\lim\limits_{x \to 1^-} f(x)$ (c) $\lim\limits_{x \to 1} f(x)$

(d) $\lim\limits_{x \to 3^+} f(x)$ (e) $\lim\limits_{x \to 3^-} f(x)$ (f) $\lim\limits_{x \to 3} f(x)$

(g) $\lim\limits_{x \to -1^+} f(x)$ (h) $\lim\limits_{x \to -1^-} f(x)$ (i) $\lim\limits_{x \to -1} f(x)$

8. Use the graph from Problem 7 to determine:

(a) $\lim\limits_{x \to 2^+} f(x)$ (b) $\lim\limits_{x \to 2^-} f(x)$ (c) $\lim\limits_{x \to 2} f(x)$

(d) $\lim\limits_{x \to 4^+} f(x)$ (e) $\lim\limits_{x \to 4^-} f(x)$ (f) $\lim\limits_{x \to 4} f(x)$

(g) $\lim\limits_{x \to -2^+} f(x)$ (h) $\lim\limits_{x \to -2^-} f(x)$ (i) $\lim\limits_{x \to -2} f(x)$

9. The Lorentz Contraction Formula in relativity theory says the length L of an object moving at v miles per second with respect to an observer is:

$$L = A \cdot \sqrt{1 - \dfrac{v^2}{c^2}}$$

where c is the speed of light (a constant).

(a) Determine the object's "rest length" ($v = 0$).

(b) Determine: $\lim\limits_{v \to c^-} L$

10. Evaluate each limit.

(a) $\lim\limits_{x \to 2^+} \lfloor x \rfloor$

(b) $\lim\limits_{x \to 2^-} \lfloor x \rfloor$

(c) $\lim\limits_{x \to -2^+} \lfloor x \rfloor$

(d) $\lim\limits_{x \to -2^-} \lfloor x \rfloor$

(e) $\lim\limits_{x \to -2.3} \lfloor x \rfloor$

(f) $\lim\limits_{x \to 3} \left\lfloor \dfrac{x}{2} \right\rfloor$

(g) $\lim\limits_{x \to 3} \dfrac{\lfloor x \rfloor}{2}$

(h) $\lim\limits_{x \to 0^+} \dfrac{\lfloor 2 + x \rfloor - \lfloor 2 \rfloor}{x}$

11. For $f(x)$ and $g(x)$ defined as:

$$f(x) = \begin{cases} 1 & \text{if } x < 1 \\ x & \text{if } 1 < x \end{cases} \qquad g(x) = \begin{cases} x & \text{if } x \neq 2 \\ 3 & \text{if } x = 2 \end{cases}$$

determine the following limits:

(a) $\lim\limits_{x \to 2} [f(x) + g(x)]$

(b) $\lim\limits_{x \to 2} \dfrac{f(x)}{g(x)}$

(c) $\lim\limits_{x \to 2} f(g(x))$

(d) $\lim\limits_{x \to 0} \dfrac{g(x)}{f(x)}$

(e) $\lim\limits_{x \to 1} \dfrac{f(x)}{g(x)}$

(f) $\lim\limits_{x \to 1} g(f(x))$

12. Give geometric interpretations for each limit and use a calculator to estimate its value.

(a) $\lim\limits_{h \to 0} \dfrac{\arctan(0 + h) - \arctan(0)}{h}$

(b) $\lim\limits_{h \to 0} \dfrac{\arctan(1 + h) - \arctan(1)}{h}$

(c) $\lim\limits_{h \to 0} \dfrac{\arctan(2 + h) - \arctan(2)}{h}$

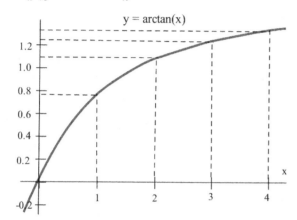

13. (a) What does $\lim\limits_{h \to 0} \dfrac{\cos(h) - 1}{h}$ represent in relation to the graph of $y = \cos(x)$? It may help to recognize that:

$$\frac{\cos(h) - 1}{h} = \frac{\cos(0 + h) - \cos(0)}{h}$$

(b) Graphically and using your calculator, determine $\lim\limits_{h \to 0} \dfrac{\cos(h) - 1}{h}$.

14. (a) What does the ratio $\dfrac{\ln(1 + h)}{h}$ represent in relation to the graph of $y = \ln(x)$? It may help to recognize that:

$$\frac{\ln(1 + h)}{h} = \frac{\ln(1 + h) - \ln(1)}{h}$$

(b) Graphically and using your calculator, determine $\lim\limits_{h \to 0} \dfrac{\ln(1 + h)}{h}$.

15. Use your calculator (to generate a table of values) to help you estimate the value of each limit.

(a) $\lim\limits_{h \to 0} \dfrac{e^h - 1}{h}$

(b) $\lim\limits_{c \to 0} \dfrac{\tan(1 + c) - \tan(1)}{c}$

(c) $\lim\limits_{t \to 0} \dfrac{g(2 + t) - g(2)}{t}$ when $g(t) = t^2 - 5$.

16. (a) For $h > 0$, find the slope of the line through the points $(h, |h|)$ and $(0, 0)$.

(b) For $h < 0$, find the slope of the line through the points $(h, |h|)$ and $(0, 0)$.

(c) Evaluate $\lim\limits_{h \to 0^-} \dfrac{|h|}{h}$, $\lim\limits_{h \to 0^+} \dfrac{|h|}{h}$ and $\lim\limits_{h \to 0} \dfrac{|h|}{h}$.

In 17–18, describe the behavior at each integer of the function $y = f(x)$ in the figure provided, using one of these phrases:

- "connected and smooth"

- "connected with a corner"

- "not connected because of a simple hole that could be plugged by adding or moving one point"

- "not connected because of a vertical jump that could not be plugged by moving one point"

17.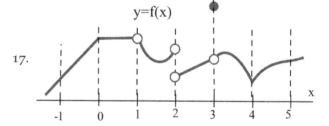

18.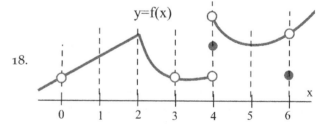

19. Use the list method to show that $\lim\limits_{x \to 2} \dfrac{|x - 2|}{x - 2}$ does not exist .

20. Show that $\lim\limits_{x \to 0} \sin\left(\dfrac{1}{x}\right)$ does not exist. (Suggestion: Let $f(x) = \sin\left(\frac{1}{x}\right)$ and let $a_n = \frac{1}{n\pi}$ so that $f(a_n) = \sin\left(\frac{1}{a_n}\right) = \sin(n\pi) = 0$ for every n. Then pick $b_n = \frac{1}{2n\pi + \frac{\pi}{2}}$ so that $f(b_n) = \sin\left(\frac{1}{b_n}\right) = \sin(2n\pi + \frac{\pi}{2}) = \sin(\frac{\pi}{2}) = 1$ for all n.)

In Problems 21–26, use the Squeezing Theorem to help evaluate each limit.

21. $\displaystyle\lim_{x\to0} x^2 \cos\left(\frac{1}{x^2}\right)$

22. $\displaystyle\lim_{x\to0} \sqrt[3]{x}\sin\left(\frac{1}{x^3}\right)$

23. $\displaystyle\lim_{x\to0} 3 + x^2 \sin\left(\frac{1}{x}\right)$

24. $\displaystyle\lim_{x\to1^-} \sqrt{1 - x^2}\cos\left(\frac{1}{x-1}\right)$

25. $\displaystyle\lim_{x\to0} x^2 \cdot \left\lfloor \frac{1}{x^2}\right\rfloor$

26. $\displaystyle\lim_{x\to0} (-1)^{\lfloor\frac{1}{x}\rfloor}(1 - \cos(x))$

27. This problem outlines the steps of a proof that $\displaystyle\lim_{\theta\to0^+}\frac{\sin(\theta)}{\theta} = 1$. Refer to the margin figure, assume that $0 < \theta < \frac{\pi}{2}$, and justify why each statement must be true.

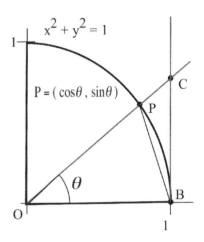

(a) Area of $\triangle OPB = \frac{1}{2}(\text{base})(\text{height}) = \frac{1}{2}\sin(\theta)$

(b) $\dfrac{\text{area of the sector (the pie shaped region) } OPB}{\text{area of the whole circle}} = \dfrac{\theta}{2\pi}$

(c) area of the sector $OPB = \pi \cdot \dfrac{\theta}{2\pi} = \dfrac{\theta}{2}$

(d) The line L through the points $(0,0)$ and $P = (\cos(\theta), \sin(\theta))$ has slope $m = \dfrac{\sin(\theta)}{\cos(\theta)}$, so $C = \left(1, \dfrac{\sin(\theta)}{\cos(\theta)}\right)$

(e) area of $\triangle OCB = \dfrac{1}{2}(\text{base})(\text{height}) = \dfrac{1}{2}(1)\dfrac{\sin(\theta)}{\cos(\theta)}$

(f) area of $\triangle OPB <$ area of sector $OPB <$ area of $\triangle OCB$

(g) $\dfrac{1}{2}\sin(\theta) < \dfrac{\theta}{2} < \dfrac{1}{2}(1)\dfrac{\sin(\theta)}{\cos(\theta)} \Rightarrow \sin(\theta) < \theta < \dfrac{\sin(\theta)}{\cos(\theta)}$

(h) $1 < \dfrac{\theta}{\sin(\theta)} < \dfrac{1}{\cos(\theta)} \Rightarrow 1 > \dfrac{\sin(\theta)}{\theta} > \cos(\theta)$

(i) $\displaystyle\lim_{\theta\to0^+} 1 = 1$ and $\displaystyle\lim_{\theta\to0^+} \cos(\theta) = 1$.

(j) $\displaystyle\lim_{\theta\to0^+} \dfrac{\sin(\theta)}{\theta} = 1$

1.2 Practice Answers

1. (a) -10 (b) 24 (c) $\frac{3}{2}$ (d) 0 (e) 0 (f) $\frac{5}{4}$ (g) -64 (h) 2

2. (a) 39 (b) $-\frac{3}{5}$ (c) $\frac{2}{3}$ 3. (a) 0 (b) 2 (c) 3 (d) 1

4. (a) slope of the line tangent to the graph of g at the point $(1, g(1))$; estimated slope ≈ -2

   (b) slope of the line tangent to the graph of g at the point $(3, g(3))$; estimated slope ≈ 0

   (c) slope of the line tangent to the graph of g at the point $(0, g(0))$; estimated slope ≈ 1

5. $\displaystyle\lim_{x\to1}\left[x^2 + 2\right] = 3$ and $\displaystyle\lim_{x\to1}[2x + 1] = 3$ so $\displaystyle\lim_{x\to1} f(x) = 3$

6. $\displaystyle\lim_{x\to0}\cos(x) = 1$ and $\displaystyle\lim_{x\to0} 1 = 1$ so $\displaystyle\lim_{x\to0}\dfrac{\sin(x)}{x} = 1$

1.3 Continuous Functions

In Section 1.2 we saw a few "nice" functions whose limits as $x \to a$ simply involved substituting a into the function: $\lim_{x \to a} f(x) = f(a)$. Functions whose limits have this substitution property are called **continuous functions** and such functions possess a number of other useful properties.

In this section we will examine what it means graphically for a function to be continuous (or not continuous), state some properties of continuous functions, and look at a few applications of these properties—including a way to solve horrible equations such as $\sin(x) = \dfrac{2x + 1}{x - 2}$.

Definition of a Continuous Function

We begin by formally stating the definition of this new concept.

> **Definition of Continuity at a Point:**
>
> A function f is **continuous** at $x = a$ if and only if
> $$\lim_{x \to a} f(x) = f(a).$$

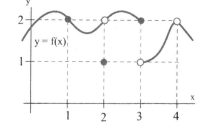

a	$f(a)$	$\lim\limits_{x \to a} f(x)$
1	2	2
2	1	2
3	2	DNE
4	undefined	2

The graph in the margin illustrates some of the different ways a function can behave at and near a point, and the accompanying table contains some numerical information about the example function f and its behavior. We can conclude from the information in the table that f is continuous at 1 because $\lim_{x \to 1} f(x) = 2 = f(1)$.

We can also conclude that f is not continuous at 2 or 3 or 4, because $\lim_{x \to 2} f(x) \neq f(2)$, $\lim_{x \to 3} f(x) \neq f(3)$ and $\lim_{x \to 4} f(x) \neq f(4)$.

Graphical Meaning of Continuity

When x is close to 1, the values of $f(x)$ are close to the value $f(1)$, and the graph of f does not have a hole or break at $x = 1$. The graph of f is "connected" at $x = 1$ and can be drawn without lifting your pencil. At $x = 2$ and $x = 4$ the graph of f has "holes," and at $x = 3$ the graph has a "break." The function f is also continuous at 1.7 (why?) and at every point shown **except** at 2, 3 and 4.

> **Informally**, we can say:
>
> - A function is **continuous** at a point if the graph of the function is **connected** there.
>
> - A function is **not continuous** at a point if its graph has a **hole** or **break** at that point.

Sometimes the definition of "continuous" (the substitution condition for limits) is easier to use if we chop it into several smaller pieces and then check whether or not our function satisfies each piece.

---

f is continuous at a if and only if:

  (i) f is defined at a

  (ii) the limit of $f(x)$, as $x \to a$, exists
      (so the left limit and right limits exist and are equal)

  (iii) the value of f at a equals the value of the limit as $x \to a$:

$$\lim_{x \to a} f(x) = f(a)$$

If f satisfies conditions (i), (ii) and (iii), then f is continuous at a. If f does not satisfy one or more of the three conditions at a, then f is not continuous at a.

For $f(x)$ in the figure on the previous page, all three conditions are satisfied for $a = 1$, so f is continuous at 1. For $a = 2$, conditions (i) and (ii) are satisfied but not (iii), so f is not continuous at 2. For $a = 3$, condition (i) is satisfied but (ii) is violated, so f is not continuous at 3. For $a = 4$, condition (i) is violated, so f is not continuous at 4.

A function is **continuous on an interval** if it is continuous at every point in the interval.

A function f is **continuous from the left** at a if $\lim\limits_{x \to a^-} f(x) = f(a)$ and is **continuous from the right** at a if $\lim\limits_{x \to a^+} f(x) = f(a)$.

Example 1. Is the function

$$f(x) = \begin{cases} x+1 & \text{if } x \leq 1 \\ 2 & \text{if } 1 < x \leq 2 \\ \frac{1}{x-3} & \text{if } x > 2 \end{cases}$$

continuous at $x = 1$? At $x = 2$? At $x = 3$?

Solution. We could answer these questions by examining the graph of $f(x)$, but let's try them without the benefit of a graph. At $x = 1$, $f(1) = 2$ and the left and right limits are equal:

$$\lim_{x \to 1^-} f(x) = \lim_{x \to 1^-} [x+1] = 2 = \lim_{x \to 1^+} 2 = \lim_{x \to 1^+} f(x)$$

and their common limit matches the value of the function at $x = 1$:

$$\lim_{x \to 1} f(x) = 2 = f(1)$$

so f is continuous at 1.

At $x = 2$, $f(2) = 2$, but the left and right limits are not equal:

$$\lim_{x \to 2^-} f(x) = \lim_{x \to 1^-} 2 = 2 \neq -1 = \lim_{x \to 2^+} \frac{1}{x-3} = \lim_{x \to 2^+} f(x)$$

so f fails condition (ii), hence is not continuous at 2. We can, however, say that f is continuous from the left (but not from the right) at 2.

At $x = 3$, $f(3) = \frac{1}{0}$, which is undefined, so f is not continuous at 3 because it fails condition (i). ◄

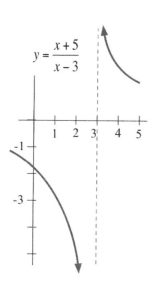

$y = \dfrac{x+5}{x-3}$

Example 2. Where is $f(x) = 3x^2 - 2x$ continuous?

Solution. By the Substitution Theorem for Polynomial and Rational Functions, $\lim_{x \to a} P(x) = P(a)$ for any polynomial $P(x)$ at any point a, so every polynomial is continuous everywhere. In particular, $f(x) = 3x^2 - 2x$ is continuous everywhere. ◄

Example 3. Where is the function $g(x) = \dfrac{x+5}{x-3}$ continuous? Where is $h(x) = \dfrac{x^2 + 4x - 5}{x^2 - 4x + 3}$ continuous?

Solution. $g(x)$ is a rational function, so by the Substitution Theorem for Polynomial and Rational Functions it is continuous everywhere except where its denominator is 0: g is continuous everywhere except at 3. The graph of g (see margin) is "connected" everywhere except at 3, where it has a vertical asymptote.

We can write the rational function $h(x)$ as:

$$h(x) = \frac{(x-1)(x+5)}{(x-1)(x-3)}$$

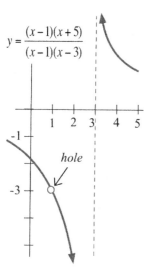

$y = \dfrac{(x-1)(x+5)}{(x-1)(x-3)}$

hole

and note that its denominator is 0 at $x = 1$ and $x = 3$, so h is continuous everywhere except 3 and 1. The graph of h (see margin) is "connected" everywhere except at 3, where it has a vertical asymptote, and 1, where it has a hole: $f(1) = \frac{0}{0}$ is undefined. ◄

Example 4. Where is $f(x) = \lfloor x \rfloor$ continuous?

Solution. The graph of $y = \lfloor x \rfloor$ seems to be "connected" except at each integer, where there is a "jump" (see margin).

If a is an integer, then $\lim_{x \to a^-} \lfloor x \rfloor = a - 1$ and $\lim_{x \to a^+} \lfloor x \rfloor = a$ so $\lim_{x \to a} \lfloor x \rfloor$ is undefined, and $\lfloor x \rfloor$ is not continuous at $x = a$.

If a is *not* an integer, then the left and right limits of $\lfloor x \rfloor$, as $x \to a$, both equal $\lfloor a \rfloor$ so: $\lim_{x \to a} \lfloor x \rfloor = a = \lfloor a \rfloor$, hence $\lfloor x \rfloor$ is continuous at $x = a$.

Summarizing: $\lfloor x \rfloor$ is continuous everywhere except at the integers. In fact, $f(x) = \lfloor x \rfloor$ is continuous from the right everywhere and is continuous from the left everywhere except at the integers. ◄

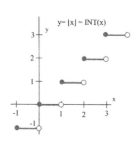

$y = \lfloor x \rfloor = \text{INT}(x)$

Practice 1. Where is $f(x) = \dfrac{|x|}{x}$ continuous?

Why Do We Care Whether a Function Is Continuous?

There are several reasons for us to examine continuous functions and their properties:

- Many applications in engineering, the sciences and business are continuous or are modeled by continuous functions or by pieces of continuous functions.

- Continuous functions share a number of useful properties that do not necessarily hold true if the function is not continuous. If a result is true of all continuous functions and we have a continuous function, then the result is true for our function. This can save us from having to show, one by one, that each result is true for each particular function we use. Some of these properties are given in the remainder of this section.

- Differential calculus has been called the study of **continuous** change, and many of the results of calculus are guaranteed to be true only for continuous functions. If you look ahead into Chapters 2 and 3, you will see that many of the theorems have the form "If f is continuous and (some additional hypothesis), then (some conclusion)."

Combinations of Continuous Functions

Not only are most of the basic functions we will encounter continuous at most points, so are basic combinations of those functions.

Theorem:

If $f(x)$ and $g(x)$ are continuous at a
  and k is any constant

then the elementary combinations of f and g

- $k \cdot f(x)$

- $f(x) + g(x)$

- $f(x) - g(x)$

- $f(x) \cdot g(x)$

- $\dfrac{f(x)}{g(x)}$ (as long as $g(a) \neq 0$)

are continuous at a.

The continuity of a function is defined using limits, and all of these results about simple combinations of continuous functions follow from the results about combinations of limits in the Main Limit Theorem.

Our hypothesis is that f and g are both continuous at a, so we can assume that

$$\lim_{x \to a} f(x) = f(a) \qquad \text{and} \qquad \lim_{x \to a} g(x) = g(a)$$

and then use the appropriate part of the Main Limit Theorem.

For example,

$$\lim_{x \to a} [f(x) + g(x)] = \lim_{x \to a} f(x) + \lim_{x \to a} g(x) = f(a) + g(a)$$

so $f + g$ is continuous at a.

Practice 2. Prove: If f and g are continuous at a, then $k \cdot f$ and $f - g$ are continuous at a (where k a constant).

Composition of Continuous Functions:

If $\quad g(x)$ is continuous at a and
$\quad\quad f(x)$ is continuous at $g(a)$

then $\quad \lim_{x \to a} f(g(x)) = f(\lim_{x \to a} g(x)) = f(g(a))$
$\quad\quad$ so $f \circ g(x) = f(g(x))$ is continuous at a.

The proof of this result involves some technical details, but just formalizes the following line of reasoning:

The hypothesis that "g is continuous at a" means that if x is close to a then $g(x)$ will be close to $g(a)$. Similarly, "f is continuous at $g(a)$" means that if $g(x)$ is close to $g(a)$ then $f(g(x)) = f \circ g(x)$ will be close to $f(g(a)) = f \circ g(a)$. Finally, we can conclude that if x is close to a, then $g(x)$ is close to $g(a)$ so $f \circ g(x)$ is close to $f \circ g(a)$ and therefore $f \circ g$ is continuous at $x = a$.

The next theorem presents an alternate version of the limit condition for continuity, which we will use occasionally in the future.

Theorem:

$$\lim_{x \to a} f(x) = f(a) \quad \text{if and only if} \quad \lim_{h \to 0} f(a + h) = f(a)$$

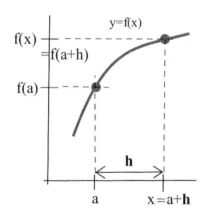

Proof. Let's define a new variable h by $h = x - a$ so that $x = a + h$ (see margin figure). Then $x \to a$ if and only if $h = x - a \to 0$, so $\lim_{x \to a} f(x) = \lim_{h \to 0} f(a + h)$ and therefore $\lim_{x \to a} f(x) = f(a)$ if and only if $\lim_{h \to 0} f(a + h) = f(a)$. $\qquad \square$

We can restate the result of this theorem as:

A function f is continuous at a if and only if $\lim_{h \to 0} f(a + h) = f(a)$.

Which Functions Are Continuous?

Fortunately, the functions we encounter most often are either continuous everywhere or continuous everywhere except at a few places.

> **Theorem**: The following functions are continuous everywhere
>
> (a) polynomials   (b) $\sin(x)$ and $\cos(x)$ (c) $|x|$

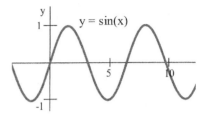

Proof. (a) This follows from the Substitution Theorem for Polynomial and Rational Functions and the definition of continuity.

(b) The graph of $y = \sin(x)$ (see margin) clearly has no holes or breaks, so it is reasonable to think that $\sin(x)$ is continuous everywhere. Justifying this algebraically, for every real number a:

$$\lim_{h \to 0} \sin(a + h) = \lim_{h \to 0} [\sin(a)\cos(h) + \cos(a)\sin(h)]$$

$$= \lim_{h \to 0} \sin(a) \cdot \lim_{h \to 0} \cos(h) + \lim_{h \to 0} \cos(a) \cdot \lim_{h \to 0} \sin(h)$$

$$= \sin(a) \cdot 1 + \cos(a) \cdot 0 = \sin(a)$$

Recall the angle addition formula for $\sin(\theta)$ and the results from Section 1.2 that $\lim_{h \to 0} \cos(h) = 1$ and $\lim_{h \to 0} \sin(h) = 0$.

so $f(x) = \sin(x)$ is continuous at every point. The justification for $f(x) = \cos(x)$ is similar.

(c) For $f(x) = |x|$, when $x > 0$, then $|x| = x$ and its graph (see margin) is a straight line and is continuous because x is a polynomial. When $x < 0$, then $|x| = -x$ and it is also continuous. The only questionable point is the "corner" on the graph when $x = 0$, but the graph there is only bent, not broken:

$$\lim_{h \to 0^+} |0 + h| = \lim_{h \to 0^+} h = 0$$

and:

$$\lim_{h \to 0^-} |0 + h| = \lim_{h \to 0^-} -h = 0$$

so:

$$\lim_{h \to 0} |0 + h| = 0 = |0|$$

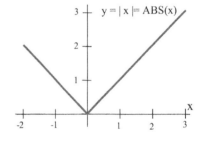

and $f(x) = |x|$ is also continuous at 0.   □

> **A continuous function can have corners but not holes or breaks.**

Even functions that fail to be continuous at some points are often continuous most places:

- A rational function is continuous **except** where the denominator is 0.

- The trig functions $\tan(x)$, $\cot(x)$, $\sec(x)$ and $\csc(x)$ are continuous **except** where they are undefined.

- The greatest integer function $\lfloor x \rfloor$ is continuous **except** at each integer.

- But the "holey" function

$$h(x) = \begin{cases} 2 & \text{if } x \text{ is a rational number} \\ 1 & \text{if } x \text{ is an irrational number} \end{cases}$$

is **discontinuous everywhere**.

Intermediate Value Property of Continuous Functions

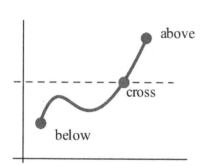

Because the graph of a continuous function is connected and does not have any holes or breaks in it, the values of the function can not "skip" or "jump over" a horizontal line (see margin figure). If one value of the continuous function is below the line and another value of the function is above the line, then **somewhere** the graph will cross the line. The next theorem makes this statement more precise. The result seems obvious, but its proof is technically difficult and is not given here.

> **Intermediate Value Theorem for Continuous Functions**:
>
> If f is continuous on the interval $[a, b]$
>   and V is any value between $f(a)$ and $f(b)$
>
> then   there is a number c between a and b so that
> $f(c) = V$. (That is, f actually takes on each
>   intermediate value between $f(a)$ and $f(b)$.)

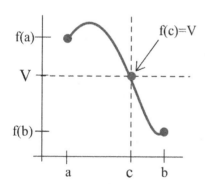

If the graph of f connects the points $(a, f(a))$ and $(b, f(b))$ and V is any number between $f(a)$ and $f(b)$, then the graph of f must cross the horizontal line $y = V$ somewhere between $x = a$ and $x = b$ (see margin figure). Since f is continuous, its graph cannot "hop" over the line $y = V$.

We often take this theorem for granted in some common situations:

- If a child's temperature rose from $98.6°$F to $101.3°$F, then there was an instant when the child's temperature was exactly $100°$F. (In fact, every temperature between $98.6°$F and $101.3°$F occurred at some instant.)

- If you dove to pick up a shell 25 feet below the surface of a lagoon, then at some instant in time you were 17 feet below the surface. (Actually, you want to be at 17 feet twice. Why?)

- If you started driving from a stop (velocity = 0) and accelerated to a velocity of 30 kilometers per hour, then there was an instant when your velocity was exactly 10 kilometers per hour.

But we cannot apply the Intermediate Value Theorem if the function is not continuous:

- In 1987 it cost 22¢ to mail a first-class letter inside the United States, and in 1990 it cost 25¢ to mail the same letter, but we cannot conclude that there was a time when it cost 23¢ or 24¢ to send the letter. (Postal rates did not increase in a continuous fashion. They jumped directly from 22¢ to 25¢.)

- Prices, taxes and rates of pay change in jumps — discrete steps — without taking on the intermediate values.

The Intermediate Value Theorem (IVT) is an example of an "existence theorem": it concludes that something exists (a number c so that $f(c) = V$). But like many existence theorems, it does not tell us how to find the the thing that exists (the value of c) and is of no use in actually finding those numbers or objects.

Bisection Algorithm for Approximating Roots

The IVT can help us finds roots of functions and solve equations. If f is continuous on $[a, b]$ and $f(a)$ and $f(b)$ have opposite signs (one is positive and one is negative), then 0 is an intermediate value between $f(a)$ and $f(b)$ so f will have a root c between $x = a$ and $x = b$ where $f(c) = 0$.

While the IVT does not tell us how to find c, it lays the groundwork for a method commonly used to approximate the roots of continuous functions.

Bisection Algorithm for Finding a Root of $f(x)$

1. Find two values of x (call them a and b) so that $f(a)$ and $f(b)$ have opposite signs. (The IVT will then guarantee that $f(x)$ has a root between a and b.)

2. Calculate the midpoint (or **bisection point**) of the interval $[a, b]$, using the formula $m = \dfrac{a+b}{2}$, and evaluate $f(m)$.

3. (a) If $f(m) = 0$, then m is a root of f and we are done.

   (b) If $f(m) \neq 0$, then $f(m)$ has the sign opposite $f(a)$ or $f(b)$:
      i. if $f(a)$ and $f(m)$ have opposite signs, then f has a root in $[a, m]$ so put $b = m$
      ii. if $f(b)$ and $f(m)$ have opposite signs, then f has a root in $[m, b]$ so put $a = m$

4. Repeat steps 2 and 3 until a root is found exactly or is approximated closely enough.

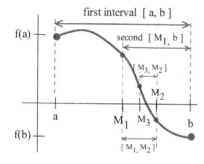

first interval [a, b]

f(a)

second [M₁, b]

[M₃, M₂]

M₂

a   M₁ M₃   b

f(b)

[M₁, M₂]'

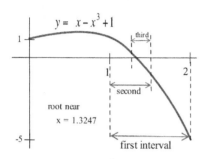

$y = x - x^3 + 1$

1

third

1   2

second

root near
x = 1.3247

-5

first interval

The length of the interval known to contain a root is cut in half each time through steps 2 and 3, so the Bisection Algorithm quickly "squeezes" in on a root (see margin figure).

The steps of the Bisection Algorithm can be done "by hand," but it is tedious to do very many of them that way. Computers are very good with this type of tedious repetition, and the algorithm is simple to program.

Example 5. Find a root of $f(x) = -x^3 + x + 1$.

Solution. $f(0) = 1$ and $f(1) = 1$ so we cannot conclude that f has a root between 0 and 1. $f(1) = 1$ and $f(2) = -5$ have opposite signs, so by the IVT (this function is a polynomial, so it is continuous everywhere and the IVT applies) we know that there is a number c between 1 and 2 such that $f(c) = 0$ (see figure). The midpoint of the interval $[1, 2]$ is $m = \frac{1+2}{2} = \frac{3}{2} = 1.5$ and $f\left(\frac{3}{2}\right) = -\frac{7}{8}$ so f changes sign between 1 and 1.5 and we can be sure that there is a root between 1 and 1.5. If we repeat the operation for the interval $[1, 1.5]$, the midpoint is $m = \frac{1+1.5}{2} = 1.25$, and $f(1.25) = \frac{19}{64} > 0$ so f changes sign between 1.25 and 1.5 and we know f has a root between 1.25 and 1.5.

Repeating this procedure a few more times, we get:

a	b	$m = \frac{b+a}{2}$	$f(a)$	$f(b)$	$f(m)$	root between	
1	2		1	−5		1	2
1	2	1.5	1	−5	−0.875	1	1.5
1	1.5	1.25	1	−0.875	0.2969	1.25	1.5
1.25	1.5	1.375	0.2969	−0.875	−0.2246	1.25	1.375
1.25	1.375	1.3125	0.2969	−0.2246	0.0515	1.3125	1.375
1.3125	1.375	1.34375					

If we continue the table, the interval containing the root will squeeze around the value 1.324718.   ◀

The Bisection Algorithm has one major drawback: there are some roots it does not find. The algorithm requires that the function take on both positive and negative values near the root so that the graph actually crosses the x-axis. The function $f(x) = x^2 - 6x + 9 = (x - 3)^2$ has the root $x = 3$ but is never negative (see margin figure). We cannot find two starting points a and b so that $f(a)$ and $f(b)$ have opposite signs, so we cannot use the Bisection Algorithm to find the root $x = 3$. In Chapter 2 we will see another method — Newton's Method — that does find roots of this type.

The Bisection Algorithm requires that we supply two starting x-values, a and b, at which the function has opposite signs. These values can often be found with a little "trial and error," or we can examine the graph of the function to help pick the two values.

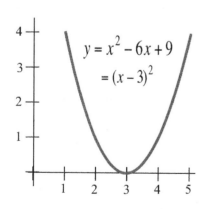

4

3

$y = x^2 - 6x + 9$

$= (x - 3)^2$

2

1

1   2   3   4   5

Finally, the Bisection Algorithm can also be used to solve equations, because the solution of any equation can always be transformed into an equivalent problem of finding roots by moving everything to one side of the equal sign. For example, the problem of solving the equation $x^3 = x + 1$ can be transformed into the equivalent problem of solving $x^3 - x - 1 = 0$ or of finding the roots of $f(x) = x^3 - x - 1$, which is equivalent to the problem we solved in the previous example.

Example 6. Find all solutions of $\sin(x) = \dfrac{2x+1}{x-2}$ (with x in radians.)

Solution. We can convert this problem of solving an equation to the problem of finding the roots of

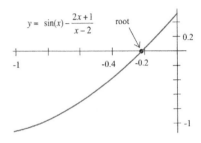

$$f(x) = \sin(x) - \frac{2x+1}{x-2} = 0$$

The function $f(x)$ is continuous everywhere except at $x = 2$, and the graph of $f(x)$ (in the margin) can help us find two starting values for the Bisection Algorithm. The graph shows that $f(-1)$ is negative and $f(0)$ is positive, and we know $f(x)$ is continuous on the interval $[-1, 0]$. Using the algorithm with the starting interval $[-1, 0]$, we know that a root is contained in the shrinking intervals $[-0.5, 0]$, $[-0.25, 0]$, $[-0.25, -0.125]$, …, $[-0.238281, -0.236328]$, …, $[-0.237176, -0.237177]$ so the root is approximately -0.237177.

We might notice that $f(0) = 0.5 > 0$ while $f(\pi) = 0 - \frac{2\pi+1}{\pi-2} \approx -6.38 < 0$. Why is it wrong to conclude that $f(x)$ has another root between $x = 0$ and $x = \pi$?   ◀

1.3 Problems

1. At which points is the function in the graph below discontinuous?

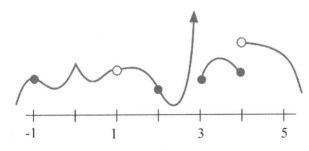

2. At which points is the function in the graph below discontinuous?

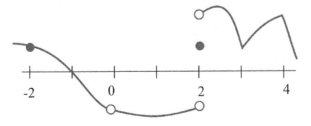

3. Find at least one point at which each function is not continuous and state which of the three conditions in the definition of continuity is violated at that point.

(a) $\dfrac{x+5}{x-3}$

(b) $\dfrac{x^2+x-6}{x-2}$

(c) $\sqrt{\cos(x)}$

(d) $\lfloor x^2 \rfloor$

(e) $\dfrac{x}{\sin(x)}$

(f) $\dfrac{x}{x}$

(g) $\ln(x^2)$

(h) $\dfrac{\pi}{x^2-6x+9}$

(i) $\tan(x)$

4. Which *two* of the following functions are not continuous? Use appropriate theorems to justify that each of the other functions is continuous.

(a) $\dfrac{7}{\sqrt{2+\sin(x)}}$

(b) $\cos^2(x^5-7x+\pi)$

(c) $\dfrac{x^2-5}{1+\cos^2(x)}$

(d) $\dfrac{x^2-5}{1+\cos(x)}$

(e) $\lfloor 3+0.5\sin(x) \rfloor$

(f) $\lfloor 0.3\sin(x)+1.5 \rfloor$

(g) $\sqrt{\cos(\sin(x))}$

(h) $\sqrt{x^2-6x+10}$

(i) $\sqrt[3]{\cos(x)}$

(j) $2^{\sin(x)}$

(k) $1-3^{-x}$

5. A continuous function f has the values:

x	0	1	2	3	4	5
$f(x)$	5	3	-2	-1	3	-2

(a) f has at least _____ roots between 0 and 5.

(b) $f(x)=4$ in at least _____ places between $x=0$ and $x=5$.

(c) $f(x)=2$ in at least _____ places between $x=0$ and $x=5$.

(d) $f(x)=3$ in at least _____ places between $x=0$ and $x=5$.

(e) Is it possible for $f(x)$ to equal 7 for some x-value(s) between 0 and 5?

6. A continuous function g has the values:

x	1	2	3	4	5	6	7
$g(x)$	-3	1	4	-1	3	-2	-1

(a) g has at least _____ roots between 1 and 5.

(b) $g(x)=3.2$ in at least _____ places between $x=1$ and $x=7$.

(c) $g(x)=-0.7$ in at least _____ places between $x=3$ and $x=7$.

(d) $g(x)=1.3$ in at least _____ places between $x=2$ and $x=6$.

(e) Is it possible for $g(x)$ to equal π for some x-value(s) between 5 and 6?

7. This problem asks you to verify that the Intermediate Value Theorem is true for some particular functions, intervals and intermediate values. In each problem you are given a function f, an interval $[a,b]$ and a value V. Verify that V is between $f(a)$ and $f(b)$ and find a value of c in the given interval so that $f(c)=V$.

(a) $f(x)=x^2$ on $[0,3]$, $V=2$

(b) $f(x)=x^2$ on $[-1,2]$, $V=3$

(c) $f(x)=\sin(x)$ on $\left[0,\frac{\pi}{2}\right]$, $V=\frac{1}{2}$

(d) $f(x)=x$ on $[0,1]$, $V=\frac{1}{3}$

(e) $f(x)=x^2-x$ on $[2,5]$, $V=4$

(f) $f(x)=\ln(x)$ on $[1,10]$, $V=2$

8. Two students claim that they both started with the points $x=1$ and $x=9$ and applied the Bisection Algorithm to the function graphed below. The first student says that the algorithm converged to the root near $x=8$, but the second claims that the algorithm will converge to the root near $x=4$. Who is correct?

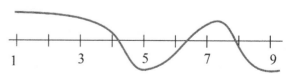

9. Two students claim that they both started with the points $x = 0$ and $x = 5$ and applied the Bisection Algorithm to the function graphed below. The first student says that the algorithm converged to the root labeled A, but the second claims that the algorithm will converge to the root labeled B. Who is correct?

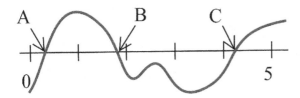

10. If you apply the Bisection Algorithm to the function graphed below, which root does the algorithm find if you use:

    (a) starting points 0 and 9?

    (b) starting points 1 and 5?

    (c) starting points 3 and 5?

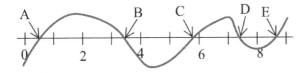

11. If you apply the Bisection Algorithm to the function graphed below, which root does the algorithm find if you use:

    (a) starting points 3 and 7?

    (b) starting points 5 and 6?

    (c) starting points 1 and 6?

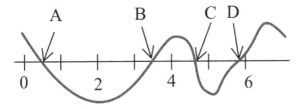

In 12–17, use the IVT to verify each function has a root in the given interval(s). Then use the Bisection Algorithm to narrow the location of that root to an interval of length less than or equal to 0.1.

12. $f(x) = x^2 - 2$ on $[0,3]$

13. $g(x) = x^3 - 3x^2 + 3$ on $[-1,0]$, $[1,2]$, $[2,4]$

14. $h(t) = t^5 - 3t + 1$ on $[1,3]$

15. $r(x) = 5 - 2^x$ on $[1,3]$

16. $s(x) = \sin(2x) - \cos(x)$ on $[0, \pi]$

17. $p(t) = t^3 + 3t + 1$ on $[-1,1]$

18. Explain what is wrong with this reasoning: If $f(x) = \frac{1}{x}$ then

$$f(-1) = -1 < 0 \quad \text{and} \quad f(1) = 1 > 0$$

so f must have a root between $x = -1$ and $x = 1$.

19. Each of the following statements is false for some functions. For each statement, sketch the graph of a counterexample.

    (a) If $f(3) = 5$ and $f(7) = -3$, then f has a root between $x = 3$ and $x = 7$.

    (b) If f has a root between $x = 2$ and $x = 5$, then $f(2)$ and $f(5)$ have opposite signs.

    (c) If the graph of a function has a sharp corner, then the function is not continuous there.

20. Define $A(x)$ to be the **area** bounded by the t- and y-axes, the curve $y = f(t)$, and the vertical line $t = x$ (see figure below). It is clear that $A(1) < 2$ and $A(3) > 2$. Do you think there is a value of x between 1 and 3 so that $A(x) = 2$? If so, justify your conclusion and estimate the location of the value of x that makes $A(x) = 2$. If not, justify your conclusion.

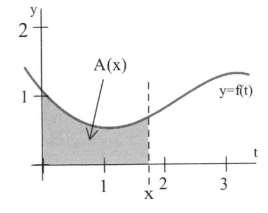

21. Define $A(x)$ to be the **area** bounded by the t- and y-axes, the curve $y = f(t)$, and the vertical line $t = x$ (see figure below).

    (a) Shade the part of the graph represented by $A(2.1) - A(2)$ and estimate the value of $\dfrac{A(2.1) - A(2)}{0.1}$.

    (b) Shade the part of the graph represented by $A(4.1) - A(4)$ and estimate the value of $\dfrac{A(4.1) - A(4)}{0.1}$.

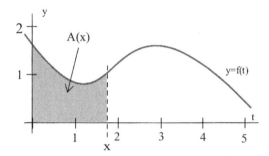

22. (a) A square sheet of paper has a straight line drawn on it from the lower-left corner to the upper-right corner. Is it possible for you to start on the left edge of the sheet and draw a "connected" line to the right edge that does not cross the diagonal line?

    (b) Prove: If f is continuous on the interval $[0, 1]$ and $0 \leq f(x) \leq 1$ for all x, then there is a number c with $0 \leq c \leq 1$ such that $f(c) = c$. (The number c is called a "fixed point" of f because the image of c is the same as c: f does not "move" c.) Hint: Define a new function $g(x) = f(x) - x$ and start by considering the values $g(0)$ and $g(1)$.

    (c) What does part (b) have to do with part (a)?

    (d) Is the theorem in part (b) true if we replace the closed interval $[0, 1]$ with the open interval $(0, 1)$?

23. A piece of string is tied in a loop and tossed onto quadrant I enclosing a single region (see figure below).

    (a) Is it always possible to find a line L passing through the origin so that L divides the region into two equal areas? (Justify your answer.)

    (b) Is it always possible to find a line L parallel to the x-axis so that L divides the region into two equal areas? (Justify your answer.)

    (c) Is it always possible to find two lines, L parallel to the x-axis and M parallel to the y-axis, so that L and M divide the region into four equal areas? (Justify your answer.)

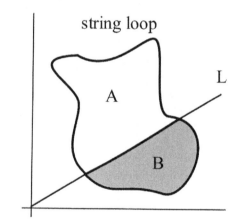

1.3 Practice Answers

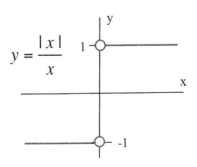

$y = \dfrac{|x|}{x}$

1. $f(x) = \dfrac{|x|}{x}$ (see margin figure) is continuous everywhere **except** at $x = 0$, where this function is not defined.

   If $a > 0$, then $\lim\limits_{x \to a} \dfrac{|x|}{x} = 1 = f(a)$ so f is continuous at a.

   If $a < 0$, then $\lim\limits_{x \to a} \dfrac{|x|}{x} = -1 = f(a)$ so f is continuous at a.

   But $f(0)$ is not defined and

   $$\lim_{x \to 0^-} \frac{|x|}{x} = -1 \neq 1 = \lim_{x \to 0^+} \frac{|x|}{x}$$

   so $\lim\limits_{x \to a} \dfrac{|x|}{x}$ does not exist.

2. (a) To prove that $k \cdot f$ is continuous at a, we need to prove that $k \cdot f$ satisfies the definition of continuity at a: $\lim\limits_{x \to a} k \cdot f(x) = k \cdot f(a)$. Using results about limits, we know

   $$\lim_{x \to a} k \cdot f(x) = k \cdot \lim_{x \to a} f(x) = k \cdot f(a)$$

   (because f is continuous at a) so $k \cdot f$ is continuous at a.

   (b) To prove that $f - g$ is continuous at a, we need to prove that $f - g$ satisfies the definition of continuity at a: $\lim\limits_{x \to a} [f(x) - g(x)] = f(a) - g(a)$. Again using information about limits:

   $$\lim_{x \to a} [f(x) - g(x)] = \lim_{x \to a} f(x) - \lim_{x \to a} g(x) = f(a) - g(a)$$

   (because f and g are both continuous at a) so $f - g$ is continuous at a.

1.4   Definition of Limit

It may seem strange that we have been using and calculating the values of limits for quite a while without having a precise definition of "limit," but the history of mathematics shows that many concepts — including limits — were successfully used before they were precisely defined or even fully understood. We have chosen to follow the historical sequence, emphasizing the intuitive and graphical meaning of limit because most students find these ideas and calculations easier than the definition.

This intuitive and graphical understanding of limit was sufficient for the first 100-plus years of the development of calculus (from Newton and Leibniz in the late 1600s to Cauchy in the early 1800s) and it is sufficient for using and understanding the results in beginning calculus.

Mathematics, however, is more than a collection of useful tools, and part of its power and beauty comes from the fact that in mathematics terms are precisely defined and results are rigorously proved. Mathematical tastes (what is mathematically beautiful, interesting, useful) change over time, but because of careful definitions and proofs, the results remain true — everywhere and forever. Textbooks seldom give all of the definitions and proofs, but it is important to mathematics that such definitions and proofs exist.

The goal of this section is to provide a precise definition of the limit of a function. The definition will not help you calculate the values of limits, but it provides a precise statement of what a limit is. The definition of limit is then used to verify the limits of some functions and prove some general results.

The Intuitive Approach

The precise ("formal") definition of limit carefully states the ideas that we have already been using graphically and intuitively. The following side-by-side columns show some of the phrases we have been using to describe limits, and those phrases — particularly the last ones — provide the basis on which to build the definition of limit.

A Particular Limit	**General Limit**
$\lim\limits_{x \to 3} 2x - 1 = 5$	$\lim\limits_{x \to a} f(x) = L$

"as the values of x approach 3, the values of $2x - 1$ approach (are arbitrarily close to) 5"

"when x is close to 3 (but not equal to 3), the value of $2x - 1$ is close to 5"

"we can guarantee that the values of $2x - 1$ are as close to 5 as we want by restricting the values of x to be sufficiently close to 3 (but not equal to 3)"

"as the values of x approach a, the values of $f(x)$ approach (are arbitrarily close to) L"

"when x is close to a (but not equal to a), the value of $f(x)$ is close to L"

"we can guarantee that the values of $f(x)$ are as close to L as we want by restricting the values of x to be sufficiently close to a (but not equal to a)"

Let's examine what the last phrase ("we can...") means for the Particular Limit in the previous discussion.

Example 1. We know $\lim\limits_{x\to3} 2x - 1 = 5$ and need to show that we can guarantee that the values of $f(x) = 2x - 1$ are as close to 5 as we want by restricting the values of x to be sufficiently close to 3.

What values of x guarantee that $f(x) = 2x - 1$ is within:

    (a) 1 unit of 5?   (b) 0.2 units of 5?   (c) E units of 5?

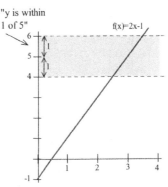
"y is within 1 of 5"

Solution. (a) "Within 1 unit of 5" means between $5 - 1 = 4$ and $5 + 1 = 6$, so the question can be rephrased as "for what values of x is $y = 2x - 1$ between 4 and 6: $4 < 2x - 1 < 6$?" We want to know which values of x ensure the values of $y = 2x - 1$ are in the the shaded band in the uppermost margin figure. The algebraic process is straightforward:

$$4 < 2x - 1 < 6 \quad\Rightarrow\quad 5 < 2x < 7 \quad\Rightarrow\quad 2.5 < x < 3.5$$

We can restate this result as follows: "If x is within 0.5 units of 3, then $y = 2x - 1$ is within 1 unit of 5." (See second margin figure) Any smaller distance also satisfies the guarantee: for example, "If x is within 0.4 units of 3, then $y = 2x - 1$ is within 1 unit of 5." (See third margin figure)

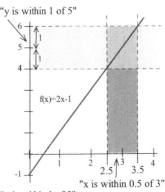
"y is within 1 of 5"
"x is within 0.5 of 3"

(b) "Within 0.2 units of 5" means between $5 - 0.2 = 4.8$ and $5 + 0.2 = 5.2$, so the question can be rephrased as "for which values of x is $y = 2x - 1$ between 4.8 and 5.2: $4.8 < 2x - 1 < 5.2$?" Solving for x, we get $5.8 < 2x < 6.2$ and $2.9 < x < 3.1$. "If x is within 0.1 units of 3, then $y = 2x - 1$ is within 0.2 units of 5." (See fourth margin figure.) Any smaller distance also satisfies the guarantee.

Rather than redoing these calculations for every possible distance from 5, we can do the work once, generally:

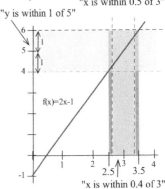
"y is within 1 of 5"
"x is within 0.4 of 3"

(c) "Within E unit of 5" means between $5 - E$ and $5 + E$, so the question becomes, "For what values of x is $y = 2x - 1$ between $5 - E$ and $5 + E$: $5 - E < 2x - 1 < 5 + E$?" Solving $5 - E < 2x - 1 < 5 + E$ for x, we get:

$$6 - E < 2x < 6 + E \quad\Rightarrow\quad 3 - \frac{E}{2} < x < 3 + \frac{E}{2}$$

"If x is within $\frac{E}{2}$ units of 3, then $y = 2x - 1$ is within E units of 5." (See last figure.) Any smaller distance also works. ◄

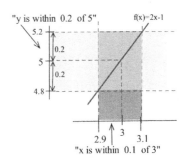
"y is within 0.2 of 5"
"x is within 0.1 of 3"

Part (c) of Example 1 illustrates the power of general solutions in mathematics. Rather than redoing similar calculations every time someone demands that $f(x) = 2x - 1$ be within some given distance of 5, we did the calculations once. And then we can quickly respond for any given distance. For the question "What values of x guarantee that $f(x) = 2x - 1$ is within 0.4, 0.1 or 0.006 units of 5?" we can answer, "If x is within 0.2 ($= \frac{0.4}{2}$), 0.05 ($= \frac{0.1}{2}$) or 0.003 ($= \frac{0.006}{2}$) units of 3."

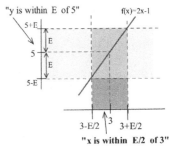
"y is within E of 5"
"x is within E/2 of 3"

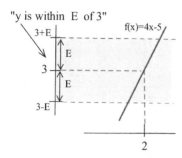

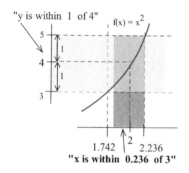

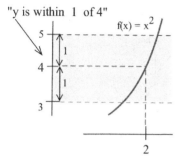

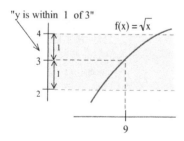

Practice 1. Knowing that $\lim\limits_{x \to 2} 4x - 5 = 3$, determine which values of x guarantee that $f(x) = 4x - 5$ is within

    (a) 1 unit of 3.   (b) 0.08 units of 3.   (c) E units of 3.

The same ideas work even if the graphs of the functions are not straight lines, but the calculations become more complicated.

Example 2. Knowing that $\lim\limits_{x \to 2} x^2 = 4$, determine which values of x guarantee that $f(x) = x^2$ is within:

    (a) 1 unit of 4.   (b) 0.2 units of 4.

State each answer in the form: "If x is within _____ units of 2, then $f(x)$ is within _____ units of 4."

Solution. (a) If x^2 is within 1 unit of 4 (and x is near 2, hence positive) then $3 < x^2 < 5$ so $\sqrt{3} < x < \sqrt{5}$ or $1.732 < x < 2.236$. The interval containing these x values extends from $2 - \sqrt{3} \approx 0.268$ units to the left of 2 to $\sqrt{5} - 2 \approx 0.236$ units to the right of 2. Because we want to specify a single distance on each side of 2, we can pick the **smaller** of the two distances, 0.236, and say: "If x is within 0.236 units of 2, then $f(x)$ is within 1 unit of 4."

(b) Similarly, if x^2 is within 0.2 units of 4 (and x is near 2, so $x > 0$) then $3.8 < x^2 < 4.2$ so $\sqrt{3.8} < x < \sqrt{4.2}$ or $1.949 < x < 2.049$. The interval containing these x values extends from $2 - \sqrt{3.8} \approx 0.051$ units to the left of 2 to $\sqrt{4.2} - 2 \approx 0.049$ units to the right of 2. Again picking the smaller of the two distances, we can say: "If x is within 0.049 units of 2, then $f(x)$ is within 1 unit of 4."   ◀

The situation in Example 2—with different distances on the left and right sides—is very common, and we **always** pick our single distance to be the **smaller** of the distances to the left and right. By using the smaller distance, we can be certain that if x is within that smaller distance on either side, then the value of $f(x)$ is within the specified distance of the value of the limit.

Practice 2. Knowing that $\lim\limits_{x \to 9} \sqrt{x} = 3$, determine which values of x guarantee that $f(x) = \sqrt{x}$ is within:

    (a) 1 unit of 3.   (b) 0.2 units of 3.

State each answer in the form: "If x is within _____ units of 9, then $f(x)$ is within _____ units of 3."

The same ideas can also be used when the function and the specified distance are given graphically, and in that case we can give the answer graphically.

Example 3. In the margin figure, $\lim_{x \to 2} f(x) = 3$. Which values of x guarantee that $y = f(x)$ is within E units (given graphically) of 3? State your answer in the form: "If x is within (*show a distance D graphically*) of 2, then $f(x)$ is within E units of 3."

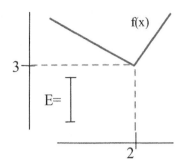

Solution. The solution process requires several steps:

(i) Use the given distance E to find the values $3 - E$ and $3 + E$ on the y-axis. (See margin.)

(ii) Sketch the horizontal band with lower edge at $y = 3 - E$ and upper edge at $y = 3 + E$.

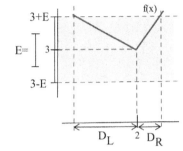

(iii) Find the first locations to the right and left of $x = 2$ where the graph of $y = f(x)$ crosses the lines $y = 3 - E$ and $y = 3 + E$, and at these locations draw vertical line segments extending to the x-axis.

(iv) On the x-axis, graphically determine the distance from 2 to the vertical line on the left (labeled D_L) and from 2 to the vertical line on the right (labeled D_R).

(v) Let the length D be the smaller of the lengths D_L and D_R.

If x is within D units of 2, then $f(x)$ is within E units of 3.   ◄

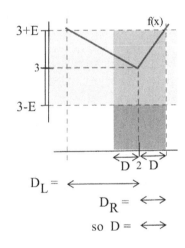

Practice 3. In the last margin figure, $\lim_{x \to 3} f(x) = 1.8$. Which values of x guarantee that $y = f(x)$ is within E units (given graphically) of 1.8?

$D_L = \longleftrightarrow$

$D_R = \longleftrightarrow$

so $D = \longleftrightarrow$

The Formal Definition of Limit

The ideas from the previous Examples and Practice problems, restated for general functions and limits, provide the basis for the definition of limit given below. The use of the lowercase Greek letters ϵ (epsilon) and δ (delta) in the definition is standard, and this definition is sometimes called the "epsilon-delta" definition of a limit.

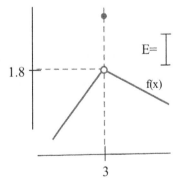

Definition of $\lim_{x \to a} f(x) = L$:
For every given number $\epsilon > 0$ there is a number $\delta > 0$ so that

if x is within δ units of a (and $x \neq a$)
then $f(x)$ is within ϵ units of L

Equivalently: $|f(x) - L| < \epsilon$ whenever $0 < |x - a| < \delta$

In this definition, ϵ represents the given distance on either side of the limiting value $y = L$, and δ is the distance on each side of the point $x = a$ on the x-axis that we have been finding in the previous

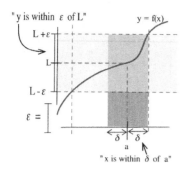

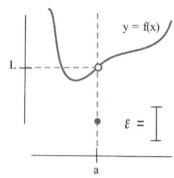

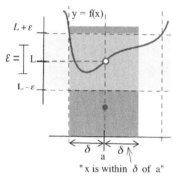

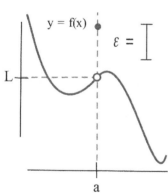

examples. This definition has the form of a "challenge and response": for any positive challenge ϵ (make $f(x)$ within ϵ of L), there is a positive response δ (start with x within δ of a and $x \neq a$).

Example 4. As seen in the second margin figure, $\lim_{x \to a} f(x) = L$, with a value for ϵ given graphically as a length. Find a length for δ that satisfies the definition of limit (so "if x is within δ of a, and $x \neq a$, then $f(x)$ is within ϵ of L").

Solution. Follow the steps outlined in Example 3. The length for δ is shown in the third margin figure, and any shorter length for δ also satisfies the definition. ◄

Practice 4. In the bottom margin figure, $\lim_{x \to a} f(x) = L$, with a value for ϵ given graphically. Find a length for δ that satisfies the definition of limit.

Example 5. Prove that $\lim_{x \to 3} 4x - 5 = 7$.

Solution. We need to show that "for every given $\epsilon > 0$ there is a $\delta > 0$ so that if x is within δ units of 3 (and $x \neq 3$) then $4x - 5$ is within ϵ units of 7."

Actually, there are two things we need to do. First, we need to find a value for δ (typically depending on ϵ) and, second, we need to show that our δ really does satisfy the "if...then..." part of the definition.

Finding δ is similar to part (c) in Example 1 and Practice 1: Assume $4x - 5$ is within ϵ units of 7 and solve for x. If $7 - \epsilon < 4x - 5 < 7 + \epsilon$ then $12 - \epsilon < 4x < 12 + \epsilon \Rightarrow 3 - \frac{\epsilon}{4} < x < 3 + \frac{\epsilon}{4}$ so x is within $\frac{\epsilon}{4}$ units of 3. Put $\delta = \frac{\epsilon}{4}$.

To show that $\delta = \frac{\epsilon}{4}$ satisfies the definition, we merely reverse the order of the steps in the previous paragraph. Assume that x is within δ units of 3. Then $3 - \delta < x < 3 + \delta$, so:

$$3 - \frac{\epsilon}{4} < x < 3 + \frac{\epsilon}{4} \quad \Rightarrow \quad 12 - \epsilon < 4x < 12 + \epsilon$$
$$\Rightarrow \quad 7 - \epsilon < 4x - 5 < 7 + \epsilon$$

so we can conclude that $f(x) = 4x - 5$ is within ϵ units of 7. This formally verifies that $\lim_{x \to 3} 4x - 5 = 7$. ◄

Practice 5. Prove that $\lim_{x \to 4} 5x + 3 = 23$.

The method used to prove the values of the limits for these particular linear functions can also be used to prove the following general result about the limits of linear functions.

Theorem: $\lim_{x \to a} mx + b = ma + b$

Proof. Let $f(x) = mx + b$.

Case 1: $m = 0$. Then $f(x) = 0x + b = b$ is simply a constant function, and any value for $\delta > 0$ satisfies the definition. Given any value of $\epsilon > 0$, let $\delta = 1$ (any positive value for δ works). If x is is within 1 unit of a, then $f(x) - f(a) = b - b = 0 < \epsilon$, so we have shown that for any $\epsilon > 0$ there is a $\delta > 0$ that satisfies the limit definition.

Case 2: $m \neq 0$. For any $\epsilon > 0$, put $\delta = \frac{\epsilon}{|m|} > 0$. If x is within $\delta = \frac{\epsilon}{|m|}$ of a then

$$a - \frac{\epsilon}{|m|} < x < a + \frac{\epsilon}{|m|} \Rightarrow \frac{\epsilon}{|m|} < x - a < \frac{\epsilon}{|m|} \Rightarrow |x - a| < \frac{\epsilon}{|m|}$$

Then the distance between $f(x)$ and $L = ma + b$ is:

$$|f(x) - L| = |(mx + b) - (ma + b)| = |mx - ma|$$
$$= |m| \cdot |x - a| < |m| \frac{\epsilon}{|m|} = \epsilon$$

so $f(x)$ is within ϵ of $L = ma + b$.

In each case, we have shown that "given any $\epsilon > 0$, there is a $\delta > 0$" that satisfies the rest of the limit definition. □

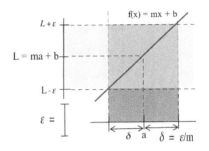

If there is even a single value of ϵ for which there is no δ, then we say that the limit "**does not exist.**"

Example 6. With $f(x)$ defined as:

$$f(x) = \begin{cases} 2 & \text{if } x < 1 \\ 4 & \text{if } x > 1 \end{cases}$$

use the limit definition to prove that $\lim\limits_{x \to 1} f(x)$ does not exist.

Solution. One common proof technique in mathematics is called "proof by contradiction" and that is the method we use here:

- We assume that the limit does exist and equals some number L.
- We show that this assumption leads to a contradiction
- We conclude that the assumption must have been false.

We therefore conclude that the limit does not exist.

First, assume that the limit exists: $\lim\limits_{x \to 1} f(x) = L$ for some value for L. Let $\epsilon = \frac{1}{2}$. Then, because we are assuming that the limit exists, there is a $\delta > 0$ so that if x is within δ of 1 then $f(x)$ is within ϵ of L.

Next, let x_1 be between 1 and $1 + \delta$. Then $x_1 > 1$ so $f(x_1) = 4$. Also, x_1 is within δ of 1 so $f(x_1) = 4$ is within $\frac{1}{2}$ of L, which means that L is between 3.5 and 4.5: $3.5 < L < 4.5$.

Let x_2 be between 1 and $1 - \delta$. Then $x_2 < 1$, so $f(x_2) = 2$. Also, x_2 is within δ of 1 so $f(x_2) = 2$ is within $\frac{1}{2}$ of L, which means that L is between 1.5 and 2.5: $1.5 < L < 2.5$.

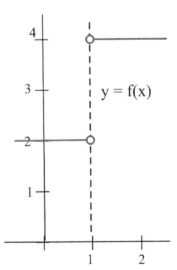

The definition says "for every ϵ" so we can certainly pick $\frac{1}{2}$ as our ϵ value; why we chose this particular value for ϵ shows up later in the proof.

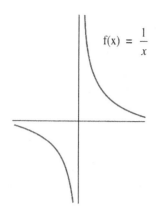

$f(x) = \dfrac{1}{x}$

There are rigorous proofs of all of the other limit properties in the Main Limit Theorem, but they are somewhat more complicated than the proofs given here.

These inequalities provide the contradiction we hoped to find. There is no value L that satisfies both $3.5 < L < 4.5$ **and** $1.5 < L < 2.5$, so our assumption must be false: $f(x)$ does not have a limit as $x \to 1$.   ◄

Practice 6. Use the limit definition to prove that $\lim\limits_{x \to 0} \dfrac{1}{x}$ does not exist.

Proofs of Two Limit Theorems

We conclude with proofs of two parts of the Main Limit Theorem so you can see how such proofs proceed — you have already used these theorems to evaluate limits.

> **Theorem:**
>
> If $\quad \lim\limits_{x \to a} f(x) = L$
>
> then $\quad \lim\limits_{x \to a} k \cdot f(x) = kL$

Proof. Case $k = 0$: The theorem is true but not very interesting:

$$\lim_{x \to a} k \cdot f(x) = \lim_{x \to a} 0 \cdot f(x) = \lim_{x \to a} 0 = 0 = 0 \cdot L = kL$$

Case $k \neq 0$: Because $\lim\limits_{x \to a} f(x) = L$, then, by the definition, for every $\epsilon > 0$ there is a $\delta > 0$ so that $|f(x) - L| < \epsilon$ whenever $|x - a| < \delta$. For any $\epsilon > 0$, we know $\frac{\epsilon}{|k|} > 0$, so pick a value of δ that satisfies $|f(x) - L| < \frac{\epsilon}{|k|}$ whenever $|x - a| < \delta$.

When $|x - a| < \delta$ ("x is within δ of a") then $|f(x) - L| < \frac{\epsilon}{|k|}$ ("$f(x)$ is within $\frac{\epsilon}{|k|}$ of L") so $|k| \cdot |f(x) - L| < \epsilon \quad \Rightarrow \quad |k \cdot f(x) - k \cdot L| < \epsilon$ (that is, $k \cdot f(x)$ is within ϵ of $k \cdot L$).   □

> **Theorem:**
>
> If $\quad \lim\limits_{x \to a} f(x) = L \quad$ and $\quad \lim\limits_{x \to a} g(x) = M$
>
> then $\quad \lim\limits_{x \to a} [f(x) + g(x)] = L + M.$

Proof. Given any $\epsilon > 0$, we know $\frac{\epsilon}{2} > 0$, so there is a number $\delta_f > 0$ such that when $|x - a| < \delta_f$ then $|f(x) - L| < \frac{\epsilon}{2}$ ("if x is within δ_f of a, then $f(x)$ is within $\frac{\epsilon}{2}$ of L").

Likewise, there is a number $\delta_g > 0$ such that when $|x - a| < \delta_g$ then $|g(x) - M| < \frac{\epsilon}{2}$ ("if x is within δ_g of a, then $g(x)$ is within $\frac{\epsilon}{2}$ of M").

Let δ be the smaller of δ_f and δ_g. If $|x - a| < \delta$ then $|f(x) - L| < \frac{\epsilon}{2}$ and $|g(x) - M| < \frac{\epsilon}{2}$ so:

Here we use the "triangle inequality":

$$|a + b| \leq |a| + |b|$$

$$|(f(x) + g(x)) - (L + M))| = |(f(x) - L) + (g(x) - M)|$$

$$\leq |f(x) - L| + |g(x) - M| < \frac{\epsilon}{2} + \frac{\epsilon}{2} = \epsilon$$

so $f(x) + g(x)$ is within ϵ of $L + M$ whenever x is within δ of a.   □

1.4 Problems

In Problems 1–4, state each answer in the form "If x is within _____ units of..."

1. Knowing that $\lim\limits_{x\to 3} 2x + 1 = 7$, what values of x guarantee that $f(x) = 2x + 1$ is within:

(a) 1 unit of 7?   (b) 0.6 units of 7?

(c) 0.04 units of 7?   (d) ϵ units of 7?

2. Knowing that $\lim\limits_{x\to 1} 3x + 2 = 5$, what values of x guarantee that $f(x) = 3x + 2$ is within:

(a) 1 unit of 5?   (b) 0.6 units of 5?

(c) 0.09 units of 5?   (d) ϵ units of 5?

3. Knowing that $\lim\limits_{x\to 2} 4x - 3 = 5$, what values of x guarantee that $f(x) = 4x - 3$ is within:

(a) 1 unit of 5?   (b) 0.4 units of 5?

(c) 0.08 units of 5?   (d) ϵ units of 5?

4. Knowing that $\lim\limits_{x\to 1} 5x - 3 = 2$, what values of x guarantee that $f(x) = 5x - 3$ is within:

(a) 1 unit of 2?   (b) 0.5 units of 2?

(c) 0.01 units of 2?   (d) ϵ units of 2?

5. For Problems 1–4, list the slope of each function f and the δ (as a function of ϵ). For these linear functions f, how is δ related to the slope?

6. You have been asked to cut two boards (exactly the same length after the cut) and place them end to end. If the combined length must be within 0.06 inches of 30 inches, then each board must be within how many inches of 15?

7. You have been asked to cut three boards (exactly the same length after the cut) and place them end to end. If the combined length must be within 0.06 inches of 30 inches, then each board must be within how many inches of 10?

8. Knowing that $\lim\limits_{x\to 3} x^2 = 9$, what values of x guarantee that $f(x) = x^2$ is within:

(a) 1 unit of 9?   (b) 0.2 units of 9?

9. Knowing that $\lim\limits_{x\to 2} x^3 = 8$, what values of x guarantee that $f(x) = x^3$ is within:

(a) 0.5 units of 8?   (b) 0.05 units of 8?

10. Knowing that $\lim\limits_{x\to 16} \sqrt{x} = 4$, what values of x guarantee that $f(x) = \sqrt{x}$ is within:

(a) 1 unit of 4?   (b) 0.1 units of 4?

11. Knowing that $\lim\limits_{x\to 3} \sqrt{1+x} = 2$, what values of x guarantee that $f(x) = \sqrt{1+x}$ is within:

(a) 1 unit of 2?   (b) 0.0002 units of 2?

12. You must cut four pieces of wire (all the same length) and form them into a square. If the area of the square must be within 0.06 in^2 of 100 in^2, then each piece of wire must be within how many inches of 10 in?

13. You need to cut four pieces of wire (all the same length) and form them into a square. If the area of the square must be within 0.06 in^2 of 25 in^2, then each piece of wire must be within how many inches of 5 in?

Problems 14–17 give $\lim\limits_{x\to a} f(x) = L$, the function f and a value for ϵ graphically. Find a length for δ that satisfies the limit definition for the given function and value of ϵ.

14.

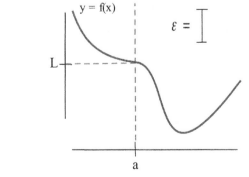

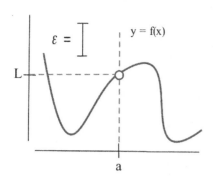

15.

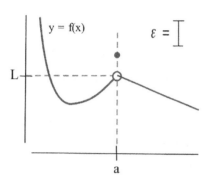

16.

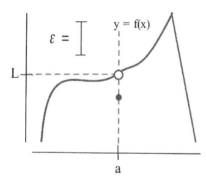

17.

18. Redo each of Problems 14–17 taking a new value of ϵ to be half the value of ϵ given in the problem.

In Problems 19–22, use the limit definition to prove that the given limit does not exist. (Find a value for $\epsilon > 0$ for which there is no δ that satisfies the definition.)

19. With $f(x)$ defined as:

$$f(x) = \begin{cases} 4 & \text{if } x < 2 \\ 3 & \text{if } x > 2 \end{cases}$$

show that $\lim\limits_{x \to 2} f(x)$ does not exist.

20. Show that $\lim\limits_{x \to 3} \lfloor x \rfloor$ does not exist.

21. With $f(x)$ defined as:

$$f(x) = \begin{cases} x & \text{if } x < 2 \\ 6 - x & \text{if } x > 2 \end{cases}$$

show that $\lim\limits_{x \to 2} f(x)$ does not exist.

22. With $f(x)$ defined as:

$$f(x) = \begin{cases} x + 1 & \text{if } x < 2 \\ x^2 & \text{if } x > 2 \end{cases}$$

show that $\lim\limits_{x \to 2} f(x)$ does not exist.

23. Prove: If $\lim\limits_{x \to a} f(x) = L$ and $\lim\limits_{x \to a} g(x) = M$ then $\lim\limits_{x \to a} [f(x) - g(x)] = L - M$.

1.4 Practice Answers

1. (a) $3 - 1 < 4x - 5 < 3 + 1 \Rightarrow 7 < 4x < 9 \Rightarrow 1.75 < x < 2.25$: "$x$ within $\frac{1}{4}$ unit of 2."

   (b) $3 - 0.08 < 4x - 5 < 3 + 0.08 \Rightarrow 7.92 < 4x < 8.08 \Rightarrow 1.98 < x < 2.02$: "$x$ within 0.02 units of 2."

   (c) $3 - E < 4x - 5 < 3 + E \Rightarrow 8 - E < 4x < 8 + E \Rightarrow 2 - \frac{E}{4} < x < 2 + \frac{E}{4}$: "$x$ within $\frac{E}{4}$ units of 2."

2. "Within 1 unit of 3": If $2 < \sqrt{x} < 4$, then $4 < x < 16$, which extends from 5 units to the left of 9 to 7 units to right of 9. Using the smaller of these two distances from 9: "If x is within 5 units of 9, then $\sqrt{x}$ is within 1 unit of 3."

"Within 0.2 units of 3": If $2.8 < \sqrt{x} < 3.2$, then $7.84 < x < 10.24$, which extends from 1.16 units to the left of 9 to 1.24 units to the right of 9. "If x is within 1.16 units of 9, then x is within 0.2 units of 3.

3.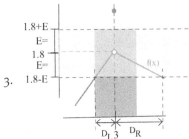

D = smaller of D_L and D_R = ⟷

4.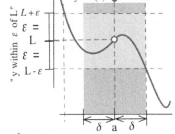

δ = ⟷ "x within δ of a"

5. Given any $\epsilon > 0$, take $\delta = \frac{\epsilon}{5}$. If x is within $\delta = \frac{\epsilon}{5}$ of 4, then $4 - \frac{\epsilon}{5} < x < 4 + \frac{\epsilon}{5}$ so:

$$-\frac{\epsilon}{5} < x - 4 < \frac{\epsilon}{5} \Rightarrow -\epsilon < 5x - 20 < \epsilon \Rightarrow -\epsilon < (5x + 3) - 23 < \epsilon$$

so, finally, $f(x) = 5x + 3$ is within ϵ of $L = 23$.

We have shown that "for any $\epsilon > 0$, there is a $\delta > 0$ (namely $\delta = \frac{\epsilon}{5}$)" so that the rest of the definition is satisfied.

6. Using "proof by contradiction" as in the solution to Example 6:

- Assume that the limit exists: $\lim_{x \to 0} \frac{1}{x} = L$ for some value of L. Let $\epsilon = 1$. Since we're assuming that the limit exists, there is a $\delta > 0$ so that if x is within δ of 0 then $f(x) = \frac{1}{x}$ is within $\epsilon = 1$ of L.

- Let x_1 be between 0 and $0 + \delta$ and also require that $x_1 < \frac{1}{2}$. Then $0 < x_1 < \frac{1}{2}$ so $f(x_1) = \frac{1}{x_1} > 2$. Because x_1 is within δ of 0, $f(x_1) > 2$ is within $\epsilon = 1$ of L, so $L > 2 - \epsilon = 1$: that is, $1 < L$. Let x_2 be between 0 and $0 - \delta$ and also require $x_2 > -\frac{1}{2}$. Then $0 > x_2 > \frac{1}{2}$ so $f(x_2) = \frac{1}{x_2} < -2$. Since x_2 is within δ of 0, $f(x_2) < -2$ is within $\epsilon = 1$ of L, so $L < -2 + \epsilon = -1 \Rightarrow -1 > L$.

- The two inequalities derived above provide the contradiction we were hoping to find. There is no value L that satisfies **both** $1 < L$ and $L < -1$, so we can conclude that our assumption was false and that $f(x) = \frac{1}{x}$ does not have a limit as $x \to 0$.

This is a much more sophisticated (= harder) problem.

The definition says "for every ϵ" so we can pick $\epsilon = 1$. For this particular limit, the definition fails for every $\epsilon > 0$.

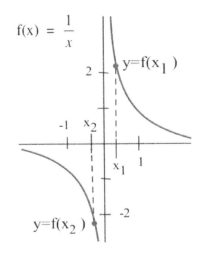

2

The Derivative

The two previous chapters have laid the foundation for the study of calculus. They provided a review of some material you will need and started to emphasize the various ways we will view and use functions: functions given by graphs, equations and tables of values.

Chapter 2 will focus on the idea of tangent lines. We will develop a definition for the derivative of a function and calculate derivatives of some functions using this definition. Then we will examine some of the properties of derivatives, see some relatively easy ways to calculate the derivatives, and begin to look at some ways we can use them.

2.0 Introduction to Derivatives

This section begins with a very graphical approach to slopes of tangent lines. It then examines the problem of finding the slopes of the tangent lines for a single function, $y = x^2$, in some detail — and illustrates how these slopes can help us solve fairly sophisticated problems.

Slopes of Tangent Lines: Graphically

The figure in the margin shows the graph of a function $y = f(x)$. We can use the information in the graph to fill in the table:

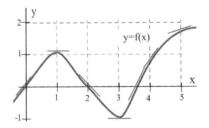

x	$y = f(x)$	$m(x)$
0	0	1
1	1	0
2	0	-1
3	-1	0
4	1	1
5	2	$\frac{1}{2}$

where $m(x)$ is the (estimated) **slope** of the line tangent to the graph of $y = f(x)$ at the point (x, y). We can estimate the values of $m(x)$ at some non-integer values of x as well: $m(0.5) \approx 0.5$ and $m(1.3) \approx -0.3$,

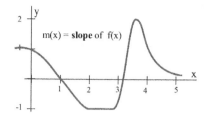

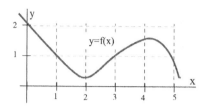

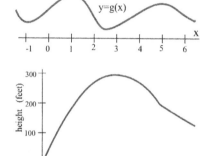

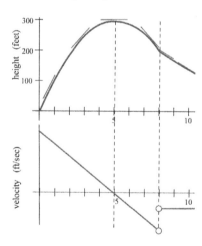

for example. We can even say something about the behavior of $m(x)$ over entire intervals: if $0 < x < 1$, then $m(x)$ is positive, for example.

The values of $m(x)$ definitely depend on the values of x (the slope varies as x varies, and there is at most one slope associated with each value of x) so $m(x)$ is a function of x. We can use the results in the table to help sketch a graph of the function $m(x)$ (see top margin figure).

Practice 1. A graph of $y = f(x)$ appears in the margin. Set up a table of (estimated) values for x and $m(x)$, the slope of the line tangent to the graph of $y = f(x)$ at the point (x, y), and then sketch a graph of the function $m(x)$.

In some applications, we need to know where the graph of a function $f(x)$ has horizontal tangent lines (that is, where the slope of the tangent line equals 0). The slopes of the lines tangent to graph of $y = f(x)$ in Practice 1 are 0 when $x = 2$ or $x \approx 4.25$.

Practice 2. At what values of x does the graph of $y = g(x)$ (in the margin) have horizontal tangent lines?

Example 1. The graph of the height of a rocket at time t appears in the margin. Sketch a graph of the **velocity** of the rocket at time t. (Remember that instantaneous velocity corresponds to the **slope of the line tangent** to the graph of position or height function.)

Solution. The penultimate margin figure shows some sample tangent line segments, while the bottom margin figure shows the velocity of the rocket.   ◄

Practice 3. The graph below shows the temperature during a summer day in Chicago. Sketch a graph of the **rate** at which the temperature is changing at each moment in time. (As with instantaneous velocity, the instantaneous rate of change for the temperature corresponds to the slope of the line tangent to the temperature graph.)

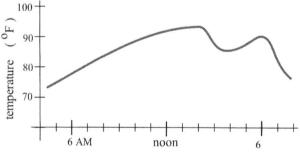

The function $m(x)$, the slope of the line tangent to the graph of $y = f(x)$ at $(x, f(x))$, is called the **derivative** of $f(x)$.

We used the idea of the slope of the tangent line all throughout Chapter 1. In Section 2.1, we will formally define the derivative of a function and begin to examine some of its properties, but first let's see what we can do when we have a formula for $f(x)$.

Tangents to $y = x^2$

When we have a formula for a function, we can determine the slope of the tangent line at a point $(x, f(x))$ by calculating the slope of the secant line through the points $(x, f(x))$ and $(x + h, f(x + h))$:

$$m_{\text{sec}} = \frac{f(x + h) - f(x)}{(x + h) - (x)}$$

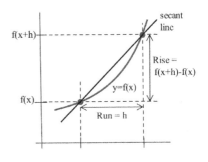

and then taking the limit of m_{sec} as h approaches 0:

$$m_{\text{tan}} = \lim_{h \to 0} m_{\text{sec}} = \lim_{h \to 0} \frac{f(x + h) - f(x)}{(x + h) - (x)}$$

Example 2. Find the slope of the line tangent to the graph of the function $y = f(x) = x^2$ at the point $(2, 4)$.

Solution. In this example, $x = 2$, so $x + h = 2 + h$ and $f(x + h) = f(2 + h) = (2 + h)^2$. The slope of the tangent line at $(2, 4)$ is

$$m_{\text{tan}} = \lim_{h \to 0} m_{\text{sec}} = \lim_{h \to 0} \frac{f(2 + h) - f(2)}{(2 + h) - (2)}$$
$$= \lim_{h \to 0} \frac{(2 + h)^2 - 2^2}{h} = \lim_{h \to 0} \frac{4 + 4h + h^2 - 4}{h}$$
$$= \lim_{h \to 0} \frac{4h + h^2}{h} = \lim_{h \to 0} [4 + h] = 4$$

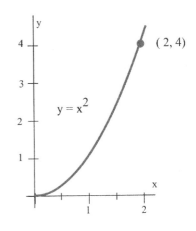

The line tangent to $y = x^2$ at the point $(2, 4)$ has slope 4. ◀

We can use the point-slope formula for a line to find an equation of this tangent line:

$$y - y_0 = m(x - x_0) \Rightarrow y - 4 = 4(x - 2) \Rightarrow y = 4x - 4$$

Practice 4. Use the method of Example 2 to show that the **slope** of the line tangent to the graph of $y = f(x) = x^2$ at the point $(1, 1)$ is $m_{\text{tan}} = 2$. Also find the values of m_{tan} at $(0, 0)$ and $(-1, 1)$.

It is possible to compute the slopes of the tangent lines one point at a time, as we have been doing, but that is not very efficient. You should have noticed in Practice 4 that the algebra for each point was very similar, so let's do all the work just once, for an arbitrary point $(x, f(x)) = (x, x^2)$ and then use the general result to find the slopes at the particular points we're interested in.

The slope of the line tangent to the graph of $y = f(x) = x^2$ at the arbitrary point (x, x^2) is:

$$m_{\tan} = \lim_{h \to 0} m_{\sec} = \lim_{h \to 0} \frac{f(x+h) - f(x)}{(x+h) - (x)}$$

$$= \lim_{h \to 0} \frac{(x+h)^2 - x^2}{h} = \lim_{h \to 0} \frac{x^2 + 2xh + h^2 - x^2}{h}$$

$$= \lim_{h \to 0} \frac{2xh + h^2}{h} = \lim_{h \to 0} [2x + h] = 2x$$

The slope of the line tangent to the graph of $y = f(x) = x^2$ at the point (x, x^2) is $m_{\tan} = 2x$. We can use this general result at any value of x without going through all of the calculations again. The slope of the line tangent to $y = f(x) = x^2$ at the point $(4, 16)$ is $m_{\tan} = 2(4) = 8$ and the slope at (p, p^2) is $m_{\tan} = 2(p) = 2p$. The value of x determines the location of our point on the curve, (x, x^2), as well as the slope of the line tangent to the curve at that point, $m_{\tan} = 2x$. The slope $m_{\tan} = 2x$ is a **function** of x and is called the **derivative** of $y = x^2$.

Simply knowing that the slope of the line tangent to the graph of $y = x^2$ is $m_{\tan} = 2x$ at a point (x, y) can help us quickly find an equation of the line tangent to the graph of $y = x^2$ at any point and answer a number of difficult-sounding questions.

Example 3. Find equations of the lines tangent to $y = x^2$ at the points $(3, 9)$ and (p, p^2).

Solution. At $(3, 9)$, the slope of the tangent line is $2x = 2(3) = 6$, and the equation of the line is $y - 9 = 6(x - 3) \Rightarrow y = 6x - 9$.

At (p, p^2), the slope of the tangent line is $2x = 2(p) = 2p$, and the equation of the line is $y - p^2 = 2p(x - p) \Rightarrow y = 2px - p^2$.   ◀

Example 4. A rocket has been programmed to follow the path $y = x^2$ in space (from left to right along the curve, as seen in the margin figure), but an emergency has arisen and the crew must return to their base, which is located at coordinates $(3, 5)$. At what point on the path $y = x^2$ should the captain turn off the engines so that the ship will coast along a path tangent to the curve to return to the base?

Solution. You might spend a few minutes trying to solve this problem without using the relation $m_{\tan} = 2x$, but the problem is much easier if we do use that result.

Let's assume that the captain turns off the engine at the point (p, q) on the curve $y = x^2$ and then try to determine what values p and q must have so that the resulting tangent line to the curve will go through the point $(3, 5)$. The point (p, q) is on the curve $y = x^2$, so $q = p^2$ and the equation of the tangent line, found in Example 3, must then be $y = 2px - p^2$.

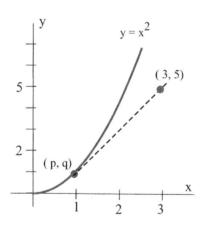

To find the value of p so that the tangent line will go through the point $(3,5)$, we can substitute the values $x = 3$ and $y = 5$ into the equation of the tangent line and solve for p:

$$y = 2px - p^2 \;\Rightarrow\; 5 = 2p(3) - p^2 \;\Rightarrow\; p^2 - 6p + 5 = 0$$
$$\Rightarrow\; (p-1)(p-5) = 0$$

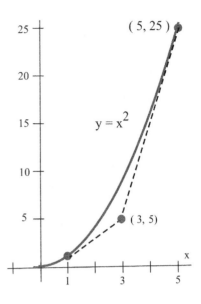

The only solutions are $p = 1$ and $p = 5$, so the only possible points are $(1,1)$ and $(5,25)$. You can verify that the tangent lines to $y = x^2$ at $(1,1)$ and $(5,25)$ both go through the point $(3,5)$. Because the ship is moving from left to right along the curve, the captain should turn off the engines at the point $(1,1)$. (Why not at $(5,25)$?) ◀

Practice 5. Verify that if the rocket engines in Example 4 are shut off at $(2,4)$, then the rocket will go through the point $(3,8)$.

2.0 Problems

1. Use the function $f(x)$ graphed below to fill in the table and then graph $m(x)$, the estimated slope of the tangent line to $y = f(x)$ at the point (x,y).

x	$f(x)$	$m(x)$	x	$f(x)$	$m(x)$
0.0			2.5		
0.5			3.0		
1.0			3.5		
1.5			4.0		
2.0					

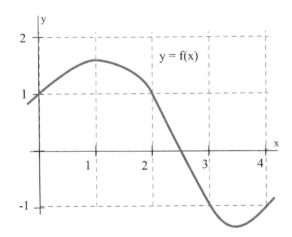

2. Use the function $g(x)$ graphed below to fill in the table and then graph $m(x)$, the estimated slope of the tangent line to $y = g(x)$ at the point (x,y).

x	$g(x)$	$m(x)$	x	$g(x)$	$m(x)$
0.0			2.5		
0.5			3.0		
1.0			3.5		
1.5			4.0		
2.0					

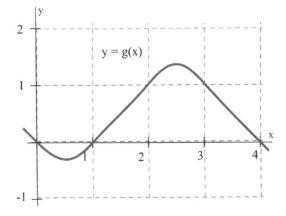

3. (a) At what values of x does the graph of f (shown below) have a horizontal tangent line?

   (b) At what value(s) of x is the value of f the largest? Smallest?

   (c) Sketch a graph of $m(x)$, the slope of the line tangent to the graph of f at the point $(x, f(x))$.

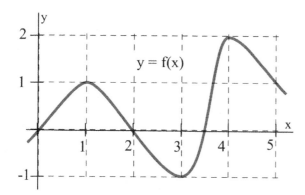

4. (a) At what values of x does the graph of g (shown below) have a horizontal tangent line?

   (b) At what value(s) of x is the value of g the largest? Smallest?

   (c) Sketch a graph of $m(x)$, the slope of the line tangent to the graph of g at the point $(x, g(x))$.

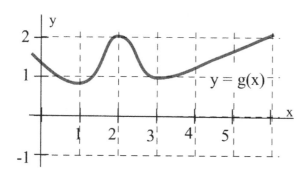

5. (a) Sketch the graph of $f(x) = \sin(x)$ on the interval $-3 \le x \le 10$.

   (b) Sketch a graph of $m(x)$, the slope of the line tangent to the graph of $\sin(x)$ at the point $(x, \sin(x))$.

   (c) Your graph in part (b) should look familiar. What function is it?

6. Match the situation descriptions with the corresponding time-velocity graphs shown below.

   (a) A car quickly leaving from a stop sign.

   (b) A car sedately leaving from a stop sign.

   (c) A student bouncing on a trampoline.

   (d) A ball thrown straight up.

   (e) A student confidently striding across campus to take a calculus test.

   (f) An unprepared student walking across campus to take a calculus test.

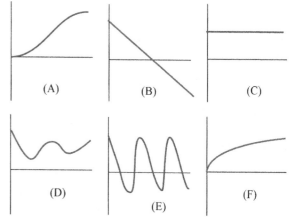

Problems 7–10 assume that a rocket is following the path $y = x^2$, from left to right.

7. At what point should the engine be turned off in order to coast along the tangent line to a base at $(5, 16)$?

8. At $(3, -7)$?   9. At $(1, 3)$?

10. Which points in the plane can not be reached by the rocket? Why not?

In Problems 11–16, perform these steps:

(a) Calculate and simplify:

$$m_{\text{sec}} = \frac{f(x+h) - f(x)}{(x+h) - (x)}$$

(b) Determine $m_{\text{tan}} = \lim_{h \to 0} m_{\text{sec}}$.

(c) Evaluate m_{tan} at $x = 2$.

(d) Find an equation of the line tangent to the graph of f at $(2, f(2))$.

11. $f(x) = 3x - 7$ 12. $f(x) = 2 - 7x$

13. $f(x) = ax + b$ where a and b are constants

14. $f(x) = x^2 + 3x$ 15. $f(x) = 8 - 3x^2$

16. $f(x) = ax^2 + bx + c$ where a, b and c are constants

In Problems 17–18, use the result:

$$f(x) = ax^2 + bx + c \Rightarrow m_{\tan} = 2ax + b$$

17. Given $f(x) = x^2 + 2x$, at which point(s) $(p, f(p))$ does the line tangent to the graph at that point also go through the point $(3, 6)$?

18. (a) If $a \neq 0$, then what is the shape of the graph of $y = f(x) = ax^2 + bx + c$?

   (b) At what value(s) of x is the line tangent to the graph of $f(x)$ horizontal?

2.0 Practice Answers

1. Approximate values of $m(x)$ appear in the table in the margin; the margin figure shows a graph of $m(x)$.

2. The tangent lines to the graph of g are horizontal (slope $= 0$) when $x \approx -1, 1, 2.5$ and 5.

3. The figure below shows a graph of the approximate rate of temperature change (slope).

x	$f(x)$	$m(x)$
0	2	-1
1	1	-1
2	$\frac{1}{3}$	0
3	1	1
4	$\frac{3}{2}$	$\frac{1}{2}$
5	1	-2

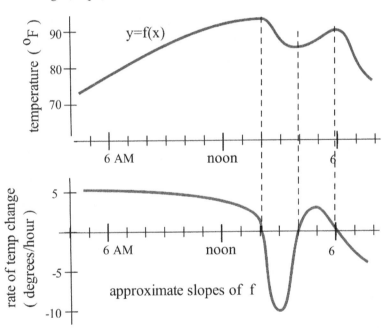

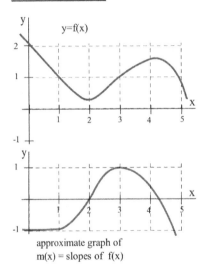

approximate graph of $m(x) =$ slopes of $f(x)$

4. At $(1, 1)$, the slope of the tangent line is:

$$m_{\tan} = \lim_{h \to 0} m_{\sec} = \lim_{h \to 0} \frac{f(1 + h) - f(1)}{(1 + h) - (1)}$$
$$= \lim_{h \to 0} \frac{(1 + h)^2 - 1^2}{h} = \lim_{h \to 0} \frac{1 + 2h + h^2 - 1}{h}$$
$$= \lim_{h \to 0} \frac{2h + h^2}{h} = \lim_{h \to 0} [2 + h] = 2$$

so the line tangent to $y = x^2$ at the point $(1,1)$ has slope 2. At $(0,0)$:

$$m_{\text{tan}} = \lim_{h \to 0} m_{\text{sec}} = \lim_{h \to 0} \frac{f(0+h) - f(1)}{(0+h) - (0)}$$

$$= \lim_{h \to 0} \frac{(0+h)^2 - 0^2}{h} = \lim_{h \to 0} \frac{h^2}{h} = \lim_{h \to 0} h = 0$$

so the line tangent to $y = x^2$ at $(0,0)$ has slope 0. At $(-1,1)$:

$$m_{\text{tan}} = \lim_{h \to 0} m_{\text{sec}} = \lim_{h \to 0} \frac{f(-1+h) - f(-1)}{(-1+h) - (-1)}$$

$$= \lim_{h \to 0} \frac{(-1+h)^2 - (-1)^2}{h} = \lim_{h \to 0} \frac{1 - 2h + h^2 - 1}{h}$$

$$= \lim_{h \to 0} \frac{-2h + h^2}{h} = \lim_{h \to 0} [-2 + h] = -2$$

so the line tangent to $y = x^2$ at the point $(-1,1)$ has slope -2.

5. From Example 4 we know the slope of the tangent line is $m_{\text{tan}} = 2x$, so the slope of the tangent line at $(2,4)$ is $m_{\text{tan}} = 2x = 2(2) = 4$. The tangent line has slope 4 and goes through the point $(2,4)$, so an equation of the tangent line (using $y - y_0 = m(x - x_0)$) is $y - 4 = 4(x - 2)$ or $y = 4x - 4$. The point $(3,8)$ satisfies the equation $y = 4x - 4$, so the point $(3,8)$ lies on the tangent line.

2.1 The Definition of Derivative

The graphical idea of a **slope of a tangent line** is very useful, but for some purposes we need a more algebraic definition of the **derivative of a function**. We will use this definition to calculate the derivatives of several functions and see that these results agree with our graphical understanding. We will also look at several different interpretations for the derivative, and derive a theorem that will allow us to easily and quickly determine the derivative of any fixed power of x.

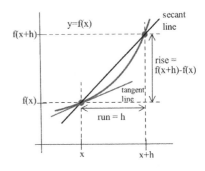

In the previous section we found the slope of the tangent line to the graph of the function $f(x) = x^2$ at an arbitrary point $(x, f(x))$ by calculating the slope of the secant line through the points $(x, f(x))$ and $(x + h, f(x + h))$:

$$m_{\text{sec}} = \frac{f(x + h) - f(x)}{(x + h) - (x)}$$

and then taking the limit of m_{sec} as h approached 0 (see margin). That approach to calculating slopes of tangent lines motivates the definition of the derivative of a function.

> **Definition of the Derivative:**
> The derivative of a function f is a new function,
> f' (pronounced "eff prime"), whose value at x is:
>
> $$f'(x) = \lim_{h \to 0} \frac{f(x + h) - f(x)}{h}$$
>
> if this limit exists and is finite.

This is **the** definition of differential calculus, and you must know it and understand what it says. The rest of this chapter and all of Chapter 3 are built on this definition, as is much of what appears in later chapters. It is remarkable that such a simple idea (the slope of a tangent line) and such a simple definition (for the derivative f') will lead to so many important ideas and applications.

Notation

There are three commonly used notations for the derivative of $y = f(x)$:

- $f'(x)$ emphasizes that the derivative is a function related to f

- $D(f)$ emphasizes that we perform an operation on f to get f'

- $\dfrac{df}{dx}$ emphasizes that the derivative is the limit of $\dfrac{\Delta f}{\Delta x} = \dfrac{f(x + h) - f(x)}{h}$

We will use all three notations so that you can become accustomed to working with each of them.

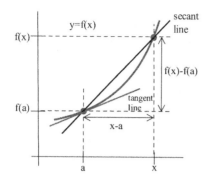

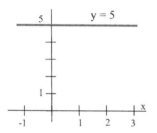

The function $f'(x)$ gives the slope of the tangent line to the graph of $y = f(x)$ at the point $(x, f(x))$, or the instantaneous rate of change of the function f at the point $(x, f(x))$.

If, in the margin figure, we let x be the point $a + h$, then $h = x - a$. As $h \to 0$, we see that $x \to a$ and:

$$f'(a) = \lim_{h \to 0} \frac{f(a + h) - f(a)}{h} = \lim_{x \to a} \frac{f(x) - f(a)}{x - a}$$

We will use whichever of these two forms is more convenient algebraically in a particular situation.

Calculating Some Derivatives Using the Definition

Fortunately, we will soon have some quick and easy ways to calculate most derivatives, but first we will need to use the definition to determine the derivatives of a few basic functions. In Section 2.2, we will use those results and some properties of derivatives to calculate derivatives of combinations of the basic functions. Let's begin by using the graphs and then the definition to find a few derivatives.

Example 1. Graph $y = f(x) = 5$ and estimate the **slope** of the tangent line at each point on the graph. Then use the definition of the derivative to calculate the exact slope of the tangent line at each point. Your graphical estimate and the exact result from the definition should agree.

Solution. The graph of $y = f(x) = 5$ is a horizontal line (see margin), which has slope 0, so we should expect that its tangent line will also have slope 0.

Using the definition: With $f(x) = 5$, then $f(x + h) = 5$ no matter what h is, so:

$$\mathbf{D}(f(x)) = \lim_{h \to 0} \frac{f(x + h) - f(x)}{h} = \lim_{h \to 0} \frac{5 - 5}{h} = \lim_{h \to 0} 0 = 0$$

and this agrees with our graphical estimate of the derivative.   ◀

Using similar steps, it is easy to show that the derivative of *any* constant function is 0.

> **Theorem:** If $f(x) = k$, then $f'(x) = 0$.

Practice 1. Graph $y = f(x) = 7x$ and estimate the slope of the tangent line at each point on the graph. Then use the definition of the derivative to calculate the exact slope of the tangent line at each point.

Example 2. Describe the derivative of $y = f(x) = 5x^3$ graphically and compute it using the definition. Find an equation of the line tangent to $y = 5x^3$ at the point $(1, 5)$.

Solution. It appears from the graph of $y = f(x) = 5x^3$ (see margin) that $f(x)$ is increasing, so the slopes of the tangent lines are positive except perhaps at $x = 0$, where the graph seems to flatten out.

With $f(x) = 5x^3$ we have:

$$f(x + h) = 5(x + h)^3 = 5(x^3 + 3x^2h + 3xh^2 + h^3)$$

and using this last expression in the definition of the derivative:

$$f'(x) = \lim_{h \to 0} \frac{f(x+h) - f(x)}{h} = \lim_{h \to 0} \frac{5(x^3 + 3x^2h + 3xh^2 + h^3) - 5x^3}{h}$$

$$= \lim_{h \to 0} \frac{15x^2h + 15xh^2 + 5h^3}{h} = \lim_{h \to 0} (15x^2 + 15xh + 5h^2) = 15x^2$$

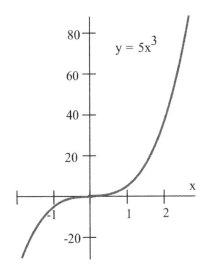

so $\mathbf{D}(5x^3) = 15x^2$, which is positive except when $x = 0$ (as we predicted from the graph).

The function $f'(x) = 15x^2$ gives the slope of the line tangent to the graph of $f(x) = 5x^3$ at the point $(x, f(x))$. At the point $(1, 5)$, the slope of the tangent line is $f'(1) = 15(1)^2 = 15$. From the point-slope formula, an equation of the tangent line to f at that point is $y - 5 = 15(x - 1)$ or $y = 15x - 10$. ◄

Practice 2. Use the definition to show that the derivative of $y = x^3$ is $\dfrac{dy}{dx} = 3x^2$. Find an equation of the line tangent to the graph of $y = x^3$ at the point $(2, 8)$.

If f has a derivative at x, we say that f is **differentiable** at x. If we have a point on the graph of a differentiable function and a slope (the derivative evaluated at the point), it is easy to write an equation of the tangent line.

Tangent Line Formula:

If $f(x)$ is differentiable at $x = a$

then   an equation of the line tangent to f at $(a, f(a))$ is:

$$y = f(a) + f'(a)(x - a)$$

Proof. The tangent line goes through the point $(a, f(a))$ with slope $f'(a)$ so, using the point-slope formula, $y - f(a) = f'(a)(x - a)$ or $y = f(a) + f'(a)(x - a)$. ☐

Practice 3. The derivatives $\mathbf{D}(x) = 1$, $\mathbf{D}(x^2) = 2x$, $\mathbf{D}(x^3) = 3x^2$ exhibit the start of a pattern. Without using the definition of the derivative, what do you think the following derivatives will be? $\mathbf{D}(x^4)$, $\mathbf{D}(x^5)$, $\mathbf{D}(x^{43})$, $\mathbf{D}(\sqrt{x}) = \mathbf{D}(x^{\frac{1}{2}})$ and $\mathbf{D}(x^\pi)$. (Just make an intelligent "guess" based on the pattern of the previous examples.)

Before further investigating the "pattern" for the derivatives of powers of x and general properties of derivatives, let's compute the derivatives of two functions that are not powers of x: $\sin(x)$ and $|x|$.

$$\boxed{\text{Theorem: } \mathbf{D}(\sin(x)) = \cos(x)}$$

The graph of $y = f(x) = \sin(x)$ (see margin) should be very familiar to you. The graph has horizontal tangent lines (slope $= 0$) when $x = \pm\frac{\pi}{2}$ and $x = \pm\frac{3\pi}{2}$ and so on. If $0 < x < \frac{\pi}{2}$, then the slopes of the tangent lines to the graph of $y = \sin(x)$ are positive. Similarly, if $\frac{\pi}{2} < x < \frac{3\pi}{2}$, then the slopes of the tangent lines are negative. Finally, because the graph of $y = \sin(x)$ is periodic, we expect that the derivative of $y = \sin(x)$ will also be periodic. Note that the function $\cos(x)$ possesses all of those desired properties for the slope function.

Proof. With $f(x) = \sin(x)$, apply an angle addition formula to get:

$$f(x + h) = \sin(x + h) = \sin(x)\cos(h) + \cos(x)\sin(h)$$

and use this formula in the definition of the derivative:

$$f'(x) = \lim_{h \to 0} \frac{f(x + h) - f(x)}{h}$$
$$= \lim_{h \to 0} \frac{(\sin(x)\cos(h) + \cos(x)\sin(h)) - \sin(x)}{h}$$

This limit looks formidable, but just collect the terms containing $\sin(x)$:

$$\lim_{h \to 0} \frac{(\sin(x)\cos(h) - \sin(x)) + \cos(x)\sin(h)}{h}$$

so you can factor out $\sin(x)$ from the first two terms, rewriting as:

$$\lim_{h \to 0} \left[\sin(x) \cdot \frac{\cos(h) - 1}{h} + \cos(x) \cdot \frac{\sin(h)}{h} \right]$$

Now calculate the limits separately:

$$\lim_{h \to 0} \sin(x) \cdot \lim_{h \to 0} \frac{\cos(h) - 1}{h} + \lim_{h \to 0} \cos(x) \cdot \lim_{h \to 0} \frac{\sin(h)}{h}$$

The first and third limits do not depend on h, and we calculated the second and fourth limits in Section 1.2:

$$\sin(x) \cdot 0 + \cos(x) \cdot 1 = \cos(x)$$

So $\mathbf{D}(\sin(x)) = \cos(x)$ and the various properties we expected of the derivative of $y = \sin(x)$ by examining its graph are true of $\cos(x)$. □

Practice 4. Show that $\mathbf{D}(\cos(x)) = -\sin(x)$ using the definition.

You will need the angle addition formula for cosine to rewrite $\cos(x + h)$ as:

$$\cos(x) \cdot \cos(h) - \sin(x) \cdot \sin(h)$$

The derivative of $\cos(x)$ resembles the situation for $\sin(x)$ but differs by an important negative sign. You should memorize both of these important derivatives.

Example 3. For $y = |x|$, find $\dfrac{dy}{dx}$.

Solution. The graph of $y = f(x) = |x|$ (see margin) is a "V" shape with its vertex at the origin. When $x > 0$, the graph is just $y = |x| = x$, which is part of a line with slope $+1$, so we should expect the derivative of $|x|$ to be $+1$. When $x < 0$, the graph is $y = |x| = -x$, which is part of a line with slope -1, so we expect the derivative of $|x|$ to be -1. When $x = 0$, the graph has a corner, and we should expect the derivative of $|x|$ to be undefined at $x = 0$, as there is no single candidate for a line tangent to the graph there.

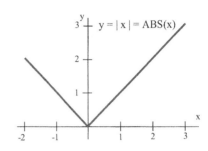

Using the definition, consider the same three cases discussed previously: $x > 0$, $x < 0$ and $x = 0$.

If $x > 0$, then, for small values of h, $x + h > 0$, so:

$$\mathbf{D}(f(x)) = \lim_{h \to 0} \frac{|x+h| - |x|}{h} = \lim_{h \to 0} \frac{x+h-x}{h} = \lim_{h \to 0} \frac{h}{h} = 1$$

If $x < 0$, then, for small values of h, $x + h < 0$, so:

$$\mathbf{D}(f(x)) = \lim_{h \to 0} \frac{|x+h| - |x|}{h} = \lim_{h \to 0} \frac{-(x+h) - (-x)}{h} = \lim_{h \to 0} \frac{-h}{h} = -1$$

When $x = 0$, the situation is a bit more complicated:

$$f'(0) = \lim_{h \to 0} \frac{|0+h| - |0|}{h} = \lim_{h \to 0} \frac{|h|}{h}$$

This is undefined, as $\displaystyle\lim_{h \to 0^+} \frac{|h|}{h} = +1$ and $\displaystyle\lim_{h \to 0^-} \frac{|h|}{h} = -1$, so:

$$\mathbf{D}(|x|) = \begin{cases} 1 & \text{if } x > 0 \\ \text{undefined} & \text{if } x = 0 \\ -1 & \text{if } x < 0 \end{cases}$$

The derivative of $|x|$ agrees with the function $\text{sgn}(x)$ defined in Chapter 0, except at $x = 0$: $\mathbf{D}(|x|)$ is undefined at $x = 0$ but $\text{sgn}(0) = 0$.

or, equivalently, $\mathbf{D}(|x|) = \dfrac{|x|}{x}$.   ◄

Practice 5. Graph $y = |x - 2|$ and $y = |2x|$ and use the *graphs* to determine $\mathbf{D}(|x - 2|)$ and $\mathbf{D}(|2x|)$.

So far we have emphasized the derivative as the slope of the line tangent to a graph. That very visual interpretation is very useful when examining the graph of a function, and we will continue to use it. Derivatives, however, are employed in a wide variety of fields and applications, and some of these fields use other interpretations. A few commonly used interpretations of the derivative follow.

Interpretations of the Derivative

General

Rate of Change The function $f'(x)$ is the rate of change of the function at x. If the units for x are years and the units for $f(x)$ are people, then the units for $\frac{df}{dx}$ are $\frac{\text{people}}{\text{year}}$, a rate of change in population.

Graphical

Slope $f'(x)$ is the slope of the line tangent to the graph of f at $(x, f(x))$.

Physical

Velocity If $f(x)$ is the position of an object at time x, then $f'(x)$ is the velocity of the object at time x. If the units for x are hours and $f(x)$ is distance, measured in miles, then the units for $f'(x) = \frac{df}{dx}$ are $\frac{\text{miles}}{\text{hour}}$, miles per hour, which is a measure of velocity.

Acceleration If $f(x)$ is the velocity of an object at time x, then $f'(x)$ is the acceleration of the object at time x. If the units for x are hours and $f(x)$ has the units $\frac{\text{miles}}{\text{hour}}$, then the units for the acceleration $f'(x) = \frac{df}{dx}$ are $\frac{\text{miles/hour}}{\text{hour}} = \frac{\text{miles}}{\text{hour}^2}$, "miles per hour per hour."

Magnification $f'(x)$ is the magnification factor of the function f for points close to x. If a and b are two points very close to x, then the distance between $f(a)$ and $f(b)$ will be close to $f'(x)$ times the original distance between a and b: $f(b) - f(a) \approx f'(x)(b - a)$.

Business

Marginal Cost If $f(x)$ is the total cost of producing x objects, then $f'(x)$ is the marginal cost, at a production level of x: (approximately) the additional cost of making one more object once we have already made x objects. If the units for x are bicycles and the units for $f(x)$ are dollars, then the units for $f'(x) = \frac{df}{dx}$ are $\frac{\text{dollars}}{\text{bicycle}}$, the cost per bicycle.

Marginal Profit If $f(x)$ is the total profit from producing and selling x objects, then $f'(x)$ is the marginal profit: the profit to be made from producing and selling one more object. If the units for x are bicycles and the units for $f(x)$ are dollars, then the units for $f'(x) = \frac{df}{dx}$ are $\frac{\text{dollars}}{\text{bicycle}}$, the profit per bicycle.

In financial contexts, the word "marginal" usually refers to the derivative or rate of change of some quantity. One of the strengths of calculus is that it provides a unity and economy of ideas among diverse applications. The vocabulary and problems may be different, but the ideas and even the notations of calculus remain useful.

Example 4. A small cork is bobbing up and down, and at time t seconds it is $h(t) = \sin(t)$ feet above the mean water level (see margin). Find the height, velocity and acceleration of the cork when $t = 2$ seconds. (Include the proper units for each answer.)

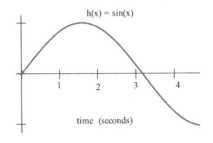

Solution. $h(t) = \sin(t)$ represents the height of the cork at any time t, so the height of the cork when $t = 2$ is $h(2) = \sin(2) \approx 0.91$ feet above the mean water level.

The velocity is the derivative of the position, so $v(t) = \frac{d}{dt}h(t) = \frac{d}{dt}\sin(t) = \cos(t)$. The derivative of position is the limit of $\frac{\Delta h}{\Delta t}$, so the units are $\frac{\text{feet}}{\text{seconds}}$. After 2 seconds, the velocity is $v(2) = \cos(2) \approx -0.42$ feet per second.

The acceleration is the derivative of the velocity, so $a(t) = \frac{d}{dt}v(t) = \frac{d}{dt}\cos(t) = -\sin(t)$. The derivative of velocity is the limit of $\frac{\Delta v}{\Delta t}$, so the units are $\frac{\text{feet/second}}{\text{seconds}}$ or $\frac{\text{feet}}{\text{second}^2}$. After 2 seconds the acceleration is $a(2) = -\sin(2) \approx -0.91\frac{\text{ft}}{\text{sec}^2}$.   ◄

Practice 6. Find the height, velocity and acceleration of the cork in the previous example after 1 second.

A Most Useful Formula: $\mathbf{D}(x^n)$

Functions that include powers of x are very common (every polynomial is a sum of terms that include powers of x) and, fortunately, it is easy to calculate the derivatives of such powers. The "pattern" emerging from the first few examples in this section is, in fact, true for all powers of x. We will only state and prove the "pattern" here for positive integer powers of x, but it is also true for other powers (as we will prove later).

> **Theorem:** If n is a positive integer, then: $\mathbf{D}(x^n) = n \cdot x^{n-1}$

This theorem is an example of the power of *generality* and *proof* in mathematics. Rather than resorting to the definition when we encounter a new exponent p in the form x^p (imagine using the definition to calculate the derivative of x^{307}), we can justify the pattern for all positive integer exponents n, and then simply apply the result for whatever exponent we have. We know, from the first examples in this section, that the theorem is true for $n = 1, 2$ and 3, but no number of *examples* would guarantee that the pattern is true for all exponents. We need a proof that what we *think* is true really *is* true.

Proof. With $f(x) = x^n$, $f(x + h) = (x + h)^n$, and in order to simplify $f(x + h) - f(x) = (x + h)^n - x^n$, we will need to expand $(x + h)^n$. However, we really only need to know the first two terms of the expansion

You may also be familiar with Pascal's triangle:

```
            1
        1       1
      1     2     1
    1     3     3     1
  1     4     6     4     1
```

Among many beautiful and amazing properties, the numbers in row n of the triangle (counting the first row as row 0) give the coefficients in the expansion of $(A + B)^n$. Notice that each entry in the interior of the triangle is the sum of the two numbers immediately above it.

and to know that all of the other terms of the expansion contain a power of h of at least 2.

The Binomial Theorem from algebra says (for $n > 3$) that:

$$(x + h)^n = x^n + n \cdot x^{n-1}h + a \cdot x^{n-2}h^2 + b \cdot x^{n-3}h^3 + \cdots + h^n$$

where a and b represent numerical coefficients. (Expand $(x + h)^n$ for a few different values of n to convince yourself of this result.) Then:

$$\mathbf{D}(f(x)) = \lim_{h \to 0} \frac{f(x + h) - f(x)}{h} = \lim_{h \to 0} \frac{(x + h)^n - x^n}{h}$$

Now expand $(x + h)^n$ to get:

$$\lim_{h \to 0} \frac{x^n + n \cdot x^{n-1}h + a \cdot x^{n-2}h^2 + b \cdot x^{n-3}h^3 + \cdots + h^n - x^n}{h}$$

Eliminating $x^n - x^n$ we get:

$$\lim_{h \to 0} \frac{n \cdot x^{n-1}h + a \cdot x^{n-2}h^2 + b \cdot x^{n-3}h^3 + \cdots + h^n}{h}$$

and we can then factor h out of the numerator:

$$\lim_{h \to 0} \frac{h(n \cdot x^{n-1} + a \cdot x^{n-2}h + b \cdot x^{n-3}h^2 + \cdots + h^{n-1})}{h}$$

and divide top and bottom by the factor h:

$$\lim_{h \to 0} \left[n \cdot x^{n-1} + a \cdot x^{n-2}h + b \cdot x^{n-3}h^2 + \cdots + h^{n-1} \right]$$

We are left with a polynomial in h and can now compute the limit by simply evaluating the polynomial at $h = 0$ to get $\mathbf{D}(x^n) = n \cdot x^{n-1}$. $\quad\square$

Practice 7. Calculate $\mathbf{D}(x^5)$, $\frac{d}{dx}(x^2)$, $\mathbf{D}(x^{100})$, $\frac{d}{dt}(t^{31})$ and $\mathbf{D}(x^0)$.

We will occasionally use the result of the theorem for the derivatives of **all** constant powers of x even though it has only been proven for positive integer powers, so far. A proof of a more general result (for all rational powers of x) appears in Section 2.9

Example 5. Find $\mathbf{D}\left(\dfrac{1}{x}\right)$ and $\dfrac{d}{dx}(\sqrt{x})$.

Solution. Rewriting the fraction using a negative exponent:

$$\mathbf{D}\left(\frac{1}{x}\right) = \mathbf{D}(x^{-1}) = -1 \cdot x^{-1-1} = -x^{-2} = -\frac{1}{x^2}$$

Rewriting the square root using a fractional exponent:

$$\frac{d}{dx}(\sqrt{x}) = \mathbf{D}(x^{\frac{1}{2}}) = \frac{1}{2} \cdot x^{\frac{1}{2}-1} = \frac{1}{2}x^{-\frac{1}{2}} = \frac{1}{2\sqrt{x}}$$

These results can also be obtained by using the definition of the derivative, but the algebra involved is slightly awkward. ◀

Practice 8. Find $\mathbf{D}(x^{\frac{3}{2}})$, $\dfrac{d}{dx}(x^{\frac{1}{3}})$, $\mathbf{D}\left(\dfrac{1}{\sqrt{x}}\right)$ and $\dfrac{d}{dt}(t^{\pi})$.

Example 6. It costs $\sqrt{x}$ hundred dollars to run a training program for x employees.

(a) How much does it cost to train 100 employees? 101 employees? If you already need to train 100 employees, how much additional money will it cost to add 1 more employee to those being trained?

(b) For $f(x) = \sqrt{x}$, calculate $f'(x)$ and evaluate f' at $x = 100$. How does $f'(100)$ compare with the last answer in part (a)?

Solution. (a) Put $f(x) = \sqrt{x} = x^{\frac{1}{2}}$ hundred dollars, the cost to train x employees. Then $f(100) = \$1000$ and $f(101) = \$1004.99$, so it costs $\$4.99$ additional to train the 101st employee. (b) $f'(x) = \frac{1}{2}x^{-\frac{1}{2}} = \frac{1}{2\sqrt{x}}$ so $f'(100) = \frac{1}{2\sqrt{100}} = \frac{1}{20}$ hundred dollars $= \$5.00$. Clearly $f'(100)$ is very close to the actual additional cost of training the 101st employee. ◄

Important Information and Results

This section contains a great deal of important information that we will continue to use throughout the rest of the course.

So it is worthwhile to collect here some of those important ideas.

Definition of Derivative: $f'(x) = \lim\limits_{h \to 0} \dfrac{f(x+h) - f(x)}{h}$

Valid if the limit exists and is finite.

Notations for the Derivative: $f'(x)$, $\mathbf{D}(f(x))$, $\frac{df}{dx}$

Tangent Line Equation: $y = f(a) + f'(a) \cdot (x - a)$

An equation of the line tangent to the graph of f at $(a, f(a))$.

Formulas:

- $\mathbf{D}(\text{constant}) = 0$

- $\mathbf{D}(x^n) = n \cdot x^{n-1}$

Proved for n = positive integer, but true for all constants n.

- $\mathbf{D}(\sin(x)) = \cos(x)$ and $\mathbf{D}(\cos(x)) = -\sin(x)$

- $\mathbf{D}(|x|) = \begin{cases} 1 & \text{if } x > 0 \\ \text{undefined} & \text{if } x = 0 \\ -1 & \text{if } x < 0 \end{cases} = \dfrac{|x|}{x}$

Interpretations of $f'(x)$:

- Slope of a line tangent to a graph

- Instantaneous rate of change of a function at a point

- Velocity or acceleration

- Magnification factor

- Marginal change

2.1 Problems

1. Match the functions f, g and h shown below with the graphs of their derivatives (show in the bottom row).

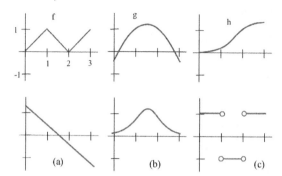

2. The figure below shows six graphs, three of which are derivatives of the other three. Match the functions with their derivatives.

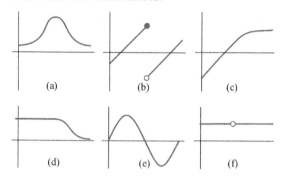

In Problems 3–6, find the slope m_{sec} of the secant line through the two given points and then calculate $m_{tan} = \lim\limits_{h\to 0} m_{sec}$.

3. $f(x) = x^2$

   (a) $(-2, 4)$, $(-2 + h, (-2 + h)^2)$

   (b) $(0.5, 0.25)$, $(0.5 + h, (0.5 + h)^2)$

4. $f(x) = 3 + x^2$

   (a) $(1, 4)$, $(1 + h, 3 + (1 + h)^2)$

   (b) $(x, 3 + x^2)$, $(x + h, 3 + (x + h)^2)$

5. $f(x) = 7x - x^2$

   (a) $(1, 6)$, $(1 + h, 7(1 + h) - (1 + h)^2)$

   (b) $(x, 7x - x^2)$, $(x + h, 7(x + h) - (x + h)^2)$

6. $f(x) = x^3 + 4x$

   (a) $(1, 5)$, $(1 + h, (1 + h)^3 + 4(1 + h))$

   (b) $(x, x^3 + 4x)$, $(x + h, (x + h)^3 + 4(x + h))$

7. Use the graph below to estimate the values of these limits. (It helps to recognize what the limit represents.)

   (a) $\lim\limits_{h\to 0} \dfrac{f(0 + h) - f(0)}{h}$ (b) $\lim\limits_{h\to 0} \dfrac{f(1 + h) - f(1)}{h}$

   (c) $\lim\limits_{w\to 0} \dfrac{f(2 + w) - 1}{w}$ (d) $\lim\limits_{h\to 0} \dfrac{f(3 + h) - f(3)}{h}$

   (e) $\lim\limits_{h\to 0} \dfrac{f(4 + h) - f(4)}{h}$ (f) $\lim\limits_{s\to 0} \dfrac{f(5 + s) - f(5)}{s}$

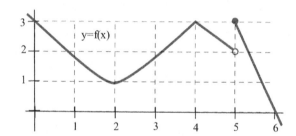

8. Use the graph below to estimate the values of these limits.

   (a) $\lim\limits_{h\to 0} \dfrac{g(0 + h) - g(0)}{h}$ (b) $\lim\limits_{h\to 0} \dfrac{g(1 + h) - g(1)}{h}$

   (c) $\lim\limits_{w\to 0} \dfrac{g(2 + w) - 2}{w}$ (d) $\lim\limits_{h\to 0} \dfrac{g(3 + h) - g(3)}{h}$

   (e) $\lim\limits_{h\to 0} \dfrac{g(4 + h) - g(4)}{h}$ (f) $\lim\limits_{s\to 0} \dfrac{g(5 + s) - g(5)}{s}$

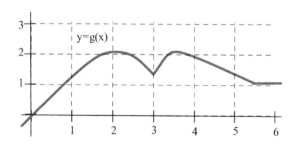

In Problems 9–12, use the definition of the derivative to calculate $f'(x)$ and then evaluate $f'(3)$.

9. $f(x) = x^2 + 8$ 10. $f(x) = 5x^2 - 2x$

11. $f(x) = 2x^3 - 5x$ 12. $f(x) = 7x^3 + x$

13. Graph $f(x) = x^2$, $g(x) = x^2 + 3$ and $h(x) = x^2 - 5$. Calculate the derivatives of f, g and h.

14. Graph $f(x) = 5x$, $g(x) = 5x + 2$ and $h(x) = 5x - 7$. Calculate the derivatives of f, g and h.

In Problems 15–18, find the slopes and equations of the lines tangent to $y = f(x)$ at the given points.

15. $f(x) = x^2 + 8$ at $(1, 9)$ and $(-2, 12)$.

16. $f(x) = 5x^2 - 2x$ at $(2, 16)$ and $(0, 0)$.

17. $f(x) = \sin(x)$ at $(\pi, 0)$ and $(\frac{\pi}{2}, 1)$.

18. $f(x) = |x + 3|$ at $(0, 3)$ and $(-3, 0)$.

19. (a) Find an equation of the line tangent to the graph of $y = x^2 + 1$ at the point $(2, 5)$.

(b) Find an equation of the line perpendicular to the graph of $y = x^2 + 1$ at $(2, 5)$.

(c) Where is the line tangent to the graph of $y = x^2 + 1$ horizontal?

(d) Find an equation of the line tangent to the graph of $y = x^2 + 1$ at the point (p, q).

(e) Find the point(s) (p, q) on the graph of $y = x^2 + 1$ so the tangent line to the curve at (p, q) goes through the point $(1, -7)$.

20. (a) Find an equation of the line tangent to the graph of $y = x^3$ at the point $(2, 8)$.

(b) Where, if ever, is the line tangent to the graph of $y = x^3$ horizontal?

(c) Find an equation of the line tangent to the graph of $y = x^3$ at the point (p, q).

(d) Find the point(s) (p, q) on the graph of $y = x^3$ so the tangent line to the curve at (p, q) goes through the point $(16, 0)$.

21. (a) Find the angle that the line tangent to $y = x^2$ at $(1, 1)$ makes with the x-axis.

(b) Find the angle that the line tangent to $y = x^3$ at $(1, 1)$ makes with the x-axis.

(c) The curves $y = x^2$ and $y = x^3$ intersect at the point $(1, 1)$. Find the angle of intersection of the two curves (actually the angle between their tangent lines) at the point $(1, 1)$.

22. The figure below shows the graph of $y = f(x)$. Sketch a graph of $y = f'(x)$.

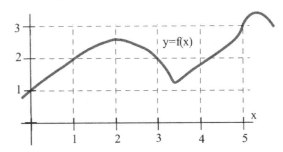

23. The figure below shows the graph of the height of an object at time t. Sketch a graph of the object's upward velocity. What are the units for each axis on the velocity graph?

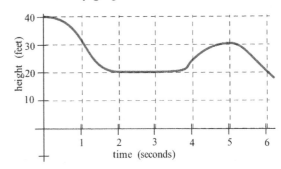

24. Fill in the table with units for $f'(x)$.

units for x	units for $f(x)$	units for $f'(x)$
hours	miles	
people	automobiles	
dollars	pancakes	
days	trout	
seconds	miles per second	
seconds	gallons	
study hours	test points	

25. A rock dropped into a deep hole will drop $d(x) = 16x^2$ feet in x seconds.

    (a) How far into the hole will the rock be after 4 seconds? After 5 seconds?

    (b) How fast will it be falling at exactly 4 seconds? After 5 seconds? After x seconds?

26. It takes $T(x) = x^2$ hours to weave x small rugs. What is the marginal production time to weave a rug? (Be sure to include the units with your answer.)

27. It costs $C(x) = \sqrt{x}$ dollars to produce x golf balls. What is the marginal production cost to make a golf ball? What is the marginal production cost when $x = 25$? When $x = 100$? (Include units.)

28. Define $A(x)$ to be the **area** bounded by the t- and y-axes, the line $y = 5$ and a vertical line at $t = x$ (see figure below).

    (a) Evaluate $A(0)$, $A(1)$, $A(2)$ and $A(3)$.

    (b) Find a formula for $A(x)$ valid for $x \geq 0$.

    (c) Determine $A'(x)$.

    (d) What does $A'(x)$ represent?

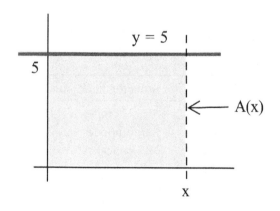

29. Define $A(x)$ to be the **area** bounded by the t-axis, the line $y = t$, and a vertical line at $t = x$ (see figure below).

    (a) Evaluate $A(0)$, $A(1)$, $A(2)$ and $A(3)$.

    (b) Find a formula for $A(x)$ valid for $x \geq 0$.

    (c) Determine $A'(x)$.

    (d) What does $A'(x)$ represent?

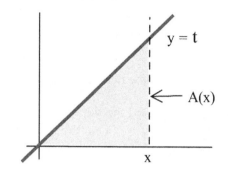

30. Compute each derivative.

    (a) $\mathbf{D}(x^{12})$

    (b) $\dfrac{d}{dx}(\sqrt[7]{x})$

    (c) $\mathbf{D}\left(\dfrac{1}{x^3}\right)$

    (d) $\dfrac{d}{dx}(x^c)$

    (e) $\mathbf{D}(|x - 2|)$

31. Compute each derivative.

    (a) $\mathbf{D}(x^9)$

    (b) $\dfrac{d}{dx}(x^{\frac{2}{3}})$

    (c) $\mathbf{D}\left(\dfrac{1}{x^4}\right)$

    (d) $\dfrac{d}{dx}(x^\pi)$

    (e) $\mathbf{D}(|x + 5|)$

    In Problems 32–37, find a function f that has the given derivative. (Each problem has several correct answers, just find one of them.)

32. $f'(x) = 4x + 3$

33. $f'(x) = 3x^2 + 8x$

34. $\mathbf{D}(f(x)) = 12x^2 - 7$

35. $f'(t) = 5\cos(t)$

36. $\frac{d}{dx}f(x) = 2x - \sin(x)$

37. $\mathbf{D}(f(x)) = x + x^2$

2.1 Practice Answers

1. The graph of $f(x) = 7x$ is a line through the origin. The slope of the line is 7. For all x:

$$m_{\tan} = \lim_{h \to 0} \frac{f(x+h) - f(x)}{h} = \lim_{h \to 0} \frac{7(x+h) - 7x}{h} = \lim_{h \to 0} \frac{7h}{h} = \lim_{h \to 0} 7 = 7$$

2. $f(x) = x^3 \Rightarrow f(x+h) = (x+h)^3 = x^3 + 3x^2h + 3xh^2 + h^3$ so:

$$f'(x) = \lim_{h \to 0} \frac{f(x+h) - f(x)}{h} = \lim_{h \to 0} \frac{x^3 + 3x^2h + 3xh^2 + h^3 - x^3}{h}$$

$$= \lim_{h \to 0} \frac{3x^2h + 3xh^2 + h^3}{h} = \lim_{h \to 0} 3x^2 + 3xh + h^2 = 3x^2$$

At the point $(2,8)$, the slope of the tangent line is $3(2)^2 = 12$ so an equation of the tangent line is $y - 8 = 12(x - 2)$ or $y = 12x - 16$.

3. $\mathbf{D}(x^4) = 4x^3$, $\mathbf{D}(x^5) = 5x^4$, $\mathbf{D}(x^{43}) = 43x^{42}$,
$\mathbf{D}(\sqrt{x}) = \mathbf{D}(x^{\frac{1}{2}}) = \frac{1}{2}x^{-\frac{1}{2}} = \frac{1}{2\sqrt{x}}$, $\mathbf{D}(x^\pi) = \pi x^{\pi - 1}$

4. Proceeding as we did to find the derivative to $\sin(x)$:

$$\mathbf{D}(\cos(x)) = \lim_{h \to 0} \frac{\cos(x+h) - \cos(x)}{h} = \lim_{h \to 0} \frac{\cos(x)\cos(h) - \sin(x)\sin(h) - \cos(x)}{h}$$

$$= \lim_{h \to 0} \left[\cos(x) \cdot \frac{\cos(h) - 1}{h} - \sin(x) \frac{\sin(h)}{h} \right] = \cos(x) \cdot 0 - \sin(x) \cdot 1 = -\sin(x)$$

5. See margin figure for the graphs of $y = |x - 2|$ and $y = |2x|$.

$$\mathbf{D}(|x - 2|) = \begin{cases} 1 & \text{if } x > 2 \\ \text{undefined} & \text{if } x = 2 \\ -1 & \text{if } x < 2 \end{cases} = \frac{|x - 2|}{x - 2}$$

$$\mathbf{D}(|2x|) = \begin{cases} 2 & \text{if } x > 0 \\ \text{undefined} & \text{if } x = 0 \\ -2 & \text{if } x < 0 \end{cases} = \frac{2|x|}{x}$$

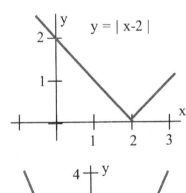

6. $h(t) = \sin(t)$ so $h(1) = \sin(1) \approx 0.84$ ft;
$v(t) = \cos(t)$ so $v(1) = \cos(1) \approx 0.54$ ft/sec;
$a(t) = -\sin(t)$ so $a(1) = -\sin(1) \approx -0.84$ ft/sec^2.

7. $\mathbf{D}(x^5) = 5x^4$, $\frac{d}{dx}(x^2) = 2x^1 = 2x$, $\mathbf{D}(x^{100}) = 100x^{99}$, $\frac{d}{dt}(t^{31}) = 31t^{30}$
and $\mathbf{D}(x^0) = 0x^{-1} = 0$ or $\mathbf{D}(x^0) = \mathbf{D}(1) = 0$

8. $\mathbf{D}(x^{\frac{3}{2}}) = \frac{3}{2}x^{\frac{1}{2}}$, $\frac{d}{dx}(x^{\frac{1}{3}}) = \frac{1}{3}x^{-\frac{2}{3}}$, $\mathbf{D}(\frac{1}{\sqrt{x}}) = \mathbf{D}(x^{-\frac{1}{2}}) = -\frac{1}{2}x^{-\frac{3}{2}}$,
$\frac{d}{dt}(t^\pi) = \pi t^{\pi - 1}$.

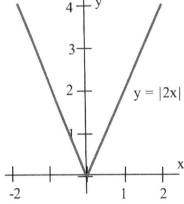

2.2 Derivatives: Properties and Formulas

This section begins with a look at which functions have derivatives. Then we'll examine how to calculate derivatives of elementary combinations of basic functions. By knowing the derivatives of some basic functions and just a few differentiation patterns, you will be able to calculate the derivatives of a tremendous variety of functions. This section contains most—but not quite all—of the general differentiation patterns you will ever need.

Which Functions Have Derivatives?

A function must be continuous in order to be differentiable.

> **Theorem:**
>
> If a function is differentiable at a point
> then it is continuous at that point.

It is vital to understand what this theorem tells us and what it does **not** tell us: If a function is differentiable at a point, then the function is automatically continuous there. If the function is continuous at a point, then the function may or may not be differentiable there.

Proof. Assume that the hypothesis (f is differentiable at the point c) is true. Then $\lim\limits_{h \to 0} \dfrac{f(c+h) - f(c)}{h}$ must exist and be equal to $f'(c)$. We want to show that f must necessarily be continuous at c, so we need to show that $\lim\limits_{h \to 0} f(c+h) = f(c)$.

It's not yet obvious why we want to do so, but we can write:

$$f(c+h) = f(c) + \frac{f(c+h) - f(c)}{h} \cdot h$$

and then compute the limit of both sides of this expression:

$$\lim_{h \to 0} f(c+h) = \lim_{h \to 0} \left(f(c) + \frac{f(c+h) - f(c)}{h} \cdot h \right)$$

$$= \lim_{h \to 0} f(c) + \lim_{h \to 0} \left(\frac{f(c+h) - f(c)}{h} \cdot h \right)$$

$$= \lim_{h \to 0} f(c) + \lim_{h \to 0} \left(\frac{f(c+h) - f(c)}{h} \right) \cdot \lim_{h \to 0} h$$

$$= f(c) + f'(c) \cdot 0 = f(c)$$

Therefore f is continuous at c. □

We often use the contrapositive form of this theorem, which tells us about some functions that do **not** have derivatives.

> **Contrapositive Form of the Theorem:**
>
> If f is not continuous at a point
> then f is not differentiable at that point.

Example 1. Show that $f(x) = \lfloor x \rfloor$ is not continuous and not differentiable at $x = 2$ (see margin figure).

Solution. The one-sided limits $\lim\limits_{x \to 2^+} \lfloor x \rfloor = 2$ and $\lim\limits_{x \to 2^-} \lfloor x \rfloor = 1$ have different values, so $\lim\limits_{x \to 2} \lfloor x \rfloor$ does not exist. Therefore $f(x) = \lfloor x \rfloor$ is not continuous at 2, and as a result it is not differentiable at 2.   ◄

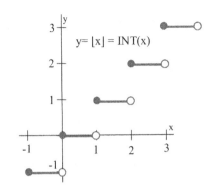

Lack of continuity implies lack of differentiability, but the next examples show that continuity is **not** enough to guarantee differentiability.

Example 2. Show that $f(x) = |x|$ is continuous but **not** differentiable at $x = 0$ (see margin figure).

Solution. We know that $\lim\limits_{x \to 0} |x| = 0 = |0|$, so f is continuous at 0, but in Section 2.1 we saw that $|x|$ was not differentiable at $x = 0$.   ◄

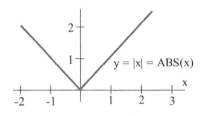

A function is not differentiable at a cusp or a "corner."

Example 3. Show that $f(x) = \sqrt[3]{x} = x^{\frac{1}{3}}$ is continuous but **not** differentiable at $x = 0$ (see margin figure).

Solution. We can verify that $\lim\limits_{x \to 0^+} \sqrt[3]{x} = \lim\limits_{x \to 0^-} \sqrt[3]{x} = 0$, so $\lim\limits_{x \to 0} \sqrt[3]{x} = 0 = \sqrt[3]{0}$ so f is continuous at 0. But $f'(x) = \frac{1}{3}x^{-\frac{2}{3}} = \frac{1}{3\sqrt[3]{x^2}}$, which is undefined at $x = 0$, so f is not differentiable at 0.   ◄

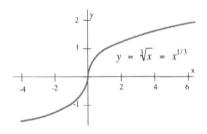

A function is not differentiable where its tangent line is vertical.

Practice 1. At which integer values of x is the graph of f in the margin figure continuous? Differentiable?

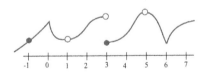

Graphically, a function is **continuous** if and only if its graph is "connected" and does not have any holes or breaks. Graphically, a function is **differentiable** if and only if it is continuous and its graph is "smooth" with no corners or vertical tangent lines.

Derivatives of Elementary Combinations of Functions

We now begin to compute derivatives of more complicated functions built from combinations of simpler functions.

Example 4. The derivative of $f(x) = x$ is $\mathbf{D}(f(x)) = 1$ and the derivative of $g(x) = 5$ is $\mathbf{D}(g(x)) = 0$. What are the derivatives of the elementary combinations: $3 \cdot f$, $f + g$, $f - g$, $f \cdot g$ and $\dfrac{f}{g}$?

Solution. The first three derivatives follow "nice" patterns:

$$\mathbf{D}(3 \cdot f(x)) = \mathbf{D}(3x) = 3 = 3 \cdot 1 = 3 \cdot \mathbf{D}(f(x))$$

$$\mathbf{D}(f(x) + g(x)) = \mathbf{D}(x + 5) = 1 = 1 + 0 = \mathbf{D}(f(x)) + \mathbf{D}(g(x))$$

$$\mathbf{D}(f(x) - g(x)) = \mathbf{D}(x - 5) = 1 = 1 - 0 = \mathbf{D}(f(x)) - \mathbf{D}(g(x))$$

yet the other two derivatives fail to follow the same "nice" patterns:

$$\mathbf{D}(f(x) \cdot g(x)) = \mathbf{D}(5x) = 5 \text{ but } \mathbf{D}(f(x)) \cdot \mathbf{D}(g(x)) = 1 \cdot 0 = 0, \text{ and}$$

$$\mathbf{D}\left(\frac{f(x)}{g(x)}\right) = \mathbf{D}\left(\frac{x}{5}\right) = \frac{1}{5} \text{ but } \frac{\mathbf{D}(f(x))}{\mathbf{D}(g(x))} = \frac{1}{0} \text{ is undefined.} \qquad \blacktriangleleft$$

The two very simple functions in the previous example show that, in general, $\mathbf{D}(f \cdot g) \neq \mathbf{D}(f) \cdot \mathbf{D}(g)$ and $\mathbf{D}\left(\dfrac{f}{g}\right) \neq \dfrac{\mathbf{D}(f)}{\mathbf{D}(g)}$.

Practice 2. For $f(x) = 6x + 8$ and $g(x) = 2$, compute the derivatives of $3 \cdot f$, $f + g$, $f - g$, $f \cdot g$ and $\dfrac{f}{g}$.

Main Differentiation Theorem:

If f and g are differentiable at x, then:

(a) **Constant Multiple Rule:**

$$\mathbf{D}(k \cdot f(x)) = k \cdot \mathbf{D}(f(x))$$

(b) **Sum Rule:**

$$\mathbf{D}(f(x) + g(x)) = \mathbf{D}(f(x)) + \mathbf{D}(g(x))$$

(c) **Difference Rule:**

$$\mathbf{D}(f(x) - g(x)) = \mathbf{D}(f(x)) - \mathbf{D}(g(x))$$

(d) **Product Rule:**

$$\mathbf{D}(f(x) \cdot g(x)) = f(x) \cdot \mathbf{D}(g(x)) + g(x) \cdot \mathbf{D}(f(x))$$

(e) **Quotient Rule:**

$$\mathbf{D}\left(\frac{f(x)}{g(x)}\right) = \frac{g(x) \cdot \mathbf{D}(f(x)) - f(x) \cdot \mathbf{D}(g(x))}{[g(x)]^2}$$

Part (e) requires that $g(x) \neq 0$.

This theorem says that the simple patterns in the previous example for constant multiples of functions and sums and differences of functions are true for all differentiable functions. It also includes the correct patterns for derivatives of products and quotients of differentiable functions.

The proofs of parts (a), (b) and (c) of this theorem are straightforward, but parts (d) and (e) require some clever algebraic manipulations. Let's look at some examples before tackling the proof.

Example 5. Recall that $\mathbf{D}(x^2) = 2x$ and $\mathbf{D}(\sin(x)) = \cos(x)$. Find $\mathbf{D}(3\sin(x))$ and $\frac{d}{dx}(5x^2 - 7\sin(x))$.

Solution. Computing $\mathbf{D}(3\sin(x))$ requires part (a) of the theorem with $k = 3$ and $f(x) = \sin(x)$ so $\mathbf{D}(3 \cdot \sin(x)) = 3 \cdot \mathbf{D}(\sin(x)) = 3\cos(x)$, while $\frac{d}{dx}(5x^2 - 7\sin(x))$ uses part (c) of the theorem with $f(x) = 5x^2$ and $g(x) = 7\sin(x)$ so:

$$\frac{d}{dx}(5x^2 - 7\sin(x)) = \frac{d}{dx}(5x^2) - \frac{d}{dx}(7\sin(x))$$
$$= 5 \cdot \frac{d}{dx}(x^2) - 7 \cdot \frac{d}{dx}(\sin(x))$$
$$= 5(2x) - 7(\cos(x))$$

which simplifies to $10x - 7\cos(x)$.   ◄

Practice 3. Find $\mathbf{D}(x^3 - 5\sin(x))$ and $\frac{d}{dx}(\sin(x) - 4x^3)$.

Practice 4. The table below gives the values of functions f and g, as well as their derivatives, at various points. Fill in the missing values for $\mathbf{D}(3 \cdot f(x))$, $\mathbf{D}(2 \cdot f(x) + g(x))$ and $\mathbf{D}(3 \cdot g(x) - f(x))$.

x	$f(x)$	$f'(x)$	$g(x)$	$g'(x)$	$\mathbf{D}(3f(x))$	$\mathbf{D}(2f(x) + g(x))$	$\mathbf{D}(3g(x) - f(x))$
0	3	−2	−4	3			
1	2	−1	1	0			
2	4	2	3	1			

Practice 5. Use the Main Differentiation Theorem to complete the table.

x	$f(x)$	$f'(x)$	$g(x)$	$g'(x)$	$\mathbf{D}(f(x) \cdot g(x))$	$\mathbf{D}\left(\dfrac{f(x)}{g(x)}\right)$	$\mathbf{D}\left(\dfrac{g(x)}{f(x)}\right)$
0	3	−2	−4	3			
1	2	−1	1	0			
2	4	2	3	1			

Example 6. Determine $\mathbf{D}(x^2 \cdot \sin(x))$ and $\frac{d}{dx}\left(\frac{x^3}{\sin(x)}\right)$.

Solution. (a) Use the Product Rule with $f(x) = x^2$ and $g(x) = \sin(x)$:

$$\mathbf{D}(x^2 \cdot \sin(x)) = \mathbf{D}(f(x) \cdot g(x)) = f(x) \cdot \mathbf{D}(g(x)) + g(x) \cdot \mathbf{D}(f(x))$$
$$= x^2 \cdot \mathbf{D}(\sin(x)) + \sin(x) \cdot \mathbf{D}(x^2)$$
$$= x^2 \cdot \cos(x) + \sin(x) \cdot 2x = x^2 \cos(x) + 2x\sin(x)$$

Many calculus students find it easier to remember the Product Rule in words: "the first function times the derivative of the second plus the second function times the derivative of the first."

(b) Use the Quotient Rule with $f(x) = x^3$ and $g(x) = \sin(x)$:

$$\frac{d}{dx}\left(\frac{x^3}{\sin(x)}\right) = \frac{d}{dx}\left(\frac{f(x)}{g(x)}\right)$$

$$= \frac{g(x) \cdot \mathbf{D}(f(x)) - f(x) \cdot \mathbf{D}(g(x))}{[g(x)]^2}$$

$$= \frac{\sin(x) \cdot \mathbf{D}(x^3) - x^3 \cdot \mathbf{D}(\sin(x))}{[\sin(x)]^2}$$

$$= \frac{\sin(x) \cdot 3x^2 - x^3 \cdot \cos(x)}{\sin^2(x)}$$

$$= \frac{3x^2 \sin(x) - x^3 \cdot \cos(x)}{\sin^2(x)}$$

The Quotient Rule in words: "the bottom times the derivative of the top minus the top times the derivative of the bottom, all over the bottom squared."

which could also be rewritten in terms of $\csc(x)$ and $\cot(x)$.   ◄

Practice 6. Find $\mathbf{D}((x^2 + 1)(7x - 3))$, $\dfrac{d}{dt}\left(\dfrac{3t - 2}{5t + 1}\right)$ and $\mathbf{D}\left(\dfrac{\cos(x)}{x}\right)$.

Now that we've seen how to use the theorem, let's prove it.

Proof. The only general fact we have about derivatives is the definition as a limit, so our proofs here will need to recast derivatives as limits and then use some results about limits. The proofs involve applications of the definition of the derivative and results about limits.

(a) Using the derivative definition and the limit laws:

$$\mathbf{D}(k \cdot f(x)) = \lim_{h \to 0} \frac{k \cdot f(x + h) - k \cdot f(x)}{h}$$

$$= \lim_{h \to 0} k \cdot \frac{f(x + h) - f(x)}{h}$$

$$= k \cdot \lim_{h \to 0} \frac{f(x + h) - f(x)}{h} = k \cdot \mathbf{D}(f(x))$$

(b) You try it (see Practice problem that follows).

(c) Once again using the derivative definition and the limit laws:

$$\mathbf{D}(f(x) - g(x)) = \lim_{h \to 0} \frac{[f(x + h) - g(x + h)] - [f(x) - g(x)]}{h}$$

$$= \lim_{h \to 0} \frac{[f(x + h) - f(x)] - [g(x + h) - g(x)]}{h}$$

$$= \lim_{h \to 0} \frac{f(x + h) - f(x)}{h} - \lim_{h \to 0} \frac{g(x + h) - g(x)}{h}$$

$$= \mathbf{D}(f(x)) - \mathbf{D}(g(x))$$

The proofs of parts (d) and (e) of the theorem are more complicated but only involve elementary techniques, used in just the right way.

Sometimes we will omit such computational proofs, but the Product and Quotient Rules are fundamental techniques you will need hundreds of times.

(d) By the hypothesis, f and g are differentiable, so:

$$\lim_{h\to 0} \frac{f(x+h) - f(x)}{h} = f'(x)$$

and:

$$\lim_{h\to 0} \frac{g(x+h) - g(x)}{h} = g'(x)$$

Also, both f and g are continuous (why?) so $\lim_{h\to 0} f(x+h) = f(x)$ and $\lim_{h\to 0} g(x+h) = g(x)$.

Let $P(x) = f(x) \cdot g(x)$. Then $P(x+h) = f(x+h) \cdot g(x+h)$ and:

$$\mathbf{D}(f(x) \cdot g(x)) = \mathbf{D}(P(x)) = \lim_{h\to 0} \frac{P(x+h) - P(x)}{h}$$

$$= \lim_{h\to 0} \frac{f(x+h) \cdot g(x+h) - f(x) \cdot g(x)}{h}$$

At this stage we need to use some cleverness to add and subtract $f(x) \cdot g(x+h)$ from the numerator (you'll see why shortly):

$$\lim_{h\to 0} \frac{f(x+h) \cdot g(x+h) + [-f(x) \cdot g(x+h) + f(x) \cdot g(x+h)] - f(x)g(x)}{h}$$

We can then split this giant fraction into two more manageable limits:

$$\lim_{h\to 0} \frac{f(x+h)g(x+h) - f(x)g(x+h)}{h} + \lim_{h\to 0} \frac{f(x)g(x+h) - f(x)g(x)}{h}$$

and then factor out a common factor from each numerator:

$$\lim_{h\to 0} g(x+h) \cdot \frac{f(x+h) - f(x)}{h} + \lim_{h\to 0} f(x) \cdot \frac{g(x+h) - g(x)}{h}$$

Taking limits of each piece (and using the continuity of $g(x)$) we get:

$$\mathbf{D}(f(x) \cdot g(x)) = g(x) \cdot f'(x) + f(x) \cdot g'(x) = g \cdot \mathbf{D}(f) + f \cdot \mathbf{D}(g)$$

The steps for a proof of the Quotient Rule appear in Problem 69.   □

Practice 7. Prove the Sum Rule: $\mathbf{D}(f(x) + g(x)) = \mathbf{D}(f(x)) + \mathbf{D}(g(x))$. (Refer to the proof of part (c) for guidance.)

Using the Differentiation Rules

You definitely need to memorize the differentiation rules, but it is vitally important that you also know **how** to use them. Sometimes it is clear that the function we want to differentiate is a sum or product of two obvious functions, but we commonly need to differentiate functions that involve several operations and functions. Memorizing the differentiation rules is only the first step in learning to use them.

Example 7. Calculate $D(x^5 + x \cdot \sin(x))$.

Solution. This function is more difficult because it involves both an addition and a multiplication. Which rule(s) should we use — or, more importantly, which rule should we use **first**?

First apply the Sum Rule to trade one derivative for two easier ones:

$$D(x^5 + x \cdot \sin(x)) = D(x^5) + D(x \cdot \sin(x))$$
$$= 5x^4 + [x \cdot D(\sin(x)) + \sin(x) \cdot D(x)]$$
$$= 5x^4 + x \cdot \cos(x) + \sin(x)$$

This last expression involves no more derivatives, so we are done. ◀

If instead of computing the derivative you were evaluating the function $x^5 + x \sin(x)$ for some particular value of x, you would:

- raise x to the 5th power

- calculate $\sin(x)$

- multiply $\sin(x)$ by x and, finally,

- **add** (sum) the values of x^5 and $x \sin(x)$

Notice that the **final** step of your **evaluation** of f indicates the **first** rule to use to calculate the **derivative** of f.

Practice 8. Which differentiation rule should you apply **first** for each of the following?

(a) $x \cdot \cos(x) - x^3 \cdot \sin(x)$ (b) $(2x - 3) \cos(x)$

(c) $2 \cos(x) - 7x^2$ (d) $\dfrac{\cos(x) + 3x}{\sqrt{x}}$

Practice 9. Calculate $D\left(\dfrac{x^2 - 5}{\sin(x)}\right)$ and $\dfrac{d}{dt}\left(\dfrac{t^2 - 5}{t \cdot \sin(t)}\right)$.

Example 8. A mass attached to a spring oscillates up and down but the motion becomes "damped" due to friction and air resistance. The height of the mass after t seconds is given by $h(t) = 5 + \dfrac{\sin(t)}{1 + t}$ (in feet). Find the height and velocity of the mass after 2 seconds.

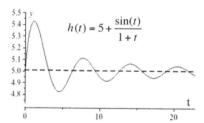

$$h(t) = 5 + \frac{\sin(t)}{1 + t}$$

Solution. The height is $h(2) = 5 + \dfrac{\sin(2)}{1 + 2} \approx 5 + \dfrac{0.909}{3} = 5.303$ feet above the ground. The velocity is $h'(2)$, so we must first compute $h'(t)$ and then evaluate the derivative at time $t = 2$:

$$h'(t) = \frac{(1 + t) \cdot \cos(t) - \sin(t) \cdot 1}{(1 + t)^2}$$

so $h'(2) = \dfrac{3 \cos(2) - \sin(2)}{9} \approx \dfrac{-2.158}{9} \approx -0.24$ feet per second. ◀

Practice 10. What are the height and velocity of the weight in the previous example after 5 seconds? What are the height and velocity of the weight be after a "long time" has passed?

Example 9. Calculate $D(x \cdot \sin(x) \cdot \cos(x))$.

Solution. Clearly we need to use the Product Rule, because the only operation in this function is multiplication. But the Product Rule deals with a product of **two** functions and here we have the product of three: x and $\sin(x)$ and $\cos(x)$. If, however, we think of our two functions as $f(x) = x \cdot \sin(x)$ and $g(x) = \cos(x)$, then we do have the product of two functions and:

$$D(x \cdot \sin(x) \cdot \cos(x)) = D(f(x) \cdot g(x))$$
$$= f(x) \cdot D(g(x)) + g(x) \cdot D(f(x))$$
$$= x \sin(x) \cdot D(\cos(x)) + \cos(x) \cdot D(x \sin(x))$$

We are not done, but we have traded one hard derivative for two easier ones. We know that $D(\cos(x)) = -\sin(x)$ and we can use the Product Rule (again) to calculate $D(x \sin(x))$. Then the last line of our calculation above becomes:

$$x \sin(x) \cdot [-\sin(x)] + \cos(x) \cdot [x D(\sin(x)) + \sin(x) D(x)]$$

and then:

$$-x \sin^2(x) + \cos(x) [x \cos(x) + \sin(x)(1)]$$

which simplifies to $-x \sin^2(x) + x \cos^2(x) + \cos(x) \sin(x)$. ◀

Evaluating a Derivative at a Point

The derivative of a function $f(x)$ is a new **function** $f'(x)$ that tells us the slope of the line tangent to the graph of f at each point x. To find the slope of the tangent line at a particular point $(c, f(c))$ on the graph of f, we should *first* calculate the derivative $f'(x)$ and *then* evaluate the function $f'(x)$ at the point $x = c$ to get the **number** $f'(c)$. If you mistakenly evaluate f first, you get a number $f(c)$, and the derivative of a constant is always equal to 0.

Example 10. Determine the slope of the line tangent to the graph of $f(x) = 3x + \sin(x)$ at $(0, f(0))$ and $(1, f(1))$.

Solution. $f'(x) = D(3x + \sin(x)) = D(3x) + D(\sin(x)) = 3 + \cos(x)$. When $x = 0$, the graph of $y = 3x + \sin(x)$ goes through the point $(0, 3(0) + \sin(0)) = (0, 0)$ with slope $f'(0) = 3 + \cos(0) = 4$. When $x = 1$, the graph goes through the point $(1, 3(1) + \sin(1)) \approx (1, 3.84)$ with slope $f'(1) = 3 + \cos(1) \approx 3.54$. ◀

Practice 11. Where do $f(x) = x^2 - 10x + 3$ and $g(x) = x^3 - 12x$ have horizontal tangent lines?

This section, like the last one, contains a great deal of important information that we will continue to use throughout the rest of the course, so we collect here some of those important results.

Important Information and Results

Differentiability and Continuity: If a function is differentiable then it must be continuous. If a function is not continuous then it cannot be differentiable. A function may be continuous at a point and not differentiable there.

Graphically: *Continuous* means "connected"; *differentiable* means "continuous, smooth and not vertical."

Differentiation Patterns:

- $[k \cdot f(x)]' = k \cdot f'(x)$

- $[f(x) + g(x)]' = f'(x) + g'(x)$

- $[f(x) - g(x)]' = f'(x) - g'(x)$

- $[f(x) \cdot g(x)]' = f(x) \cdot g'(x) + g(x) \cdot f'(x)$

- $\left[\dfrac{f(x)}{g(x)}\right]' = \dfrac{g(x) \cdot f'(x) - f(x) \cdot g'(x)}{[g(x)]^2}$

- The *final step* used to evaluate a function f indicates the *first rule* used to differentiate f.

Evaluating a derivative at a point: First differentiate and *then* evaluate.

2.2 Problems

1. Use the graph of $y = f(x)$ below to determine:

   (a) at which integers f is continuous.

   (b) at which integers f is differentiable.

2. Use the graph of $y = g(x)$ below to determine:

   (a) at which integers g is continuous.

   (b) at which integers g is differentiable.

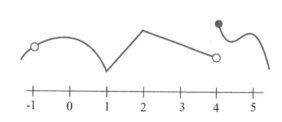

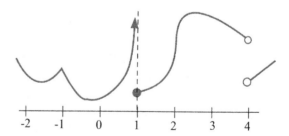

3. Use the values given in the table to determine the values of $f \cdot g$, $\mathbf{D}(f \cdot g)$, $\frac{f}{g}$ and $\mathbf{D}\left(\frac{f}{g}\right)$.

x	$f(x)$	$f'(x)$	$g(x)$	$g'(x)$	$f(x) \cdot g(x)$	$\mathbf{D}(f(x) \cdot g(x))$	$\frac{f(x)}{g(x)}$	$\mathbf{D}\left(\frac{f(x)}{g(x)}\right)$
0	2	3	1	5				
1	-3	2	5	-2				
2	0	-3	2	4				
3	1	-1	0	3				

4. Use the values given in the table to determine the values of $f \cdot g$, $\mathbf{D}(f \cdot g)$, $\frac{f}{g}$ and $\mathbf{D}\left(\frac{f}{g}\right)$.

x	$f(x)$	$f'(x)$	$g(x)$	$g'(x)$	$f(x) \cdot g(x)$	$\mathbf{D}(f(x) \cdot g(x))$	$\frac{f(x)}{g(x)}$	$\mathbf{D}\left(\frac{f(x)}{g(x)}\right)$
0	4	2	3	−3				
1	0	3	2	1				
2	−2	5	0	−1				
3	−1	−2	−3	4				

5. Use the information in the figure below to plot the values of the functions $f + g$, $f \cdot g$ and $\frac{f}{g}$ and their derivatives at $x = 1, 2$ and 3.

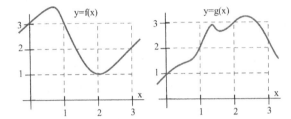

6. Use the information in the figure above to plot the values of the functions $2f$, $f - g$ and $\frac{g}{f}$ and their derivatives at $x = 1, 2$ and 3.

7. Calculate $\mathbf{D}((x - 5)(3x + 7))$ by:

   (a) using the Product Rule.

   (b) expanding and then differentiating.

   Verify that both methods give the same result.

8. Calculate $\mathbf{D}\left(\dfrac{x^3 - 3x + 2}{\sqrt{x}}\right)$ by:

   (a) using the Quotient Rule.

   (b) rewriting and then differentiating.

   Verify that both methods give the same result.

In Problems 9–26, compute each derivative.

9. $\dfrac{d}{dx}\left(19x^3 - 7\right)$

10. $\dfrac{d}{dt}\left(5\cos(t) + \dfrac{\pi}{2}\right)$

11. $\mathbf{D}(\sin(x) + \cos(x))$

12. $\mathbf{D}(7\sin(x) - 3\cos(x))$

13. $\mathbf{D}(x^2 \cdot \cos(x))$

14. $\mathbf{D}(\sqrt{x} \cdot \sin(x))$

15. $\mathbf{D}(\sin^2(x))$

16. $\mathbf{D}(\cos^2(x))$

17. $\dfrac{d}{dx}\left(\dfrac{\cos(x)}{x^2}\right)$

18. $\dfrac{d}{dt}\left(\dfrac{\sin(t)}{t^3}\right)$

19. $\dfrac{d}{dx}\left(\dfrac{1}{1 + x^2}\right)$

20. $\dfrac{d}{dt}\left(\dfrac{t}{1 + t^3}\right)$

21. $\dfrac{d}{d\theta}\left(\dfrac{1}{\cos(\theta)}\right)$

22. $\dfrac{d}{d\theta}\left(\dfrac{1}{\sin(\theta)}\right)$

23. $\dfrac{d}{d\theta}\left(\dfrac{\sin(\theta)}{\cos(\theta)}\right)$

24. $\dfrac{d}{d\theta}\left(\dfrac{\cos(\theta)}{\sin(\theta)}\right)$

25. $\mathbf{D}\left(8x^5 - 3x^4 + 2x^3 + 7x^2 - 12x + 147\right)$

26. (a) $\mathbf{D}(\sin(x))$ (b) $\mathbf{D}(\sin(x) + 7)$
    (c) $\mathbf{D}(\sin(x) - 8000)$ (d) $\mathbf{D}(\sin(x) + k)$

27. Find values for the constants a, b and c so that the parabola $f(x) = ax^2 + bx + c$ has $f(0) = 0$, $f'(0) = 0$ and $f'(10) = 30$.

28. If f is a differentiable function, how are the:

    (a) graphs of $y = f(x)$ and $y = f(x) + k$ related?

    (b) derivatives of $f(x)$ and $f(x) + k$ related?

29. If f and g are differentiable functions that always differ by a constant ($f(x) - g(x) = k$ for all x) then what can you conclude about their graphs? Their derivatives?

30. If f and g are differentiable functions whose sum is a constant ($f(x) + g(x) = k$ for all x) then what can you conclude about their graphs? Their derivatives?

31. If the product of f and g is a constant (that is, $f(x) \cdot g(x) = k$ for all x) then how are $\dfrac{\mathbf{D}(f(x))}{f(x)}$ and $\dfrac{\mathbf{D}(g(x))}{g(x)}$ related?

32. If the quotient of f and g is a constant ($\dfrac{f(x)}{g(x)} = k$ for all x) then how are $g \cdot f'$ and $f \cdot g'$ related?

In Problems 33–40:

(a) calculate $f'(1)$

(b) determine where $f'(x) = 0$.

33. $f(x) = x^2 - 5x + 13$

34. $f(x) = 5x^2 - 40x + 73$

35. $f(x) = 3x - 2\cos(x)$

36. $f(x) = |x + 2|$

37. $f(x) = x^3 + 9x^2 + 6$

38. $f(x) = x^3 + 3x^2 + 3x - 1$

39. $f(x) = x^3 + 2x^2 + 2x - 1$

40. $f(x) = \dfrac{7x}{x^2 + 4}$

41. $f(x) = x \cdot \sin(x)$ and $0 \le x \le 5$. (You may need to use the Bisection Algorithm or the "trace" option on a calculator to approximate where $f'(x) = 0$.)

42. $f(x) = Ax^2 + Bx + C$, where B and C are constants and $A \ne 0$ is constant.

43. $f(x) = x^3 + Ax^2 + Bx + C$ with constants A, B and C. Can you find conditions on the constants A, B and C that will guarantee that the graph of $y = f(x)$ has two distinct "turning points"? (Here a "turning point" means a place where the curve changes from increasing to decreasing or from decreasing to increasing, like the vertex of a parabola.)

In 44–51, where are the functions differentiable?

44. $f(x) = |x| \cos(x)$ 45. $f(x) = \tan(x)$

46. $f(x) = \dfrac{x - 5}{x + 3}$ 47. $f(x) = \dfrac{x^2 + x}{x^2 - 3x}$

48. $f(x) = |x^2 - 4|$ 49. $f(x) = |x^3 - 1|$

50. $f(x) = \begin{cases} 0 & \text{if } x < 0 \\ \sin(x) & \text{if } x \ge 0 \end{cases}$

51. $f(x) = \begin{cases} x & \text{if } x < 0 \\ \sin(x) & \text{if } x \ge 0 \end{cases}$

52. For what value(s) of A is

$$f(x) = \begin{cases} Ax - 4 & \text{if } x < 2 \\ x^2 + x & \text{if } x \ge 2 \end{cases}$$

differentiable at $x = 2$?

53. For what values of A and B is

$$f(x) = \begin{cases} Ax + B & \text{if } x < 1 \\ x^2 + x & \text{if } x \ge 1 \end{cases}$$

differentiable at $x = 1$?

54. An arrow shot straight up from ground level (get out of the way!) with an initial velocity of 128 feet per second will be at height $h(x) = -16x^2 + 128x$ feet after x seconds (see figure below).

(a) Determine the velocity of the arrow when $x = 0$, 1 and 2 seconds.

(b) What is the velocity of the arrow, $v(x)$, at any time x?

(c) At what time x will the velocity of the arrow be 0?

(d) What is the greatest height the arrow reaches?

(e) How long will the arrow be aloft?

(f) Use the answer for the velocity in part (b) to determine the acceleration, $a(x) = v'(x)$, at any time x.

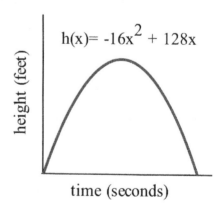

55. If an arrow is shot straight up from ground level on the moon with an initial velocity of 128 feet per second, its height will be $h(x) = -2.65x^2 + 128x$ feet after x seconds. Redo parts (a)–(e) of problem 40 using this new formula for $h(x)$.

56. In general, if an arrow is shot straight upward with an initial velocity of 128 feet per second from ground level on a planet with a constant gravitational acceleration of g feet per second^2 then its height will be $h(x) = -\frac{g}{2}x^2 + 128x$ feet after x seconds. Answer the questions in problem 40 for arrows shot on Mars and Jupiter.

object	g (ft/sec^2)	g (cm/sec^2)
Mercury	11.8	358
Venus	20.1	887
Earth	32.2	981
moon	5.3	162
Mars	12.3	374
Jupiter	85.3	2601
Saturn	36.6	1117
Uranus	34.4	1049
Neptune	43.5	1325

Source: *CRC Handbook of Chemistry and Physics*

57. If an object on Earth is propelled upward from ground level with an initial velocity of v_0 feet per second, then its height after x seconds will be $h(x) = -16x^2 + v_0 x$.

    (a) Find the object's velocity after x seconds.

    (b) Find the greatest height the object will reach.

    (c) How long will the object remain aloft?

58. In order for a 6-foot-tall basketball player to dunk the ball, the player must achieve a vertical jump of about 3 feet. Use the information in the previous problems to answer the following questions.

    (a) What is the smallest initial vertical velocity the player can have and still dunk the ball?

    (b) With the initial velocity achieved in part (a), how high would the player jump on the moon?

59. The best high jumpers in the world manage to lift their centers of mass approximately 3.75 feet.

    (a) What is the initial vertical velocity these high jumpers attain?

    (b) How long are these high jumpers in the air?

    (c) How high would they lift their centers of mass on the moon?

60. (a) Find an equation for the line L that is tangent to the curve $y = \frac{1}{x}$ at the point $(1,1)$.

    (b) Determine where L intersects the x-axis and the y-axis.

    (c) Determine the area of the region in the first quadrant bounded by L, the x-axis and the y-axis (see figure below).

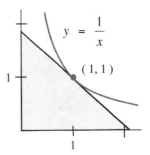

61. (a) Find an equation for the line L that is tangent to the curve $y = \frac{1}{x}$ at the point $(2, \frac{1}{2})$.

    (b) Graph $y = \frac{1}{x}$ and L and determine where L intersects the x-axis and the y-axis.

    (c) Determine the area of the region in the first quadrant bounded by L, the x-axis and the y-axis.

62. (a) Find an equation for the line L that is tangent to the curve $y = \frac{1}{x}$ at the point $(p, \frac{1}{p})$ (assuming $p \neq 0$).

    (b) Determine where L intersects the x-axis and the y-axis.

    (c) Determine the area of the region in the first quadrant bounded by L, the x-axis and the y-axis.

    (d) How does the area of the triangle in part (c) depend on the initial point $(p, \frac{1}{p})$?

63. Find values for the coefficients a, b and c so that the parabola $f(x) = ax^2 + bx + c$ goes through the point $(1,4)$ and is tangent to the line $y = 9x - 13$ at the point $(3,14)$.

64. Find values for the coefficients a, b and c so that the parabola $f(x) = ax^2 + bx + c$ goes through the point $(0,1)$ and is also tangent to the line $y = 3x - 2$ at the point $(2,4)$.

65. (a) Find a function f so that $\mathbf{D}(f(x)) = 3x^2$.

    (b) Find another function g with $\mathbf{D}(g(x)) = 3x^2$.

    (c) Can you find more functions whose derivatives are $3x^2$?

66. (a) Find a function f so that $f'(x) = 6x + \cos(x)$.

    (b) Find another function g with $g'(x) = f'(x)$.

67. The graph of $y = f'(x)$ appears below.

    (a) Assume $f(0) = 0$ and sketch a graph of $y = f(x)$.

    (b) Assume $f(0) = 1$ and graph $y = f(x)$.

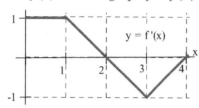

68. The graph of $y = g'(x)$ appears below. Assume that g is continuous.

    (a) Assume $g(0) = 0$ and sketch a graph of $y = g(x)$.

    (b) Assume $g(0) = 1$ and graph $y = g(x)$.

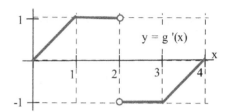

69. Assume that f and g are differentiable functions and that $g(x) \neq 0$. State why each step in the following proof of the Quotient Rule is valid.

Proof of the Quotient Rule

$$\mathbf{D}\left(\frac{f(x)}{g(x)}\right) = \lim_{h \to 0} \frac{1}{h}\left[\frac{f(x+h)}{g(x+h)} - \frac{f(x)}{g(x)}\right] = \lim_{h \to 0} \frac{1}{h}\left[\frac{f(x+h)g(x) - g(x+h)f(x)}{g(x+h)g(x)}\right]$$

$$= \lim_{h \to 0} \frac{1}{g(x+h)g(x)}\left[\frac{f(x+h)g(x) + (-f(x)g(x) + f(x)g(x)) - g(x+h)f(x)}{h}\right]$$

$$= \lim_{h \to 0} \frac{1}{g(x+h)g(x)}\left[g(x)\frac{f(x+h) - f(x)}{h} + f(x)\frac{g(x) - g(x+h)}{h}\right]$$

$$= \frac{1}{[g(x)]^2}\left[g(x) \cdot f'(x) - f(x) \cdot g'(x)\right]$$

$$= \frac{g(x) \cdot f'(x) - f(x) \cdot g'(x)}{[g(x)]^2}$$

Practice Answers

1. f is continuous at $x = -1, 0, 2, 4, 6$ and 7.

    f is differentiable at $x = -1, 2, 4,$ and 7.

2. $f(x) = 6x + 8$ and $g(x) = 2$ so $\mathbf{D}(f(x)) = 6$ and $\mathbf{D}(g(x)) = 0$.

    $\mathbf{D}(3 \cdot f(x)) = 3 \cdot \mathbf{D}(f(x)) = 3(6) = 18$

    $\mathbf{D}(f(x) + g(x)) = \mathbf{D}(f(x)) + \mathbf{D}(g(x)) = 6 + 0 = 6$

    $\mathbf{D}(f(x) - g(x)) = \mathbf{D}(f(x)) - \mathbf{D}(g(x)) = 6 - 0 = 6$

    $\mathbf{D}(f(x) \cdot g(x)) = f(x)g'(x) + g(x)f'(x) = (6x + 8)(0) + (2)(6) = 12$

    $\mathbf{D}\left(\frac{f(x)}{g(x)}\right) = \frac{g(x)f'(x) - f(x)g'(x)}{[g(x)]^2} = \frac{(2)(6) - (6x+8)(0)}{2^2} = \frac{12}{4} = 3$

3. $\mathbf{D}(x^3 - 5\sin(x)) = \mathbf{D}(x^3) - 5 \cdot \mathbf{D}(\sin(x)) = 3x^2 - 5\cos(x)$

$\dfrac{d}{dx}\left(\sin(x) - 4x^3\right) = \dfrac{d}{dx}\sin(x) - 4 \cdot \dfrac{d}{dx}x^3 = \cos(x) - 12x^2$

4.

$\mathbf{D}(3f(x))$	$\mathbf{D}(2f(x) + g(x))$	$\mathbf{D}(3g(x) - f(x))$
-6	-1	11
-3	-2	1
6	5	1

5.

$\mathbf{D}(f(x) \cdot g(x))$	$\mathbf{D}\left(\dfrac{f(x)}{g(x)}\right)$	$\mathbf{D}\left(\dfrac{g(x)}{f(x)}\right)$
$3 \cdot 3 + (-4)(-2) = 17$	$\dfrac{-4(-2)-(3)(3)}{(-4)^2} = -\dfrac{1}{16}$	$\dfrac{(3)(3)-(-4)(-2)}{3^2} = \dfrac{1}{9}$
$2 \cdot 0 + 1(-1) = -1$	$\dfrac{1(-1)-(2)(0)}{1^2} = -1$	$\dfrac{2(0)-1(-1)}{2^2} = \dfrac{1}{4}$
$4 \cdot 1 + 3 \cdot 2 = 10$	$\dfrac{3(2)-(4)(1)}{3^2} = \dfrac{2}{9}$	$\dfrac{4(1)-3(2)}{4^2} = -\dfrac{1}{8}$

6. $\mathbf{D}((x^2 + 1)(7x - 3)) = (x^2 + 1)\,\mathbf{D}(7x - 3) + (7x - 3)\,\mathbf{D}(x^2 + 1)$

$\qquad\qquad = (x^2 + 1)(7) + (7x - 3)(2x) = 21x^2 - 6x + 7$

or: $\mathbf{D}((x^2 + 1)(7x - 3)) = \mathbf{D}(7x^3 - 3x^2 + 7x) = 21x^2 - 6x + 7$

$\dfrac{d}{dt}\left(\dfrac{3t - 2}{5t + 1}\right) = \dfrac{(5t + 1)\,\mathbf{D}(3t - 2) - (3t - 2)\,\mathbf{D}(5t + 1)}{(5t + 1)^2} = \dfrac{(5t + 1)(3) - (3t - 2)(5)}{(5t + 1)^2} = \dfrac{13}{(5t + 1)^2}$

$\mathbf{D}\left(\dfrac{\cos(x)}{x}\right) = \dfrac{x\,\mathbf{D}(\cos(x)) - \cos(x)\,\mathbf{D}(x)}{x^2} = \dfrac{x(-\sin(x)) - \cos(x)(1)}{x^2} = \dfrac{-x \cdot \sin(x) - \cos(x)}{x^2}$

7. Mimicking the proof of the Difference Rule:

$$\mathbf{D}(f(x) + g(x)) = \lim_{h \to 0} \dfrac{[f(x + h) + g(x + h)] - [f(x) + g(x)]}{h}$$

$$= \lim_{h \to 0} \dfrac{[f(x + h) - f(x)] + [g(x + h) - g(x)]}{h}$$

$$= \lim_{h \to 0} \dfrac{f(x + h) - f(x)}{h} + \lim_{h \to 0} \dfrac{g(x + h) - g(x)}{h}$$

$$= \mathbf{D}(f(x)) + \mathbf{D}(g(x))$$

8. (a) difference rule (b) product rule (c) difference rule (d) quotient rule

9. $\mathbf{D}\left(\dfrac{x^2 - 5}{\sin(x)}\right) = \dfrac{\sin(x)\,\mathbf{D}(x^2 - 5) - (x^2 - 5)\,\mathbf{D}(\sin(x))}{(\sin(x))^2} = \dfrac{\sin(x)(2x) - (x^2 - 5)\cos(x)}{\sin^2(x)}$

$\dfrac{d}{dt}\left(\dfrac{t^2 - 5}{t \cdot \sin(t)}\right) = \dfrac{t \cdot \sin(t)\,\mathbf{D}(t^2 - 5) - (t^2 - 5)\,\mathbf{D}(t \cdot \sin(t))}{(t \cdot \sin(t))^2} = \dfrac{t \cdot \sin(t)(2t) - (t^2 - 5)\,[t\cos(t) + \sin(t)]}{t^2 \cdot \sin^2(t)}$

10. $h(5) = 5 + \dfrac{\sin(5)}{1+5} \approx 4.84$ ft.; $v(5) = h'(5) = \dfrac{(1+5)\cos(5) - \sin(5)}{(1+5)^2} \approx 0.074$ ft/sec.

"long time": $h(t) = 5 + \dfrac{\sin(t)}{1+t} \approx 5$ feet when t is very large;

$h'(t) = \dfrac{(1 + t)\cos(t) - \sin(t)}{(1 + t)^2} = \dfrac{\cos(t)}{1 + t} - \dfrac{\sin(t)}{(1 + t)^2} \approx 0$ ft/sec when t is very large.

11. $f'(x) = 2x - 10$ so $f'(x) = 0 \Rightarrow 2x - 10 = 0 \Rightarrow x = 5$.

$g'(x) = 3x^2 - 12$ so $g'(x) = 0 \Rightarrow 3x^2 - 12 = 0 \Rightarrow x^2 = 4 \Rightarrow x = \pm 2$.

2.3 More Differentiation Patterns

Polynomials are very useful, but they are not the only functions we need. This section uses the ideas of the two previous sections to develop techniques for differentiating **powers of functions**, and to determine the derivatives of some particular functions that occur often in applications: the **trigonometric** and **exponential** functions.

As you focus on learning how to differentiate different types and combinations of functions, it is important to remember what derivatives are and what they measure. Calculators and computers are available to calculate derivatives. Part of your job as a professional will be to decide which functions need to be differentiated and how to use the resulting derivatives. You can succeed at that only if you understand what a derivative is and what it measures.

A Power Rule for Functions: $\mathbf{D}(f^n(x))$

If we apply the Product Rule to the product of a function with itself, a pattern emerges.

$$\mathbf{D}(f^2) = \mathbf{D}(f \cdot f) = f \cdot \mathbf{D}(f) + f \cdot \mathbf{D}(f) = \qquad\qquad = 2f \cdot \mathbf{D}(f)$$
$$\mathbf{D}(f^3) = \mathbf{D}(f^2 \cdot f) = f^2 \cdot \mathbf{D}(f) + f \cdot \mathbf{D}(f^2) = f^2 \cdot \mathbf{D}(f) + f \cdot 2f \cdot \mathbf{D}(f) = 3f^2 \cdot \mathbf{D}(f)$$
$$\mathbf{D}(f^4) = \mathbf{D}(f^3 \cdot f) = f^3 \cdot \mathbf{D}(f) + f \cdot \mathbf{D}(f^3) = f^3 \cdot \mathbf{D}(f) + f \cdot 3f^2 \cdot \mathbf{D}(f) = 4f^3 \cdot \mathbf{D}(f)$$

Practice 1. What is the pattern here? What do you think the results will be for $\mathbf{D}(f^5)$ and $\mathbf{D}(f^{13})$?

We could keep differentiating higher and higher powers of $f(x)$ by writing them as products of lower powers of $f(x)$ and using the Product Rule, but the Power Rule for Functions guarantees that the pattern we just saw for the small integer powers also works for all constant powers of functions.

The Power Rule for Functions is a special case of a more general theorem, the Chain Rule, which we will examine in Section 2.4, so we will wait until then to prove the Power Rule for Functions.

> **Power Rule for Functions:**
>
> If $\quad p$ is any constant
>
> then $\quad \mathbf{D}(f^p(x)) = p \cdot f^{p-1}(x) \cdot \mathbf{D}(f(x)).$

Example 1. Use the Power Rule for Functions to find:

Remember: $\sin^2(x) = [\sin(x)]^2$

(a) $\mathbf{D}((x^3 - 5)^2)$ (b) $\dfrac{d}{dx}\left(\sqrt{2x + 3x^5}\right)$ (c) $\mathbf{D}(\sin^2(x))$

Solution. (a) To match the pattern of the Power Rule for $\mathbf{D}((x^3 - 5)^2)$, let $f(x) = x^3 - 5$ and $p = 2$. Then:

Check that you get the same answer by first expanding $(x^3 - 5)^2$ and then taking the derivative.

$$\mathbf{D}((x^3 - 5)^2) = \mathbf{D}(f^p(x)) = p \cdot f^{p-1}(x) \cdot \mathbf{D}(f(x))$$
$$= 2(x^3 - 5)^1 \, \mathbf{D}(x^3 - 5) = 2(x^3 - 5)(3x^2) = 6x^2(x^3 - 5)$$

(b) To match the pattern for $\dfrac{d}{dx}\left(\sqrt{2x+3x^5}\right) = \dfrac{d}{dx}\left((2x+3x^5)^{\frac{1}{2}}\right)$, let
$f(x) = 2x + 3x^5$ and take $p = \frac{1}{2}$. Then:

$$\frac{d}{dx}\left(\sqrt{2x+3x^5}\right) = \frac{d}{dx}(f^p(x)) = p \cdot f^{p-1}(x) \cdot \frac{d}{dx}(f(x))$$
$$= \frac{1}{2}(2x+3x^5)^{-\frac{1}{2}}\frac{d}{dx}(2x+3x^5)$$
$$= \frac{1}{2}(2x+3x^5)^{-\frac{1}{2}}(2+15x^4) = \frac{2+15x^4}{2\sqrt{2x+3x^5}}$$

(c) To match the pattern for $\mathbf{D}(\sin^2(x))$, let $f(x) = \sin(x)$ and $p = 2$:

$$\mathbf{D}(\sin^2(x)) = \mathbf{D}(f^p(x)) = p \cdot f^{p-1}(x) \cdot \mathbf{D}(f(x))$$
$$= 2\sin^1(x)\,\mathbf{D}(\sin(x)) = 2\sin(x)\cos(x)$$

We could also rewrite this last expression as $\sin(2x)$.   ◀

Practice 2. Use the Power Rule for Functions to find:

(a) $\dfrac{d}{dx}\left((2x^5 - \pi)^2\right)$ (b) $\mathbf{D}\left(\sqrt{x+7x^2}\right)$ (c) $\mathbf{D}(\cos^4(x))$

Example 2. Use calculus to show that the line tangent to the circle $x^2 + y^2 = 25$ at the point $(3,4)$ has slope $-\frac{3}{4}$.

Solution. The top half of the circle is the graph of $f(x) = \sqrt{25 - x^2}$ so:

$$f'(x) = \mathbf{D}\left((25-x^2)^{\frac{1}{2}}\right) = \frac{1}{2}(25-x^2)^{-\frac{1}{2}} \cdot \mathbf{D}(25-x^2) = \frac{-x}{\sqrt{25-x^2}}$$

and $f'(3) = \dfrac{-3}{\sqrt{25-3^2}} = -\dfrac{3}{4}$. As a check, you can verify that the slope of the radial line through the center of the circle $(0,0)$ and the point $(3,4)$ has slope $\frac{4}{3}$ and is perpendicular to the tangent line that has a slope of $-\frac{3}{4}$.   ◀

Derivatives of Trigonometric Functions

We have some general rules that apply to any elementary combination of differentiable functions, but in order to use the rules we still need to know the derivatives of some basic functions. Here we will begin to add to the list of functions whose derivatives we know.

We already know the derivatives of the sine and cosine functions, and each of the other four trigonometric functions is just a ratio involving sines or cosines. Using the Quotient Rule, we can easily differentiate the rest of the trigonometric functions.

Theorem:

$$\mathbf{D}(\tan(x)) = \sec^2(x) \qquad \mathbf{D}(\sec(x)) = \sec(x) \cdot \tan(x)$$
$$\mathbf{D}(\cot(x)) = -\csc^2(x) \qquad \mathbf{D}(\csc(x)) = -\csc(x) \cdot \cot(x)$$

Proof. From trigonometry, we know $\tan(x) = \dfrac{\sin(x)}{\cos(x)}$, $\cot(x) = \dfrac{\cos(x)}{\sin(x)}$, $\sec(x) = \dfrac{1}{\cos(x)}$ and $\csc(x) = \dfrac{1}{\sin(x)}$. From calculus, we already know $\mathbf{D}(\sin(x)) = \cos(x)$ and $\mathbf{D}(\cos(x)) = -\sin(x)$. So:

$$\begin{aligned}\mathbf{D}(\tan(x)) = \mathbf{D}\left(\frac{\sin(x)}{\cos(x)}\right) &= \frac{\cos(x) \cdot \mathbf{D}(\sin(x)) - \sin(x) \cdot \mathbf{D}(\cos(x))}{(\cos(x))^2} \\ &= \frac{\cos(x) \cdot \cos(x) - \sin(x)(-\sin(x))}{\cos^2(x)} \\ &= \frac{\cos^2(x) + \sin^2(x)}{\cos^2(x)} = \frac{1}{\cos^2(x)} = \sec^2(x) \end{aligned}$$

Similarly:

$$\begin{aligned}\mathbf{D}(\sec(x)) = \mathbf{D}\left(\frac{1}{\cos(x)}\right) &= \frac{\cos(x) \cdot \mathbf{D}(1) - 1 \cdot \mathbf{D}(\cos(x))}{(\cos(x))^2} \\ &= \frac{\cos(x) \cdot 0 - (-\sin(x))}{\cos^2(x)} \\ &= \frac{\sin(x)}{\cos^2(x)} = \frac{1}{\cos(x)} \cdot \frac{\sin(x)}{\cos(x)} = \sec(x) \cdot \tan(x) \end{aligned}$$

Instead of the Quotient Rule, we could have used the Power Rule to calculate $\mathbf{D}(\sec(x)) = \mathbf{D}((\cos(x))^{-1})$. $\qquad\square$

Practice 3. Use the Quotient Rule on $f(x) = \cot(x) = \dfrac{\cos(x)}{\sin(x)}$ to prove that $f'(x) = -\csc^2(x)$.

Practice 4. Prove that $\mathbf{D}(\csc(x)) = -\csc(x) \cdot \cot(x)$. The justification of this result is very similar to the justification for $\mathbf{D}(\sec(x))$.

Practice 5. Find: (a) $\mathbf{D}(x^5 \tan(x))$ (b) $\dfrac{d}{dt}\left(\dfrac{\sec(t)}{t}\right)$ (c) $\mathbf{D}\left(\sqrt{\cot(x) - x}\right)$

Derivatives of Exponential Functions

We can estimate the value of a derivative of an exponential function (a function of the form $f(x) = a^x$ where $a > 0$) by estimating the slope of the line tangent to the graph of such a function, or we can numerically approximate those slopes.

Example 3. Estimate the value of the derivative of $f(x) = 2^x$ at the point $(0, 2^0) = (0, 1)$ by approximating the slope of the line tangent to $f(x) = 2^x$ at that point.

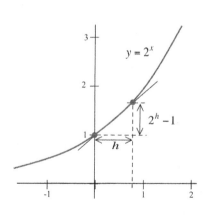

Solution. We can get estimates from the graph of $f(x) = 2^x$ by carefully graphing $y = 2^x$ for small values of x (so that x is near 0), sketching secant lines, and then measuring the slopes of the secant lines (see margin figure).

We can also estimate the slope numerically by using the definition of the derivative:

$$f'(0) = \lim_{h \to 0} \frac{f(0+h) - f(0)}{h} = \lim_{h \to 0} \frac{2^{0+h} - 2^0}{h} = \lim_{h \to 0} \frac{2^h - 1}{h}$$

and evaluating $\dfrac{2^h - 1}{h}$ for some very small values of h. From the table below we can see that $f'(0) \approx 0.693$. ◀

h	$\frac{2^h-1}{h}$	$\frac{3^h-1}{h}$	$\frac{e^h-1}{h}$
$+0.1$	0.717734625		
-0.1	0.669670084		
$+0.01$	0.695555006		
-0.01	0.690750451		
$+0.001$	0.693387463		
-0.001	0.692907009		
$\downarrow$	$\downarrow$	$\downarrow$	$\downarrow$
0	≈ 0.693	≈ 1.099	1

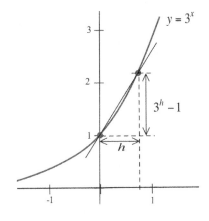

Practice 6. Fill in the table for $\dfrac{3^h - 1}{h}$ and show that the slope of the line tangent to $g(x) = 3^x$ at $(0,1)$ is approximately 1.099.

At $(0,1)$, the slope of the tangent to $y = 2^x$ is less than 1 and the slope of the tangent to $y = 3^x$ is slightly greater than 1. You might expect that there is a number b between 2 and 3 so that the slope of the tangent to $y = b^x$ is exactly 1. Indeed, there is such a number, $e \approx 2.71828182845904$, with

$$\lim_{h \to 0} \frac{e^h - 1}{h} = 1$$

The number e is irrational and plays a very important role in calculus and applications.

In fact, e is a "transcendental" number, which means that it is not the root of any polynomial equation with rational coefficients.

Don't worry—we'll tie up some of these loose ends in Chapter 7.

We have not proved that this number e with the desired limit property actually exists, but if we assume it does, then it becomes relatively straightforward to calculate $\mathbf{D}(e^x)$.

> **Theorem: $\mathbf{D}(e^x) = e^x$**

Proof. Using the definition of the derivative:

$$\mathbf{D}(e^x) = \lim_{h \to 0} \frac{e^{x+h} - e^x}{h} = \lim_{h \to 0} \frac{e^x \cdot e^h - e^x}{h}$$

$$= \lim_{h \to 0} e^x \cdot \frac{e^h - 1}{h} = \lim_{h \to 0} e^x \cdot \lim_{h \to 0} \frac{e^h - 1}{h}$$

$$= e^x \cdot 1 = e^x$$

The function $f(x) = e^x$ is its own derivative: $f'(x) = f(x)$. □

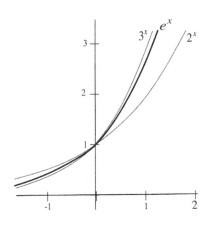

Notice that the limit property of e that we assumed was true actually says that for $f(x) = e^x$, $f'(0) = 1$. So knowing the derivative of $f(x) = e^x$ at a single point $(x = 0)$ allows us to determine its derivative at every other point.

Graphically: the **height** of $f(x) = e^x$ at any point and the **slope** of the tangent to $f(x) = e^x$ at that point are the same: as the graph gets higher, its slope gets steeper.

Example 4. Find: (a) $\dfrac{d}{dt}(t \cdot e^t)$ (b) $\mathbf{D}\left(\dfrac{e^x}{\sin(x)}\right)$ (c) $\mathbf{D}(e^{5x})$

Solution. (a) Using the Product Rule with $f(t) = t$ and $g(t) = e^t$:

$$\frac{d}{dt}(t \cdot e^t) = t \cdot \mathbf{D}(e^t) + e^t \cdot \mathbf{D}(t) = t \cdot e^t + e^t \cdot 1 = (t+1)e^t$$

(b) Using the Quotient Rule with $f(x) = e^x$ and $g(x) = \sin(x)$:

$$\mathbf{D}\left(\frac{e^x}{\sin(x)}\right) = \frac{\sin(x) \cdot \mathbf{D}(e^x) - e^x \cdot \mathbf{D}(\sin(x))}{[\sin(x)]^2}$$
$$= \frac{\sin(x) \cdot e^x - e^x(\cos(x))}{\sin^2(x)}$$

(c) Using the Power Rule for Functions with $f(x) = e^x$ and $p = 5$:

$$\mathbf{D}((e^x)^5) = 5(e^x)^4 \cdot \mathbf{D}(e^x) = 5e^{4x} \cdot e^x = 5e^{5x}$$

where we have rewritten e^{5x} as $(e^x)^5$. ◀

Practice 7. Find: (a) $\mathbf{D}(x^3 e^x)$ (b) $\mathbf{D}((e^x)^3)$.

Higher Derivatives: Derivatives of Derivatives

The derivative of a function f is a new function f' and if this new function is differentiable we can calculate the derivative of this new function to get the derivative of the derivative of f, denoted by f'' and called the **second derivative** of f.

For example, if $f(x) = x^5$ then $f'(x) = 5x^4$ and:

$$f''(x) = (f'(x))' = (5x^4)' = 20x^3$$

Definitions: Given a differentiable function f,

- the first derivative is $f'(x)$, the rate of change of f.

- the second derivative is $f''(x) = (f'(x))'$, the rate of change of f'.

- the third derivative is $f'''(x) = (f''(x))'$, the rate of change of f''.

For $y = f(x)$, we write $f'(x) = \dfrac{dy}{dx}$, so we can extend that notation to write $f''(x) = \dfrac{d}{dx}\left(\dfrac{dy}{dx}\right) = \dfrac{d^2y}{dx^2}$, $f'''(x) = \dfrac{d}{dx}\left(\dfrac{d^2y}{dx^2}\right) = \dfrac{d^3y}{dx^3}$ and so on.

Practice 8. Find f', f'' and f''' for $f(x) = 3x^7$, $f(x) = \sin(x)$ and $f(x) = x \cdot \cos(x)$.

If $f(x)$ represents the position of a particle at time x, then $v(x) = f'(x)$ will represent the velocity (rate of change of the position) of the particle and $a(x) = v'(x) = f''(x)$ will represent the acceleration (the rate of change of the velocity) of the particle.

Example 5. The height (in feet) of a particle at time t seconds is given by $t^3 - 4t^2 + 8t$. Find the height, velocity and acceleration of the particle when $t = 0$, 1 and 2 seconds.

Solution. $f(t) = t^3 - 4t^2 + 8t$ so $f(0) = 0$ feet, $f(1) = 5$ feet and $f(2) = 8$ feet. The velocity is given by $v(t) = f'(t) = 3t^2 - 8t + 8$ so $v(0) = 8$ ft/sec, $v(1) = 3$ ft/sec and $v(2) = 4$ ft/sec. At each of these times the velocity is positive and the particle is moving upward (increasing in height). The acceleration is $a(t) = 6t - 8$ so $a(0) = -8$ ft/sec^2, $a(1) = -2$ ft/sec^2 and $a(2) = 4$ ft/sec^2.   ◄

We will examine the geometric (graphical) meaning of the second derivative in the next chapter.

A Really "Bent" Function

In Section 1.2 we saw that the "holey" function

$$h(x) = \begin{cases} 2 & \text{if } x \text{ is a rational number} \\ 1 & \text{if } x \text{ is an irrational number} \end{cases}$$

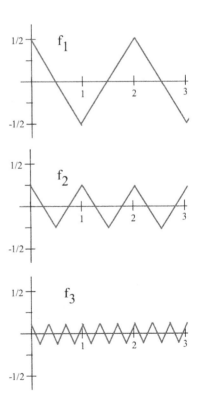

is discontinuous at every value of x, so $h(x)$ is not differentiable anywhere. We can create graphs of continuous functions that are not differentiable at several places just by putting corners at those places, but how many corners can a continuous function have? How badly can a continuous function fail to be differentiable?

In the mid-1800s, the German mathematician Karl Weierstrass surprised and even shocked the mathematical world by creating a function that was **continuous everywhere but differentiable nowhere** — a function whose graph was everywhere connected and everywhere bent! He used techniques we have not investigated yet, but we can begin to see how such a function could be built.

Start with a function f_1 (see margin) that zigzags between the values $\frac{1}{2}$ and $-\frac{1}{2}$ and has a "corner" at each integer. This starting function f_1 is continuous everywhere and is differentiable everywhere except at the integers. Next create a list of functions f_2, f_3, f_4, ..., each of which is "shorter" than the previous one but with many more "corners" than the previous one. For example, we might make f_2 zigzag between the values $\frac{1}{4}$ and $-\frac{1}{4}$ and have "corners" at $\pm\frac{1}{2}$, $\pm\frac{3}{2}$, $\pm\frac{5}{2}$, etc.; f_3 zigzag between $\frac{1}{9}$ and $-\frac{1}{9}$ and have "corners" at $\pm\frac{1}{3}$, $\pm\frac{2}{3}$, $\pm\frac{3}{3} = \pm 1$, etc.

If we add f_1 and f_2, we get a continuous function (because the sum of two continuous functions is continuous) with corners at 0, $\pm\frac{1}{2}$, ± 1, $\pm\frac{3}{2}$, ... If we then add f_3 to the previous sum, we get a new continuous function with even more corners. If we continue adding the functions in our list "indefinitely," the final result will be a continuous function that is differentiable nowhere.

We haven't developed enough mathematics here to precisely describe what it means to add an infinite number of functions together or to verify that the resulting function is nowhere differentiable—but we will. You can at least start to imagine what a strange, totally "bent" function it must be. Until Weierstrass created his "everywhere continuous, nowhere differentiable" function, most mathematicians thought a continuous function could only be "bad" in a few places. Weierstrass' function was (and is) considered "pathological," a great example of how bad something can be. The mathematician Charles Hermite expressed a reaction shared by many when they first encounter the Weierstrass function: "I turn away with fright and horror from this lamentable evil of functions which do not have derivatives."

Important Results

Power Rule for Functions: $\mathbf{D}(f^p(x)) = p \cdot f^{p-1}(x) \cdot \mathbf{D}(f(x))$

Derivatives of the Trigonometric Functions:

$$\mathbf{D}(\sin(x)) = \cos(x) \qquad \mathbf{D}(\cos(x)) = -\sin(x)$$
$$\mathbf{D}(\tan(x)) = \sec^2(x) \qquad \mathbf{D}(\cot(x)) = -\csc^2(x)$$
$$\mathbf{D}(\sec(x)) = \sec(x)\tan(x) \qquad \mathbf{D}(\csc(x)) = -\csc(x)\cot(x)$$

Derivative of the Exponential Function: $\mathbf{D}(e^x) = e^x$

2.3 Problems

1. Let $f(1) = 2$ and $f'(1) = 3$. Find the values of each of the following derivatives at $x = 1$.

   (a) $\mathbf{D}(f^2(x))$

   (b) $\mathbf{D}(f^5(x))$

   (c) $\mathbf{D}(\sqrt{f(x)})$

2. Let $f(2) = -2$ and $f'(2) = 5$. Find the values of each of the following derivatives at $x = 2$.

   (a) $\mathbf{D}(f^2(x))$

   (b) $\mathbf{D}(f^{-3}(x))$

   (c) $\mathbf{D}(\sqrt{f(x)})$

3. For $x = 1$ and $x = 3$ estimate the values of $f(x)$ (whose graph appears below), $f'(x)$ and

   (a) $\dfrac{d}{dx}\left(f^2(x)\right)$ (b) $\mathbf{D}\left(f^3(x)\right)$ (c) $\mathbf{D}\left(f^5(x)\right)$

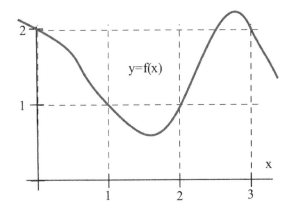

4. For $x = 0$ and $x = 2$ estimate the values of $f(x)$ (whose graph appears above), $f'(x)$ and

   (a) $\mathbf{D}\left(f^2(x)\right)$ (b) $\dfrac{d}{dx}\left(f^3(x)\right)$ (c) $\dfrac{d}{dx}\left(f^5(x)\right)$

In Problems 5–10, find $f'(x)$.

5. $f(x) = (2x - 8)^5$

6. $f(x) = (6x - x^2)^{10}$

7. $f(x) = x \cdot (3x + 7)^5$

8. $f(x) = (2x + 3)^6 \cdot (x - 2)^4$

9. $f(x) = \sqrt{x^2 + 6x - 1}$

10. $f(x) = \dfrac{x - 5}{(x + 3)^4}$

11. A mass attached to the end of a spring is at a height of $h(t) = 3 - 2\sin(t)$ feet above the floor t seconds after it is released.

    (a) Graph $h(t)$.

    (b) At what height is the mass when it is released?

    (c) How high does above the floor and how close to the floor does the mass ever get?

    (d) Determine the height, velocity and acceleration at time t. (Be sure to include the correct units.)

    (e) Why is this an unrealistic model of the motion of a mass attached to a real spring?

12. A mass attached to a spring is at a height of $h(t) = 3 - \dfrac{2\sin(t)}{1 + 0.1t^2}$ feet above the floor t seconds after it is released.

    (a) Graph $h(t)$.

    (b) At what height is the mass when it is released?

    (c) Determine the velocity of the mass at time t.

    (d) What happens to the height and the velocity of the mass a "long time" after it is released?

13. The kinetic energy K of an object of mass m and velocity v is $\frac{1}{2}mv^2$.

    (a) Find the kinetic energy of an object with mass m and height $h(t) = 5t$ feet at $t = 1$ and $t = 2$ seconds.

    (b) Find the kinetic energy of an object with mass m and height $h(t) = t^2$ feet at $t = 1$ and $t = 2$ seconds.

14. An object of mass m is attached to a spring and has height $h(t) = 3 + \sin(t)$ feet at time t seconds.

    (a) Find the height and kinetic energy of the object when $t = 1, 2$ and 3 seconds.

    (b) Find the rate of change in the kinetic energy of the object when $t = 1, 2$ and 3 seconds.

    (c) Can K ever be negative? Can $\dfrac{dK}{dt}$ ever be negative? Why?

In Problems 15–20, compute $f'(x)$.

15. $f(x) = x \cdot \sin(x)$

16. $f(x) = \sin^5(x)$

17. $f(x) = e^x - \sec(x)$

18. $f(x) = \sqrt{\cos(x) + 1}$

19. $f(x) = e^{-x} + \sin(x)$

20. $f(x) = \sqrt{x^2 - 4x + 3}$

In Problems 21–26, find an equation for the line tangent to the graph of $y = f(x)$ at the given point.

21. $f(x) = (x - 5)^7$ at $(4, -1)$

22. $f(x) = e^x$ at $(0, 1)$

23. $f(x) = \sqrt{25 - x^2}$ at $(3,4)$

24. $f(x) = \sin^3(x)$ at $(\pi, 0)$

25. $f(x) = (x - a)^5$ at $(a, 0)$

26. $f(x) = x \cdot \cos^5(x)$ at $(0,0)$

27. (a) Find an equation for the line tangent to $f(x) = e^x$ at the point $(3, e^3)$.

    (b) Where will this tangent line intersect the x-axis?

    (c) Where will the tangent line to $f(x) = e^x$ at the point (p, e^p) intersect the x-axis?

In Problems 28–33, calculate f' and f''.

28. $f(x) = 7x^2 + 5x - 3$

29. $f(x) = \cos(x)$

30. $f(x) = \sin(x)$

31. $f(x) = x^2 \cdot \sin(x)$

32. $f(x) = x \cdot \sin(x)$

33. $f(x) = e^x \cdot \cos(x)$

34. Calculate the first 8 derivatives of $f(x) = \sin(x)$. What is the pattern? What is the 208th derivative of $\sin(x)$?

35. What will the second derivative of a quadratic polynomial be? The third derivative? The fourth derivative?

36. What will the third derivative of a cubic polynomial be? The fourth derivative?

37. What can you say about the n-th and $(n + 1)$-st derivatives of a polynomial of degree n?

In Problems 38–42, you are given f'. Find a function f with the given derivative.

38. $f'(x) = 4x + 2$

39. $f'(x) = 5e^x$

40. $f'(x) = 3 \cdot \sin^2(x) \cdot \cos(x)$

41. $f'(x) = 5(1 + e^x)^4 \cdot e^x$

42. $f'(x) = e^x + \sin(x)$

43. The function $f(x)$ defined as

$$f(x) = \begin{cases} x \cdot \sin(\frac{1}{x}) & \text{if } x \neq 0 \\ 0 & \text{if } x = 0 \end{cases}$$

shown below is continuous at 0 because we can show (using the Squeezing Theorem) that

$$\lim_{h \to 0} f(x) = 0 = f(0)$$

Is f differentiable at 0? To answer this question, use the definition of $f'(0)$ and consider

$$\lim_{h \to 0} \frac{f(0 + h) - f(0)}{h}$$

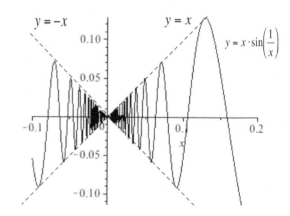

44. The function $f(x)$ defined as

$$f(x) = \begin{cases} x^2 \cdot \sin(\frac{1}{x}) & \text{if } x \neq 0 \\ 0 & \text{if } x = 0 \end{cases}$$

(shown at the top of the next page) is continuous at 0 because we can show (using the Squeezing Theorem) that

$$\lim_{h \to 0} f(x) = 0 = f(0)$$

Is f differentiable at 0? To answer this question, use the definition of $f'(0)$ and consider

$$\lim_{h \to 0} \frac{f(0 + h) - f(0)}{h}$$

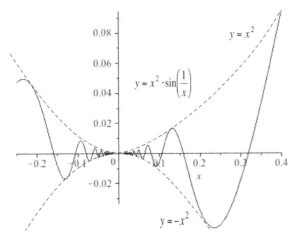

The number e appears in a variety of unusual situations. Problems 45–48 illustrate a few of these.

45. Use your calculator to examine the values of $f(x) = \left(1 + \dfrac{1}{x}\right)^x$ when x is relatively large (for example, $x = 100$, 1000 and $10{,}000$. Try some other large values for x. If x is large, the value of $f(x)$ is close to what number?

46. If you put \$1 into a bank account that pays 1% interest per year and compounds the interest x times a year, then after one year you will have $\left(1 + \dfrac{0.01}{x}\right)^x$ dollars in the account.

   (a) How much money will you have after one year if the bank calculates the interest once a year?

   (b) How much money will you have after one year if the bank calculates the interest twice a year?

   (c) How much money will you have after one year if the bank calculates the interest 365 times a year?

   (d) How does your answer to part (c) compare with $e^{0.01}$?

47. Define $n!$ to be the product of all positive integers from 1 through n. For example, $2! = 1 \cdot 2 = 2$, $3! = 1 \cdot 2 \cdot 3 = 6$ and $4! = 1 \cdot 2 \cdot 3 \cdot 4 = 24$.

   (a) Calculate the value of the sums:

$$s_1 = 1 + \frac{1}{1!}$$
$$s_2 = 1 + \frac{1}{1!} + \frac{1}{2!}$$
$$s_3 = 1 + \frac{1}{1!} + \frac{1}{2!} + \frac{1}{3!}$$
$$s_4 = 1 + \frac{1}{1!} + \frac{1}{2!} + \frac{1}{3!} + \frac{1}{4!}$$
$$s_5 = 1 + \frac{1}{1!} + \frac{1}{2!} + \frac{1}{3!} + \frac{1}{4!} + \frac{1}{5!}$$
$$s_6 = 1 + \frac{1}{1!} + \frac{1}{2!} + \frac{1}{3!} + \frac{1}{4!} + \frac{1}{5!} + \frac{1}{6!}$$

   (b) What value do the sums in part (a) seem to be approaching?

   (c) Calculate s_7 and s_8.

48. If it is late at night and you are tired of studying calculus, try the following experiment with a friend. Take the 2 through 10 of hearts from a regular deck of cards and shuffle these nine cards well. Have your friend do the same with the 2 through 10 of spades. Now compare your cards one at a time. If there is a match, for example you both play a 5, then the game is over and you win. If you make it through the entire nine cards with no match, then your friend wins. If you play the game **many times**, then the ratio:

$$\frac{\text{total number of games played}}{\text{number of times your friend wins}}$$

will be approximately equal to e.

2.3 Practice Answers

1. The pattern is $\mathbf{D}(f^n(x)) = n \cdot f^{n-1}(x) \cdot \mathbf{D}(f(x))$:
   $$\mathbf{D}(f^5(x)) = 5f^4(x) \cdot \mathbf{D}(f(x)) \text{ and } \mathbf{D}(f^{13}(x)) = 13f^{12}(x) \cdot \mathbf{D}(f(x))$$

2. $\dfrac{d}{dx}(2x^5 - \pi)^2 = 2(2x^5 - \pi)^1 \mathbf{D}(2x^5 - \pi) = 2(2x^5 - \pi)(10x^4) = 40x^9 - 20\pi x^4$

   $$\mathbf{D}\left((x + 7x^2)^{\frac{1}{2}}\right) = \frac{1}{2}(x + 7x^2)^{-\frac{1}{2}} \mathbf{D}(x + 7x^2) = \frac{1 + 14x}{2\sqrt{x + 7x^2}}$$

   $$\mathbf{D}\left((\cos(x))^4\right) = 4(\cos(x))^3 \mathbf{D}(\cos(x)) = 4(\cos(x))^3(-\sin(x)) = -4\cos^3(x)\sin(x)$$

3. Mimicking the proof for the derivative of $\tan(x)$:

$$\mathbf{D}\left(\frac{\cos(x)}{\sin(x)}\right) = \frac{\sin(x) \cdot \mathbf{D}(\cos(x)) - \cos(x) \cdot \mathbf{D}(\sin(x))}{(\sin(x))^2}$$

$$= \frac{\sin(x)(-\sin(x)) - \cos(x)(\cos(x))}{\sin^2(x)}$$

$$= \frac{-(\sin^2(x) + \cos^2(x))}{\sin^2(x)} = \frac{-1}{\sin^2(x)} = -\csc^2(x)$$

4. Mimicking the proof for the derivative of $\sec(x)$:

$$\mathbf{D}(\csc(x)) = \mathbf{D}\left(\frac{1}{\sin(x)}\right) = \frac{\sin(x) \cdot \mathbf{D}(1) - 1 \cdot \mathbf{D}(\sin(x)}{\sin^2(x)}$$

$$= \frac{\sin(x) \cdot 0 - \cos(x)}{\sin^2(x)} = -\frac{1}{\sin(x)} \cdot \frac{\cos(x)}{\sin(x)} = -\cot(x)\csc(x)$$

5. $\mathbf{D}(x^5 \cdot \tan(x)) = x^5 \, \mathbf{D}(\tan(x)) + \tan(x) \, \mathbf{D}(x^5) = x^5 \sec^2(x) + \tan(x)(5x^4)$

$$\frac{d}{dt}\left(\frac{\sec(t)}{t}\right) = \frac{t \, \mathbf{D}(\sec(t)) - \sec(t) \, \mathbf{D}(t)}{t^2} = \frac{t \sec(t)\tan(t) - \sec(t)}{t^2}$$

$$\mathbf{D}\left((\cot(x) - x)^{\frac{1}{2}}\right) = \frac{1}{2}(\cot(x) - x)^{-\frac{1}{2}} \, \mathbf{D}(\cot(x) - x)$$

$$= \frac{1}{2}(\cot(x) - x)^{-\frac{1}{2}}(-\csc^2(x) - 1) = \frac{-\csc^2(x) - 1}{2\sqrt{\cot(x) - x}}$$

6. Filling in values for both 3^x and e^x:

h	$\frac{2^h - 1}{h}$	$\frac{3^h - 1}{h}$	$\frac{e^h - 1}{h}$
$+0.1$	0.717734625	1.161231740	1.0517091808
-0.1	0.669670084	1.040415402	0.9516258196
$+0.01$	0.695555006	1.104669194	1.0050167084
-0.01	0.690750451	1.092599583	0.9950166251
$+0.001$	0.693387463	1.099215984	1.0005001667
-0.001	0.692907009	1.098009035	0.9995001666
$\downarrow$	$\downarrow$	$\downarrow$	$\downarrow$
0	≈ 0.693	≈ 1.0986	1

7. $\mathbf{D}(x^3 e^x) = x^3 \, \mathbf{D}(e^x) + e^x \, \mathbf{D}(x^3) = x^3 e^x + e^x \cdot 3x^2 = x^2 e^x(x + 3)$
$\mathbf{D}\left((e^x)^3\right) = 3(e^x)^2 \, \mathbf{D}(e^x) = 3e^{2x} \cdot e^x = 3e^{3x}$

8. $f(x) = 3x^7 \Rightarrow f'(x) = 21x^6 \Rightarrow f''(x) = 126x^5 \Rightarrow f'''(x) = 630x^4$
$f(x) = \sin(x) \Rightarrow f'(x) = \cos(x) \Rightarrow f''(x) = -\sin(x)$
$\qquad\qquad \Rightarrow f'''(x) = -\cos(x)$
$f(x) = x \cdot \cos(x) \Rightarrow f'(x) = -x\sin(x) + \cos(x)$
$\Rightarrow f''(x) = -x\cos(x) - 2\sin(x) \Rightarrow f'''(x) = x\sin(x) - 3\cos(x)$

2.4 The Chain Rule

The Chain Rule is the **most important and most often used** of the differentiation patterns. It enables us to differentiate **composites** of functions such as $y = \sin(x^2)$. It is a powerful tool for determining the derivatives of some **new functions** such as logarithms and inverse trigonometric functions. And it leads to important **applications** in a variety of fields. You will need the Chain Rule hundreds of times in this course. Practice with it now will save you time—and points—later. Fortunately, with practice, the Chain Rule is also easy to use. We already know how to differentiate the composition of some functions.

Example 1. For $f(x) = 5x - 4$ and $g(x) = 2x + 1$, find $f \circ g(x)$ and $\mathbf{D}(f \circ g(x))$.

Solution. Writing $f \circ g(x) = f(g(x)) = 5(2x + 1) - 4 = 10x + 1$, we can compute that $\mathbf{D}(f \circ g(x)) = \mathbf{D}(10x + 1) = 10$. ◄

Practice 1. For $f(x) = 5x - 4$ and $g(x) = x^2$, find $f \circ g(x)$, $\mathbf{D}(f \circ g(x))$, $g \circ f(x)$ and $\mathbf{D}(g \circ f(x))$.

Some compositions, however, are still very difficult to differentiate. We know the derivatives of $g(x) = x^2$ and $h(x) = \sin(x)$, and we know how to differentiate certain combinations of these functions, such as $x^2 + \sin(x)$, $x^2 \cdot \sin(x)$ and even $\sin^2(x) = (\sin(x))^2$. But the derivative of the simple composition $f(x) = h \circ g(x) = \sin(x^2)$ is hard—until we know the Chain Rule.

To see just how difficult, try using the definition of derivative on it.

Example 2. (a) Suppose amplifier Y doubles the strength of the output signal from amplifier U, and U triples the strength of the original signal x. How does the final signal out of Y compare with the original signal x?

(b) Suppose y changes twice as fast as u, and u changes three times as fast as x. How does the rate of change of y compare with the rate of change of x?

Solution. In each case we are comparing the result of a composition, and the answer to each question is 6, the product of the two amplifications or rates of change. In part (a), we have that:

$$\frac{\text{signal out of } Y}{\text{signal } x} = \frac{\text{signal out of } Y}{\text{signal out of } U} \cdot \frac{\text{signal out of } U}{\text{signal } x} = 2 \cdot 3 = 6$$

In part (b):

$$\frac{\Delta y}{\Delta x} = \frac{\Delta y}{\Delta u} \cdot \frac{\Delta u}{\Delta x} = 2 \cdot 3 = 6$$

These examples are simple cases of the Chain Rule for differentiating a composition of functions. ◄

The Chain Rule

We can express the chain rule using more than one type of notation. Each will be useful in various situations.

Chain Rule (Leibniz notation form):

If y is a differentiable function of u and

       u is a differentiable function of x

then y is a differentiable function of x and

$$\frac{dy}{dx} = \frac{dy}{du} \cdot \frac{du}{dx}.$$

Idea for a proof. If $\Delta u \neq 0$ then:

$$\frac{dy}{dx} = \lim_{\Delta x \to 0} \frac{\Delta y}{\Delta x} = \lim_{\Delta x \to 0} \frac{\Delta y}{\Delta u} \cdot \frac{\Delta u}{\Delta x} = \left(\lim_{\Delta x \to 0} \frac{\Delta y}{\Delta u} \right) \left(\lim_{\Delta x \to 0} \frac{\Delta u}{\Delta x} \right)$$

$$= \left(\lim_{\Delta u \to 0} \frac{\Delta y}{\Delta u} \right) \left(\lim_{\Delta x \to 0} \frac{\Delta u}{\Delta x} \right) = \frac{dy}{du} \cdot \frac{du}{dx}$$

The key step here is to argue that $\Delta x \to 0$ implies $\Delta u \to 0$, which follows from the continuity of u as as function of x.

Although this nice short argument gets to the heart of why the Chain Rule works, it is not quite valid. If $\frac{du}{dx} \neq 0$, then it is possible to show that $\Delta u \neq 0$ for all "very small" values of Δx, and the "idea for a proof" becomes a real proof. There are, however, functions for which $\Delta u = 0$ for infinitely many small values of Δx (no matter how close to 0 we restrict Δx) and this creates problems with the simple argument outlined above.

The symbol $\frac{dy}{du}$ is a single symbol, as is $\frac{du}{dx}$, so we cannot eliminate du from the product $\frac{dy}{du} \frac{du}{dx}$ in the Chain Rule by "cancelling" du as we can with Δu in the fractions $\frac{\Delta y}{\Delta u} \cdot \frac{\Delta u}{\Delta x}$. It is, however, perfectly fine to use the *idea* of cancelling du to help you remember the proper statement of the Chain Rule.

> A justification that holds true for **all** cases is more complicated and provides no new conceptual insight. Problem 84 at the end of this section guides you through a rigorous proof of the Chain Rule.

Example 3. Write $y = \cos(x^2 + 3)$ as $y = \cos(u)$ with $u = x^2 + 3$ and find $\frac{dy}{dx}$.

Solution. $y = \cos(u) \Rightarrow \frac{dy}{du} = -\sin(u)$ and $u = x^2 + 3 \Rightarrow \frac{du}{dx} = 2x$. Using the Chain Rule:

$$\frac{dy}{dx} = \frac{dy}{du} \cdot \frac{du}{dx} = -\sin(u) \cdot 2x = -2x \cdot \sin(x^2 + 3)$$

Notice that in the last step we have eliminated the intermediate variable u to express the derivative only in terms of x.   ◀

Practice 2. Find $\frac{dy}{dx}$ for $y = \sin(4x + e^x)$.

We can also state the Chain Rule in terms of composition of functions. The notation is different, but the meaning is precisely the same.

Chain Rule (composition form):

If g is differentiable at x and
      f is differentiable at $g(x)$

then the composite $f \circ g$ is differentiable at x and
      $(f \circ g)'(x) = \mathbf{D}(f(g(x))) = f'(g(x)) \cdot g'(x).$

You may find it easier to think of the result of the composition form of the Chain Rule in words: "the derivative of the outside function (evaluated at the original inside function) times the derivative of the inside function" where f is the outside function and g is the inside function.

Example 4. Differentiate $\sin(x^2)$.

Solution. We can write the function $\sin(x^2)$ as the composition $f \circ g$ of two simple functions: $f(x) = \sin(x)$ and $g(x) = x^2$: $f \circ g(x) = f(g(x)) = f(x^2) = \sin(x^2)$. Both f and g are differentiable functions with derivatives $f'(x) = \cos(x)$ and $g'(x) = 2x$, so the Chain Rule says:

$$\mathbf{D}(\sin(x^2)) = (f \circ g)'(x) = f'(g(x)) \cdot g'(x) = \cos(g(x)) \cdot 2x$$
$$= \cos(x^2) \cdot 2x = 2x \cos(x^2)$$

Check that you get the same answer using the Leibniz notation. ◀

If you tried using the definition of derivative to calculate the derivative of this function at the beginning of this section, you can really appreciate the power of the Chain Rule for differentiating compositions of functions, even simple ones like these.

Example 5. The table below gives values for f, f', g and g' at various points. Use these values to determine $(f \circ g)(x)$ and $(f \circ g)'(x)$ at $x = -1$ and $x = 0$.

x	$f(x)$	$g(x)$	$f'(x)$	$g'(x)$	$(f \circ g)(x)$	$(f \circ g)'(x)$
-1	2	3	1	0		
0	-1	1	3	2		
1	1	0	-1	3		
2	3	-1	0	1		
3	0	2	2	-1		

Solution. $(f \circ g)(-1) = f(g(-1)) = f(3) = 0$, $(f \circ g)(0) = f(g(0)) = f(1) = 1$, $(f \circ g)'(-1) = f'(g(-1)) \cdot g'(-1) = f'(3) \cdot 0 = 2 \cdot 0 = 0$ and $(f \circ g)'(0) = f'(g(0)) \cdot g'(0) = f'(1) \cdot 2 = (-1)(2) = -2.$ ◀

Practice 3. Fill in the table in Example 5 for $(f \circ g)(x)$ and $(f \circ g)'(x)$ at $x = 1, 2$ and 3.

Neither form of the Chain Rule is inherently superior to the other — use the one you prefer or the one that appears most useful in a particular situation. The Chain Rule will be used hundreds of times in the rest of this book, and it is important that you master its usage. The time you spend now mastering and understanding how to use the Chain Rule will be paid back tenfold over the next several chapters.

Example 6. Determine $\mathbf{D}\left(e^{\cos(x)}\right)$ using each form of the Chain Rule.

Solution. Using the Leibniz notation: $y = e^u$ and $u = \cos(x)$ so we have $\frac{dy}{du} = e^u$ and $\frac{du}{dx} = -\sin(x)$. Applying the Chain Rule:

$$\frac{dy}{dx} = \frac{dy}{du} \cdot \frac{du}{dx} = e^u \cdot (-\sin(x)) = -\sin(x) \cdot e^{\cos(x)}$$

We can also write the function $e^{\cos(x)}$ as the composition of $f(x) = e^x$ with $g(x) = \cos(x)$, so the Chain Rule says:

$$\mathbf{D}(e^{\cos(x)}) = f'(g(x)) \cdot g'(x) = e^{g(x)} \cdot (-\sin(x)) = -\sin(x) \cdot e^{\cos(x)}$$

because $\mathbf{D}(e^x) = e^x$ and $\mathbf{D}(\cos(x)) = -\sin(x)$.   ◀

Practice 4. Calculate $\mathbf{D}(\sin(7x - 1))$, $\frac{d}{dx}(\sin(ax + b))$ and $\frac{d}{dt}\left(e^{3t}\right)$.

Practice 5. Use the graph of g given in the margin along with the Chain Rule to estimate $\mathbf{D}(\sin(g(x)))$ and $\mathbf{D}(g(\sin(x)))$ at $x = \pi$.

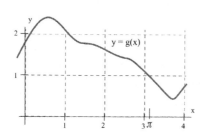

The Chain Rule is a general differentiation pattern that can be used along with other general patterns like the Product and Quotient Rules.

Example 7. Determine $\mathbf{D}\left(e^{3x} \cdot \sin(5x + 7)\right)$ and $\frac{d}{dx}(\cos(x \cdot e^x))$.

Solution. The function $e^{3x}\sin(5x + 7)$ is a product of two functions so we need the Product Rule first:

$$\mathbf{D}(e^{3x} \cdot \sin(5x + 7)) = e^{3x} \cdot \mathbf{D}(\sin(5x + 7)) + \sin(5x + 7) \cdot \mathbf{D}(e^{3x})$$
$$= e^{3x} \cdot \cos(5x + 7) \cdot 5 + \sin(5x + 7) \cdot e^{3x} \cdot 3$$
$$= 5e^{3x}\cos(5x + 7) + 3e^{3x}\sin(5x + 7)$$

The function $\cos(x \cdot e^x)$ is a composition of cosine with a product so we need the Chain Rule first:

$$\frac{d}{dx}(\cos(x \cdot e^x)) = -\sin(x \cdot e^x) \cdot \frac{d}{dx}(x \cdot e^x)$$
$$= -\sin(xe^x) \cdot \left(x \cdot \frac{d}{dx}(e^x) + e^x \cdot \frac{d}{dx}(x)\right)$$
$$= -\sin(xe^x) \cdot (xe^x + e^x)$$

We could also write this last answer as $-(x + 1)e^x \sin(e^x)$.   ◀

Sometimes we want to differentiate a composition of more than two functions. We can do so if we proceed in a careful, step-by-step way.

Example 8. Find $\mathbf{D}(\sin(\sqrt{x^3 + 1}))$.

Solution. The function $\sin(\sqrt{x^3 + 1})$ can be viewed as a composition $f \circ g$ of $f(x) = \sin(x)$ and $g(x) = \sqrt{x^3 + 1}$. Then:

$$(\sin(\sqrt{x^3 + 1}))' = f'(g(x)) \cdot g'(x) = \cos(g(x)) \cdot g'(x)$$
$$= \cos(\sqrt{x^3 + 1}) \cdot \mathbf{D}(\sqrt{x^3 + 1})$$

For the derivative of $\sqrt{x^3 + 1}$, we can use the Chain Rule again or its special case, the Power Rule:

$$\mathbf{D}(\sqrt{x^3 + 1}) = \mathbf{D}((x^3 + 1)^{\frac{1}{2}}) = \frac{1}{2}(x^3 + 1)^{-\frac{1}{2}} \cdot \mathbf{D}(x^3 + 1)$$
$$= \frac{1}{2}(x^3 + 1)^{-\frac{1}{2}} \cdot 3x^2$$

Finally, $\mathbf{D}\left(\sin(\sqrt{x^3 + 1})\right) = \cos(\sqrt{x^3 + 1}) \cdot \frac{1}{2}(x^3 + 1)^{-\frac{1}{2}} \cdot 3x^2$, which can be rewritten as $\dfrac{3x^2 \cos(\sqrt{x^3 + 1})}{2\sqrt{x^3 + 1}}$. ◄

This example was more complicated than the earlier ones, but it is just a matter of applying the Chain Rule twice, to a composition of a composition. If you proceed step by step and don't get lost in the details of the problem, these multiple applications of the Chain Rule are relatively straightforward.

We can also use the Leibniz form of the Chain Rule for a composition of more than two functions. If $y = \sin(\sqrt{x^3 + 1})$, then $y = \sin(u)$ with $u = \sqrt{w}$ and $w = x^3 + 1$. The Leibniz form of the Chain Rule says:

$$\frac{dy}{dx} = \frac{dy}{du} \cdot \frac{du}{dw} \cdot \frac{dw}{dx} = \cos(u) \cdot \frac{1}{2\sqrt{w}} \cdot 3x^2$$
$$= \cos(\sqrt{x^3 + 1}) \cdot \frac{1}{2\sqrt{x^3 + 1}} \cdot 3x^2$$

which agrees with our previous answer.

Practice 6. (a) Find $\mathbf{D}(\sin(\cos(5x)))$. (b) For $y = e^{\cos(3x)}$, find $\frac{dy}{dx}$.

The Chain Rule and Tables of Derivatives

With the Chain Rule, the derivatives of all sorts of strange and wonderful functions become available. If we know f' and g', then we also know the derivatives of their composition: $(f(g(x))' = f'(g(x)) \cdot g'(x)$.

We have begun to build a list of derivatives of "basic" functions, such as x^n, $\sin(x)$ and e^x. We will continue to add to that list later in the course, but if we peek ahead at the rest of that list—spoiler alert!—to (for example) see that $\mathbf{D}(\arctan(x)) = \dfrac{1}{1 + x^2}$, then we can use the Chain Rule to compute derivatives of compositions of those functions.

Example 9. Given that $D(\arcsin(x)) = \dfrac{1}{\sqrt{1-x^2}}$, compute the derivatives $D(\arcsin(5x))$ and $\dfrac{d}{dx}(\arcsin(e^x))$.

Solution. Write $\arcsin(5x)$ as the composition of $f(x) = \arcsin(x)$ with $g(x) = 5x$. We know $g'(x) = 5$ and $f'(x) = \dfrac{1}{\sqrt{1-x^2}}$, so we have $f'(g(x)) = \dfrac{1}{\sqrt{1-(g(x))^2}} = \dfrac{1}{\sqrt{1-25x^2}}$. Then:

$$D(\arcsin(5x)) = f'(g(x)) \cdot g'(x) = \dfrac{1}{\sqrt{1-(5x)^2}} \cdot 5 = \dfrac{5}{\sqrt{1-25x^2}}$$

We can write $y = \arcsin(e^x)$ as $y = \arcsin(u)$ with $u = e^x$, and we know that $\dfrac{dy}{du} = \dfrac{1}{\sqrt{1-u^2}}$ and $\dfrac{du}{dx} = e^x$ so:

$$\dfrac{dy}{dx} = \dfrac{dy}{du} \cdot \dfrac{du}{dx} = \dfrac{1}{\sqrt{1-u^2}} \cdot e^x = \dfrac{e^x}{\sqrt{1-e^{2x}}}$$

We can generalize this result to say that $D(\arcsin(f(x))) = \dfrac{f'(x)}{\sqrt{1-(f(x))^2}}$ or, in Leibniz notation, $\dfrac{d}{du}(\arcsin(u)) = \dfrac{1}{\sqrt{1-u^2}} \cdot \dfrac{du}{dx}$. ◀

Practice 7. Given that $D(\arctan(x)) = \dfrac{1}{1+x^2}$, compute the derivatives $D(\arctan(x^3))$ and $\dfrac{d}{dx}(\arctan(e^x))$.

Appendix D in the back of this book shows the derivative patterns for a variety of functions. You may not know much about some of the functions, but with the given differentiation patterns and the Chain Rule you should be able to calculate derivatives of compositions that involve these new functions. It is just a matter of following the pattern.

Practice 8. Use the patterns $D(\sinh(x)) = \cosh(x)$ and $D(\ln(x)) = \frac{1}{x}$ to determine:

(a) $D(\sinh(5x-7))$ (b) $\dfrac{d}{dx}\left(\ln(3+e^{2x})\right)$ (c) $D(\arcsin(1+3x))$

Example 10. If $D(F(x)) = e^x \cdot \sin(x)$, find $D(F(5x))$ and $\dfrac{d}{dt}\left(F(t^3)\right)$.

Solution. $D(F(5x)) = D(F(g(x))$ with $g(x) = 5x$ and we know that $F'(x) = e^x \cdot \sin(x)$ so:

$$D(F(5x)) = F'(g(x)) \cdot g'(x) = e^{g(x)} \cdot \sin(g(x)) \cdot 5 = e^{5x} \cdot \sin(5x) \cdot 5$$

With $y = F(u)$ and $u = t^3$ we know $\dfrac{dy}{du} = e^u \cdot \sin(u)$ and $\dfrac{du}{dt} = 3t^2$ so:

$$\dfrac{dy}{dt} = \dfrac{dy}{du} \cdot \dfrac{du}{dt} = e^u \cdot \sin(u) \cdot 3t^2 = e^{t^3} \cdot \sin(t^3) \cdot 3t^2$$

Notice that we have eliminated the intermediate variable u (which didn't appear in the original problem) from the final answer. ◀

Proof of the Power Rule For Functions

We started using the Power Rule For Functions in Section 2.3. Now we can easily prove it.

> **Power Rule For Functions:**
>
> If $\quad p$ is any constant
>
> then $\quad \mathbf{D}(f^p(x)) = p \cdot f^{p-1}(x) \cdot \mathbf{D}(f(x))$.

Proof. Write $y = f^p(x)$ as $y = u^p$ with $u = f(x)$. Then $\dfrac{dy}{du} = p \cdot u^{p-1}$ and $\dfrac{du}{dx} = f'(x)$ so:

$$\frac{dy}{dx} = \frac{dy}{du} \cdot \frac{du}{dx} = p \cdot u^{p-1} \cdot f'(x) = p \cdot f^{p-1}(x) \cdot f'(x)$$

by the Chain Rule. $\qquad\qquad\qquad\qquad\qquad\qquad\qquad\square$

2.4 Problems

In Problems 1–6 , find two functions f and g so that the given function is the composition of f and g.

1. $y = (x^3 - 7x)^5$

2. $y = \sin^4(3x - 8)$

3. $y = \sqrt{(2 + \sin(x))^5}$

4. $y = \dfrac{1}{\sqrt{x^2 + 9}}$

5. $y = |x^2 - 4|$

6. $y = \tan(\sqrt{x})$

7. For each function in Problems 1–6, write y as a function of u for some u that is a function of x.

For Problems 8–9, use the values given in this table to determine the indicated quantities:

x	$f(x)$	$g(x)$	$f'(x)$	$g'(x)$	$(f \circ g)(x)$	$(f \circ g)'(x)$
-2	2	-1	1	1		
-1	1	2	0	2		
0	-2	1	2	-1		
1	0	-2	-1	2		
2	1	0	1	-1		

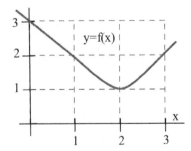

8. $(f \circ g)(x)$ and $(f \circ g)'(x)$ at $x = 1$ and $x = 2$.

9. $(f \circ g)(x)$ and $(f \circ g)'(x)$ at $x = -2, -1$ and 0.

10. Using the figure in the margin, estimate the values of $g(x)$, $g'(x)$, $(f \circ g)(x)$, $f'(g(x))$ and $(f \circ g)'(x)$ at $x = 1$.

11. Using the figure in the margin, estimate the values of $g(x)$, $g'(x)$, $(f \circ g)(x)$, $f'(g(x))$ and $(f \circ g)'(x)$ at $x = 2$.

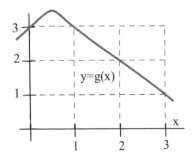

In Problems 12–22, compute the derivative.

12. $\mathbf{D}\left((x^2 + 2x + 3)^{87}\right)$

13. $\mathbf{D}\left(\left(1 - \dfrac{3}{x}\right)^4\right)$

14. $\dfrac{d}{dx}\left(x + \dfrac{1}{x}\right)^5$

15. $\mathbf{D}\left(\dfrac{5}{\sqrt{2 + \sin(x)}}\right)$

16. $\dfrac{d}{dt}\left(t \cdot \sin(3t + 2)\right)$

17. $\dfrac{d}{dx}\left(x^2 \cdot \sin(x^2 + 3)\right)$

18. $\dfrac{d}{dx}\left(\sin(2x) \cdot \cos(5x + 1)\right)$

19. $\mathbf{D}\left(\dfrac{7}{\cos(x^3 - x)}\right)$

20. $\dfrac{d}{dt}\left(\dfrac{5}{3 + e^t}\right)$

21. $\mathbf{D}\left(e^x + e^{-x}\right)$

22. $\mathbf{D}\left(e^x - e^{-x}\right)$

23. An object attached to a spring is at a height of $h(t) = 3 - \cos(2t)$ feet above the floor t seconds after it is released.

   (a) At what height was it released?
   (b) Determine its height, velocity and acceleration at any time t.
   (c) If the object has mass m, determine its kinetic energy $K = \frac{1}{2}mv^2$ and $\frac{dK}{dt}$ at any time t.

24. An employee with d days of production experience will be able to produce approximately $P(d) = 3 + 15(1 - e^{-0.2d})$ items per day.

   (a) Graph $P(d)$.
   (b) Approximately how many items will a beginning employee be able to produce each day?
   (c) How many items will a very experienced employee be able to produce each day?
   (d) What is the marginal production rate of an employee with 5 days of experience? (Include units for your answer. What does this mean?)

25. The air pressure $P(h)$, in pounds per square inch, at an altitude of h feet above sea level is approximately $P(h) = 14.7e^{-0.0000385h}$.

   (a) What is the air pressure at sea level?
   (b) What is the air pressure at 30,000 feet?
   (c) At what altitude is the air pressure 10 pounds per square inch?

(d) If you are in a balloon that is 2,000 feet above the Pacific Ocean and is rising at 500 feet per minute, how fast is the air pressure on the balloon changing?

(e) If the temperature of the gas in the balloon remained constant during this ascent, what would happen to the volume of the balloon?

Find the indicated derivatives in Problems 26–33.

26. $\mathbf{D}\left(\dfrac{(2x + 3)^2}{(5x - 7)^3}\right)$

27. $\dfrac{d}{dz}\sqrt{1 + \cos^2(z)}$

28. $\mathbf{D}\left(\sin(3x + 5)\right)$

29. $\dfrac{d}{dx}\tan(3x + 5)$

30. $\dfrac{d}{dt}\cos(7t^2)$

31. $\mathbf{D}\left(\sin(\sqrt{x + 1})\right)$

32. $\mathbf{D}\left(\sec(\sqrt{x + 1})\right)$

33. $\dfrac{d}{dx}\left(e^{\sin(x)}\right)$

In Problems 34–37, calculate $f'(x) \cdot x'(t)$ when $t = 3$ and use these values to determine the value of $\dfrac{d}{dt}\left(f(x(t))\right)$ when $t = 3$.

34. $f(x) = \cos(x)$, $x = t^2 - t + 5$

35. $f(x) = \sqrt{x}$, $x = 2 + \dfrac{21}{t}$

36. $f(x) = e^x$, $x = \sin(t)$

37. $f(x) = \tan^3(x)$, $x = 8$

In 38–43, find a function that has the given function as its derivative. (You are given a function $f'(x)$ and are asked to find a corresponding function $f(x)$.)

38. $f'(x) = (3x + 1)^4$

39. $f'(x) = (7x - 13)^{10}$

40. $f'(x) = \sqrt{3x - 4}$

41. $f'(x) = \sin(2x - 3)$

42. $f'(x) = 6e^{3x}$

43. $f'(x) = \cos(x)e^{\sin(x)}$

If two functions are equal, then their derivatives are also equal. In 44–47, differentiate each side of the trigonometric identity to get a new identity.

44. $\sin^2(x) = \frac{1}{2} - \frac{1}{2}\cos(2x)$

45. $\cos(2x) = \cos^2(x) - \sin^2(x)$

46. $\sin(2x) = 2\sin(x) \cdot \cos(x)$

47. $\sin(3x) = 3\sin(x) - 4\sin^3(x)$

Derivatives of Families of Functions

So far we have emphasized derivatives of particular functions, but sometimes we want to investigate the derivatives of a whole family of functions all at once. In 48–71, A, B, C and D represent constants and the given formulas describe families of functions.

For Problems 48–65, calculate $y' = \frac{dy}{dx}$.

48. $y = Ax^3 - B$

49. $y = Ax^3 + Bx^2 + C$

50. $y = \sin(Ax + B)$

51. $y = \sin(Ax^2 + B)$

52. $y = Ax^3 + \cos(Bx)$

53. $y = \sqrt{A + Bx^2}$

54. $y = \sqrt{A - Bx^2}$

55. $y = A - \cos(Bx)$

56. $y = \cos(Ax + B)$

57. $y = \cos(Ax^2 + B)$

58. $y = A \cdot e^{Bx}$

59. $y = x \cdot e^{Bx}$

60. $y = e^{Ax} + e^{-Ax}$

61. $y = e^{Ax} - e^{-Ax}$

62. $y = \dfrac{\sin(Ax)}{x}$

63. $y = \dfrac{Ax}{\sin(Bx)}$

64. $y = \dfrac{1}{Ax + B}$

65. $y = \dfrac{Ax + B}{Cx + D}$

In 66–71, (a) find y' (b) find the value(s) of x so that $y' = 0$ and (c) find y''. Typically your answer in part (b) will contain A's, B's and (sometimes) C's.

66. $y = Ax^2 + Bx + C$

67. $y = Ax(B - x) = ABx - Ax^2$

68. $y = Ax(B - x^2) = ABx - Ax^3$

69. $y = Ax^2(B - x) = ABx^2 - Ax^3$

70. $y = Ax^2 + Bx$

71. $y = Ax^3 + Bx^2 + C$

In Problems 72–83, use the differentiation patterns
$\mathbf{D}(\arctan(x)) = \dfrac{1}{1 + x^2}$, $\mathbf{D}(\arcsin(x)) = \dfrac{1}{\sqrt{1 - x^2}}$
and $\mathbf{D}(\ln(x)) = \dfrac{1}{x}$. We have not derived the derivatives for these functions (yet), but if you are handed the derivative pattern then you should be able to use that pattern to compute derivatives of associated composite functions.

72. $\mathbf{D}\left(\arctan(7x)\right)$

73. $\mathbf{D}\left(\arctan(x^2)\right)$

74. $\dfrac{d}{dt}\left(\arctan(\ln(t))\right)$

75. $\dfrac{d}{dx}\left(\arctan(e^x)\right)$

76. $\dfrac{d}{dw}\left(\arcsin(4w)\right)$

77. $\dfrac{d}{dx}\left(\arcsin(x^3)\right)$

78. $\mathbf{D}\left(\arcsin(\ln(x))\right)$

79. $\mathbf{D}\left(\arcsin(e^t)\right)$

80. $\mathbf{D}\left(\ln(3x + 1)\right)$

81. $\mathbf{D}\left(\ln(\sin(x))\right)$

82. $\dfrac{d}{dx}\left(\ln(\arctan(x))\right)$

83. $\dfrac{d}{ds}\left(\ln(e^s)\right)$

84. To prove the Chain Rule, assume $g(x)$ is differentiable at $x = a$ and $f(x)$ is differentiable at $x = g(a)$. We need to show that

$$\lim_{x \to a} \frac{f(g(x)) - f(g(a))}{x - a}$$

exists and is equal to $f'(g(a)) \cdot g'(a)$. To do this, define a new function F as:

$$F(y) = \begin{cases} \frac{f(y) - f(g(a))}{y - g(a)} & \text{if } y \neq g(a) \\ f'(g(a)) & \text{if } y = g(a) \end{cases}$$

and justify each of the following statements.

(a) $F(y)$ is continuous at $y = g(a)$ because:

$$\lim_{y \to g(a)} F(y) = \lim_{y \to g(a)} \frac{f(y) - f(g(a))}{y - g(a)} = F(g(a))$$

(b) By considering separately the cases $g(x) = g(a)$ and $g(x) \neq g(a)$:

$$\frac{f(g(x)) - f(g(a))}{x - a} = F(g(x)) \cdot \frac{g(x) - g(a)}{x - a}$$

for all $x \neq a$.

(c) $\lim\limits_{x \to a} \dfrac{f(g(x)) - f(g(a))}{x - a} = \lim\limits_{x \to a} F(g(x)) \cdot \dfrac{g(x) - g(a)}{x - a}$

(d) $\lim\limits_{x \to a} F(g(x)) \cdot \dfrac{g(x) - g(a)}{x - a} = F(g(a)) \cdot g'(a)$

(e) $\lim\limits_{x \to a} \dfrac{f(g(x)) - f(g(a))}{x - a} = f'(g(a)) \cdot g'(a)$

(f) $(f \circ g)'(a) = f'(g(a)) \cdot g'(a)$

2.4 Practice Answers

1. $f(x) = 5x - 4$ and $g(x) = x^2 \Rightarrow f'(x) = 5$ and $g'(x) = 2x$, so
$f \circ g(x) = f(g(x)) = f(x^2)) = 5x^2 - 4$ and $\mathbf{D}(5x^2 - 4) = 10x$ or:

$$\mathbf{D}(f \circ g(x)) = f'(g(x)) \cdot g'(x) = 5 \cdot 2x = 10x$$

$g \circ f(x) = g(f(x)) = g(5x - 4) = (5x - 4)^2 = 25x^2 - 40x + 16$ and
$\mathbf{D}(25x^2 - 40x + 16) = 50x - 40$ or:

$$\mathbf{D}(g \circ f(x)) = g'(f(x)) \cdot f'(x) = 2(5x - 4) \cdot 5 = 50x - 40$$

2. $\dfrac{d}{dx}\left(\sin(4x + e^x)\right) = \cos(4x + e^x) \cdot \mathbf{D}(4x + e^x) = \cos(4x + e^x) \cdot (4 + e^x)$

3. To fill in the last column, compute:

$$f'(g(1)) \cdot g'(1) = f'(0) \cdot 3 = (3)(3) = 9$$
$$f'(g(2)) \cdot g'(2) = f'(-1) \cdot 1 = (1)(1) = 1$$
$$f'(g(3)) \cdot g'(3) = f'(2) \cdot (-1) = (0)(-1) = 0$$

x	$f(x)$	$g(x)$	$f'(x)$	$g'(x)$	$(f \circ g)(x)$	$(f \circ g)'(x)$
1	1	0	-1	3	-1	9
2	3	-1	0	1	2	3
3	0	2	2	-1	3	0

4. $\mathbf{D}\left(\sin(7x - 1)\right) = \cos(7x - 1) \cdot \mathbf{D}(7x - 1) = 7 \cdot \cos(7x - 1)$
$\dfrac{d}{dx}\left(\sin(ax + b)\right) = \cos(ax + b) \cdot \mathbf{D}(ax + b) = a \cdot \cos(ax + b)$
$\dfrac{d}{dt}\left(e^{3t}\right) = e^{3t} \cdot \dfrac{d}{dt}(3t) = 3 \cdot e^{3t}$

5. $\mathbf{D}\left(\sin(g(x))\right) = \cos(g(x)) \cdot g'(x)$. At $x = \pi$, $\cos(g(\pi)) \cdot g'(\pi) \approx \cos(0.86) \cdot (-1) \approx -0.65$. $\mathbf{D}\left(g(\sin(x))\right) = g'(\sin(x)) \cdot \cos(x)$. At $x = \pi$, $g'(\sin(\pi)) \cdot \cos(\pi) = g'(0) \cdot (-1) \approx -2$

6. $\mathbf{D}\left(\sin(\cos(5x))\right) = \cos(\cos(5x)) \cdot \mathbf{D}(\cos(5x))$
$\qquad = \cos(\cos(5x)) \cdot (-\sin(5x)) \cdot \mathbf{D}(5x) = -5 \cdot \sin(5x) \cdot \cos(\cos(5x))$
$\dfrac{d}{dx}\left(e^{\cos(3x)}\right) = e^{\cos(3x)} \cdot \mathbf{D}(\cos(3x)) = e^{\cos(3x)}(-\sin(3x)) \mathbf{D}(3x)$
$\qquad = -3 \cdot \sin(3x) \cdot e^{\cos(3x)}$

7. $\mathbf{D}\left(\arctan(x^3)\right) = \dfrac{1}{1 + (x^3)^2} \cdot \mathbf{D}(x^3) = \dfrac{3x^2}{1 + x^6}$
$\dfrac{d}{dx}\left(\arctan(e^x)\right) = \dfrac{1}{1 + (e^x)^2} \cdot \mathbf{D}(e^x) = \dfrac{e^x}{1 + e^{2x}}$

8. $\mathbf{D}(\sinh(5x - 7)) = \cosh(5x - 7) \cdot \mathbf{D}(5x - 7) = 5 \cdot \cosh(5x - 7)$
$\dfrac{d}{dx}\left(\ln(3 + e^{2x})\right) = \dfrac{1}{3 + e^{2x}} \cdot \mathbf{D}(3 + e^{2x}) = \dfrac{2e^{2x}}{3 + e^{2x}}$
$\mathbf{D}(\arcsin(1 + 3x)) = \dfrac{1}{\sqrt{1 - (1 + 3x)^2}} \cdot \mathbf{D}(1 + 3x) = \dfrac{3}{\sqrt{1 - (1 + 3x)^2}}$

2.5 Applications of the Chain Rule

The Chain Rule can help us determine the derivatives of logarithmic functions like $f(x) = \ln(x)$ and general exponential functions like $f(x) = a^x$. We will also use it to answer some applied questions and to find slopes of graphs given by parametric equations.

Derivatives of Logarithms

You know from precalculus that the natural logarithm $\ln(x)$ is defined as the inverse of the exponential function e^x: $e^{\ln(x)} = x$ for $x > 0$. We can use this identity along with the Chain Rule to determine the derivative of the natural logarithm.

$$\mathbf{D}(\ln(x)) = \frac{1}{x} \quad \text{and} \quad \mathbf{D}\left(\ln(g(x))\right) = \frac{g'(x)}{g(x)}$$

Proof. We know that $\mathbf{D}(e^u) = e^u$, so using the Chain Rule we have $\mathbf{D}\left(e^{f(x)}\right) = e^{f(x)} \cdot f'(x)$. Differentiating each side of the identity $e^{\ln(x)} = x$, we get:

$$\mathbf{D}\left(e^{\ln(x)}\right) = \mathbf{D}(x) \Rightarrow e^{\ln(x)} \cdot \mathbf{D}(\ln(x)) = 1$$

$$\Rightarrow x \cdot \mathbf{D}(\ln(x)) = 1 \Rightarrow \mathbf{D}(\ln(x)) = \frac{1}{x}$$

The function $\ln(g(x))$ is the composition of $f(x) = \ln(x)$ with $g(x)$ so the Chain Rule says:

$$\mathbf{D}\left(\ln(g(x))\right) = \mathbf{D}\left(f(g(x))\right) = f'(g(x)) \cdot g'(x) = \frac{1}{g(x)} \cdot g'(x) = \frac{g'(x)}{g(x)}$$

You can remember the differentiation pattern for the the natural logarithm in words as: "one over the inside times the the derivative of the inside."

Graph $f(x) = \ln(x)$ along with $f'(x) = \frac{1}{x}$ and compare the behavior of the function at various points with the values of its derivative at those points. Does $y = \frac{1}{x}$ possess the properties you would expect to see from the derivative of $f(x) = \ln(x)$? □

Example 1. Find $\mathbf{D}(\ln(\sin(x)))$ and $\mathbf{D}(\ln(x^2 + 3))$.

Solution. Using the pattern $\mathbf{D}(\ln(g(x))) = \frac{g'(x)}{g(x)}$ with $g(x) = \sin(x)$:

$$\mathbf{D}(\ln(\sin(x))) = \frac{g'(x)}{g(x)} = \frac{\mathbf{D}(\sin(x))}{\sin(x)} = \frac{\cos(x)}{\sin(x)} = \cot(x)$$

With $g(x) = x^2 + 3$, $\mathbf{D}(\ln(x^2 + 3)) = \frac{g'(x)}{g(x)} = \frac{2x}{x^2 + 3}$. ◀

We can use the Change of Base Formula from precalculus to rewrite any logarithm as a natural logarithm, and then we can differentiate the resulting natural logarithm.

> **Change of Base Formula for Logarithms:**
>
> $$\log_a(x) = \frac{\log_b(x)}{\log_b(a)} \text{ for all positive } a, b \text{ and } x.$$

Your calculator likely has two logarithm buttons: **ln** for the natural logarithm (base e) and **log** for the common logarithm (base 10). Be careful, however, as more advanced mathematics texts (as well as the Web site Wolfram|Alpha) use log for the (base e) natural logarithm.

Example 2. Use the Change of Base formula and your calculator to find $\log_\pi(7)$ and $\log_2(8)$.

Solution. $\log_\pi(7) = \dfrac{\ln(7)}{\ln(\pi)} \approx \dfrac{1.946}{1.145} \approx 1.700$. (Check that $\pi^{1.7} \approx 7$.)

Likewise, $\log_2(8) = \dfrac{\ln(8)}{\ln(2)} = 3$. ◀

Practice 1. Find the values of $\log_9 20$, $\log_3 20$ and $\log_\pi e$.

Putting $b = e$ in the Change of Base Formula, $\log_a(x) = \dfrac{\log_e(x)}{\log_e(a)} = \dfrac{\ln(x)}{\ln(a)}$, so any logarithm can be written as a natural logarithm divided by a constant. This makes any logarithmic function easy to differentiate.

> $$\mathbf{D}\left(\log_a(x)\right) = \frac{1}{x \ln(a)} \quad \text{and} \quad \mathbf{D}\left(\log_a(f(x))\right) = \frac{f'(x)}{f(x)} \cdot \frac{1}{\ln(a)}$$

Proof. $\mathbf{D}\left(\log_a(x)\right) = \mathbf{D}\left(\dfrac{\ln x}{\ln a}\right) = \dfrac{1}{\ln(a)} \cdot \mathbf{D}(\ln x) = \dfrac{1}{\ln(a)} \cdot \dfrac{1}{x} = \dfrac{1}{x \ln(a)}$.
The second differentiation formula follows from the Chain Rule. □

Practice 2. Calculate $\mathbf{D}\left(\log_{10}(\sin(x))\right)$ and $\mathbf{D}\left(\log_\pi(e^x)\right)$.

The number e might seem like an "unnatural" base for a natural logarithm, but of all the possible bases, the logarithm with base e has the nicest and easiest derivative. The natural logarithm is even related to the distribution of prime numbers. In 1896, the mathematicians Hadamard and Vallée-Poussin proved the following conjecture of Gauss (the Prime Number Theorem): For large values of N,

$$\text{number of primes less than } N \approx \frac{N}{\ln(N)}$$

Derivative of a^x

Once we know the derivative of e^x and the Chain Rule, it is relatively easy to determine the derivative of a^x for any $a > 0$.

> $$\mathbf{D}(a^x) = a^x \cdot \ln(a) \text{ for } a > 0.$$

Proof. If $a > 0$, then $a^x > 0$ and $a^x = e^{\ln(a^x)} = e^{x \cdot \ln(a)}$, so we have:

$$\mathbf{D}(a^x) = \mathbf{D}\left(e^{\ln(a^x)}\right) = \mathbf{D}\left(e^{x \cdot \ln(a)}\right) = e^{x \cdot \ln(a)} \cdot \mathbf{D}(x \cdot \ln(a)) = a^x \cdot \ln(a).$$

$\square$

Example 3. Calculate $\mathbf{D}(7^x)$ and $\dfrac{d}{dt}\left(2^{\sin(t)}\right)$.

Solution. $\mathbf{D}(7^x) = 7^x \cdot \ln(7) \approx (1.95)7^x$. We can write $y = 2^{\sin(t)}$ as $y = 2^u$ with $u = \sin(t)$. Using the Chain Rule: $\dfrac{dy}{dt} = \dfrac{dy}{du} \cdot \dfrac{du}{dt} = 2^u \cdot \ln(2) \cos(t) = 2^{\sin(t)} \cdot \ln(2) \cdot \cos(t)$. ◀

Practice 3. Calculate $\mathbf{D}\left(\sin(2^x)\right)$ and $\dfrac{d}{dt}\left(3^{t^2}\right)$.

Some Applied Problems

Let's examine some applications involving more complicated functions.

Example 4. A ball at the end of a rubber band (see margin) is oscillating up and down, and its height (in feet) above the floor at time t seconds is $h(t) = 5 + 2\sin\left(\dfrac{t}{2}\right)$ (with t in radians).

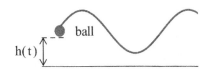
ball
h(t)

(a) How fast is the ball traveling after 2 seconds? After 4 seconds? After 60 seconds?

(b) Is the ball moving up or down after 2 seconds? After 4 seconds? After 60 seconds?

(c) Is the vertical velocity of the ball ever 0?

Solution. (a) $v(t) = h'(t) = \mathbf{D}\left(5 + 2\sin\left(\dfrac{t}{2}\right)\right) = 2\cos\left(\dfrac{t}{2}\right) \cdot \dfrac{1}{2}$ so $v(t) = \cos\left(\dfrac{t}{2}\right)$ feet/second: $v(2) = \cos\left(\dfrac{2}{2}\right) \approx 0.540$ ft/s, $v(4) = \cos\left(\dfrac{4}{2}\right) \approx -0.416$ ft/s, and $v(60) = \cos\left(\dfrac{60}{2}\right) \approx 0.154$ ft/s.

(b) The ball is moving up at $t = 2$ and $t = 60$, down when $t = 4$.

(c) $v(t) = \cos\left(\dfrac{t}{2}\right) = 0$ when $\dfrac{t}{2} = \dfrac{\pi}{2} \pm k \cdot \pi \Rightarrow t = \pi \pm 2\pi k$ for any integer k. ◀

Example 5. If 2,400 people now have a disease, and the number of people with the disease appears to double every 3 years, then the number of people expected to have the disease in t years is $y = 2400 \cdot 2^{\frac{t}{3}}$.

(a) How many people are expected to have the disease in 2 years?

(b) When are 50,000 people expected to have the disease?

(c) How fast is the number of people with the disease growing now? How fast is it expected to be growing 2 years from now?

Solution. (a) In 2 years, $y = 2400 \cdot 2^{\frac{2}{3}} \approx 3{,}810$ people.

(b) We know $y = 50000$ and need to solve $50000 = 2400 \cdot 2^{\frac{t}{3}}$ for t. Taking logarithms of each side of the equation: $\ln(50000) = \ln\left(2400 \cdot 2^{\frac{2}{3}}\right) = \ln(2400) + \frac{t}{3} \cdot \ln(2)$ so $10.819 \approx 7.783 + 0.231t$ and $t \approx 13.14$ years. We expect 50,000 people to have the disease about 13 years from now.

(c) This question asks for $\dfrac{dy}{dt}$ when $t = 0$ and $t = 2$.

$$\frac{dy}{dt} = \frac{d}{dt}\left(2400 \cdot 2^{\frac{t}{3}}\right) = 2400 \cdot 2^{\frac{t}{3}} \cdot \ln(2) \cdot \frac{1}{3} \approx 554.5 \cdot 2^{\frac{t}{3}}$$

Now, at $t = 0$, the rate of growth of the disease is approximately $554.5 \cdot 2^{0} \approx 554.5$ people/year. In 2 years, the rate of growth will be approximately $554.5 \cdot 2^{\frac{2}{3}} \approx 880$ people/year. ◀

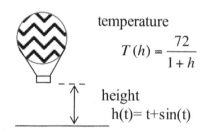

temperature

$T(h) = \dfrac{72}{1+h}$

height

$h(t) = t + \sin(t)$

Example 6. You are riding in a balloon, and at time t (in minutes) you are $h(t) = t + \sin(t)$ thousand feet above sea level. If the temperature at an elevation h is $T(h) = \dfrac{72}{1+h}$ degrees Fahrenheit, then how fast is the temperature changing when $t = 5$ minutes?

Solution. As t changes, your elevation will change. And, as your elevation changes, so will the temperature. It is not difficult to write the temperature as a function of time, and then we could calculate $\dfrac{dT}{dt} = T'(t)$ and evaluate $T'(5)$. Or we could use the Chain Rule:

$$\frac{dT}{dt} = \frac{dT}{dh} \cdot \frac{dh}{dt} = -\frac{72}{(1+h)^2} \cdot (1 + \cos(t))$$

At $t = 5$, $h(5) = 5 + \sin(5) \approx 4.04$ so $T'(5) \approx -\frac{72}{(1+4.04)^2} \cdot (1 + 0.284) \approx -3.64\,^\circ$/minute. ◀

Practice 4. Write the temperature T in the previous example as a function of the variable t alone and then differentiate T to determine the value of $\dfrac{dT}{dt}$ when $t = 5$ minutes.

Example 7. A scientist has determined that, under optimum conditions, an initial population of 40 bacteria will grow "exponentially" to $f(t) = 40 \cdot e^{\frac{t}{5}}$ bacteria after t hours.

(a) Graph $y = f(t)$ for $0 \le t \le 15$. Calculate $f(0)$, $f(5)$ and $f(10)$.

(b) How fast is the population increasing at time t? (Find $f'(t)$.)

(c) Show that the rate of population increase, $f'(t)$, is proportional to the population, $f(t)$, at any time t. (Show $f'(t) = K \cdot f(t)$ for some constant K.)

Solution. (a) The graph of $y = f(t)$ appears in the margin. $f(0) = 40 \cdot e^{\frac{0}{5}} = 40$ bacteria, $f(5) = 40 \cdot e^{\frac{5}{5}} = 40e \approx 109$ bacteria and $f(10) = 40 \cdot e^{\frac{10}{5}} \approx 296$ bacteria.

(b) $f'(t) = \frac{d}{dt}\left(f(t) \right) = \frac{d}{dt}\left(40 \cdot e^{\frac{t}{5}} \right) = 40 \cdot e^{\frac{t}{5}} \cdot \frac{d}{dt}\left(\frac{t}{5} \right) = 40 \cdot e^{\frac{t}{5}} \cdot \frac{1}{5} = 8 \cdot e^{\frac{t}{5}}$ bacteria/hour.

(c) $f'(t) = 8 \cdot e^{\frac{t}{5}} = \frac{1}{5} \cdot 40 e^{\frac{t}{5}} = \frac{1}{5} f(t)$ so $f'(t) = K \cdot f(t)$ with $K = \frac{1}{5}$. The rate of change of the population is proportional to its size.   ◄

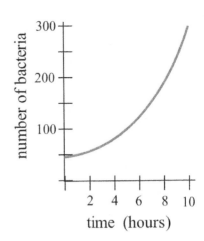

Parametric Equations

Suppose a robot has been programmed to move in the xy-plane so at time t its x-coordinate will be $\sin(t)$ and its y-coordinate will be t^2. Both x and y are functions of the independent parameter t: $x(t) = \sin(t)$ and $y(t) = t^2$. The path of the robot (see margin) can be found by plotting $(x, y) = (x(t), y(t))$ for lots of values of t.

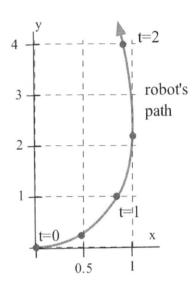

t	$x(t) = \sin(t)$	$y(t) = t^2$	point
0	0	0	$(0,0)$
0.5	0.48	0.25	$(0.48, 0.25)$
1.0	0.84	1	$(0.84, 1)$
1.5	1.00	2.25	$(1, 2.25)$
2.0	0.91	4	$(0.91, 4)$

Typically we know $x(t)$ and $y(t)$ and need to find $\dfrac{dy}{dx}$, the slope of the tangent line to the graph of $(x(t), y(t))$. The Chain Rule says:

$$\frac{dy}{dt} = \frac{dy}{dx} \cdot \frac{dx}{dt}$$

so , algebraically solving for $\dfrac{dy}{dx}$, we get:

$$\frac{dy}{dx} = \frac{\frac{dy}{dt}}{\frac{dx}{dt}}$$

If we can calculate $\dfrac{dy}{dt}$ and $\dfrac{dx}{dt}$, the derivatives of y and x with respect to the parameter t, then we can determine $\dfrac{dy}{dx}$, the rate of change of y with respect to x.

> If $x = x(t)$ and $y = y(t)$ are differentiable
> with respect to t and $\dfrac{dx}{dt} \neq 0$
>
> then $\dfrac{dy}{dx} = \dfrac{\frac{dy}{dt}}{\frac{dx}{dt}}$.

Example 8. Find the slope of the tangent line to the graph of $(x,y) = (\sin(t), t^2)$ when $t = 2$.

Solution. $\dfrac{dx}{dt} = \cos(t)$ and $\dfrac{dy}{dt} = 2t$. When $t = 2$, the object is at the point $(\sin(2), 2^2) \approx (0.91, 4)$ and the slope of the tangent line is:

$$\frac{dy}{dx} = \frac{\frac{dy}{dt}}{\frac{dx}{dt}} = \frac{2t}{\cos(t)} = \frac{2 \cdot 2}{\cos(2)} \approx \frac{4}{-0.42} \approx -9.61$$

Notice in the figure that the slope of the tangent line to the curve at $(0.91, 4)$ is negative and very steep. ◀

Practice 5. Graph $(x,y) = (3\cos(t), 2\sin(t))$ and find the slope of the tangent line when $t = \frac{\pi}{2}$.

When we calculated $\dfrac{dy}{dx}$, the slope of the tangent line to the graph of $(x(t), y(t))$, we used the derivatives $\dfrac{dx}{dt}$ and $\dfrac{dy}{dt}$. Each of these also has a geometric meaning: $\dfrac{dx}{dt}$ measures the rate of change of $x(t)$ with respect to t: it tells us whether the x-coordinate is increasing or decreasing as the t-variable increases (and how fast it is changing), while $\dfrac{dy}{dt}$ measures the rate of change of $y(t)$ with respect to t.

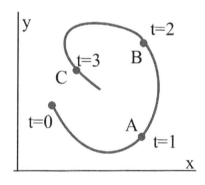

Example 9. For the parametric graph in the margin, determine whether $\dfrac{dx}{dt}, \dfrac{dy}{dt}$ and $\dfrac{dy}{dx}$ are positive or negative when $t = 2$.

Solution. As we move through the point B (where $t = 2$) in the direction of increasing values of t, we are moving to the left, so $x(t)$ is decreasing and $\dfrac{dx}{dt} < 0$. The values of $y(t)$ are increasing, so $\dfrac{dy}{dt} > 0$. Finally, the slope of the tangent line, $\dfrac{dy}{dx}$, is negative. ◀

As a check on the sign of $\dfrac{dy}{dx}$ in the previous example:

$$\frac{dy}{dx} = \frac{\frac{dy}{dt}}{\frac{dx}{dt}} = \frac{\text{positive}}{\text{negative}} = \text{negative}$$

Practice 6. For the parametric graph in the previous example, tell whether $\dfrac{dx}{dt}, \dfrac{dy}{dt}$ and $\dfrac{dy}{dx}$ are positive or negative at $t = 1$ and $t = 3$.

Speed

If we know the position of an object at any time, then we can determine its speed. The formula for speed comes from the distance formula and looks a lot like it, but involves derivatives.

> If $x = x(t)$ and $y = y(t)$ give the location of an object
> at time t and both are differentiable functions of t
> then   the speed of the object is
>
> $$\sqrt{\left(\frac{dx}{dt}\right)^2 + \left(\frac{dy}{dt}\right)^2}$$

Proof. The speed of an object is the limit, as $\Delta t \to 0$, of (see margin):

$$\frac{\text{change in position}}{\text{change in time}} = \frac{\sqrt{(\Delta x)^2 + (\Delta y)^2}}{\Delta t} = \sqrt{\frac{(\Delta x)^2 + (\Delta y)^2}{(\Delta t)^2}}$$

$$= \sqrt{\left(\frac{\Delta x}{\Delta t}\right)^2 + \left(\frac{\Delta y}{\Delta t}\right)^2} \to \sqrt{\left(\frac{dx}{dt}\right)^2 + \left(\frac{dy}{dt}\right)^2}$$

as $\Delta t \to 0$. $\square$

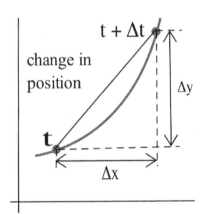

Example 10. Find the speed of the object whose location at time t is $(x, y) = (\sin(t), t^2)$ when $t = 0$ and $t = 1$.

Solution. $\dfrac{dx}{dt} = \cos(t)$ and $\dfrac{dy}{dt} = 2t$ so:

$$\text{speed} = \sqrt{(\cos(t))^2 + (2t)^2} = \sqrt{\cos^2(t) + 4t^2}$$

When $t = 0$, speed $= \sqrt{\cos^2(0) + 4(0)^2} = \sqrt{1 + 0} = 1$. When $t = 1$, speed $= \sqrt{\cos^2(1) + 4(1)^2} \approx \sqrt{0.29 + 4} \approx 2.07$.   ◀

Practice 7. Show that an object located at $(x, y) = (3\sin(t), 3\cos(t))$ at time t has a constant speed. (This object is moving on a circular path.)

Practice 8. Is the object at $(x, y) = (3\cos(t), 2\sin(t))$ at time t traveling faster at the top of the ellipse ($t = \frac{\pi}{2}$) or at the right edge ($t = 0$)?

2.5 Problems

In Problems 1–27, differentiate the given function.

1. $\ln(5x)$

2. $\ln(x^2)$

3. $\ln(x^k)$

4. $\ln(x^x) = x \cdot \ln(x)$

5. $\ln(\cos(x))$

6. $\cos(\ln(x))$

7. $\log_2(5x)$

8. $\log_2(kx)$

9. $\ln(\sin(x))$

10. $\ln(kx)$

11. $\log_2(\sin(x))$

12. $\ln(e^x)$

13. $\log_5(5^x)$

14. $\ln\left(e^{f(x)}\right)$

15. $x \cdot \ln(3x)$

16. $e^x \cdot \ln(x)$

17. $\dfrac{\ln(x)}{x}$

18. $\sqrt{x + \ln(3x)}$

19. $\ln\left(\sqrt{5x-3}\right)$ 20. $\ln(\cos(t))$

21. $\cos(\ln(w))$ 22. $\ln(ax+b)$

23. $\ln\left(\sqrt{t+1}\right)$ 24. 3^x

25. $5^{\sin(x)}$ 26. $x\cdot\ln(x)-x$

27. $\ln\left(\sec(x)+\tan(x)\right)$

28. Find the slope of the line tangent to $f(x)=\ln(x)$ at the point $(e,1)$. Find the slope of the line tangent to $g(x)=e^x$ at the point $(1,e)$. How are the slopes of f and g at these points related?

29. Find a point P on the graph of $f(x)=\ln(x)$ so the tangent line to f at P goes through the origin.

30. You are moving from left to right along the graph of $y=\ln(x)$ (see figure below).

   (a) If the x-coordinate of your location at time t seconds is $x(t)=3t+2$, then how fast is your elevation increasing?

   (b) If the x-coordinate of your location at time t seconds is $x(t)=e^t$, then how fast is your elevation increasing?

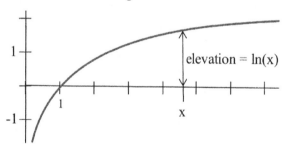

31. The percent of a population, $p(t)$, who have heard a rumor by time t is often modeled by
$$p(t)=\frac{100}{1+Ae^{-t}}=100\left(1+Ae^{-t}\right)^{-1}$$
for some positive constant A. Calculate $p'(t)$, the rate at which the rumor is spreading.

32. If we start with A atoms of a radioactive material that has a "half-life" (the time it takes for half of the material to decay) of 500 years, then the number of radioactive atoms left after t years is $r(t)=A\cdot e^{-Kt}$ where $K=\dfrac{\ln(2)}{500}$. Calculate $r'(t)$ and show that $r'(t)$ is proportional to $r(t)$ (that is, $r'(t)=b\cdot r(t)$ for some constant b).

In 33–41, find a function with the given derivative.

33. $f'(x)=\dfrac{8}{x}$ 34. $h'(x)=\dfrac{3}{3x+5}$

35. $f'(x)=\dfrac{\cos(x)}{3+\sin(x)}$ 36. $g'(x)=\dfrac{x}{1+x^2}$

37. $g'(x)=3e^{5x}$ 38. $h'(x)=e^2$

39. $f'(x)=2x\cdot e^{x^2}$ 40. $g'(x)=\cos(x)e^{\sin(x)}$

41. $h'(x)=\cot(x)=\dfrac{\cos(x)}{\sin(x)}$

42. Define $A(x)$ to be the **area** bounded between the t-axis, the graph of $y=f(t)$ and a vertical line at $t=x$ (see figure below). The area under each "hump" of f is 2 square inches.

   (a) Graph $A(x)$ for $0\le x\le 9$.

   (b) Graph $A'(x)$ for $0\le x\le 9$.

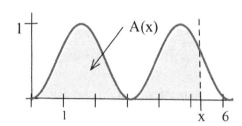

Problems 43–48 involve parametric equations.

43. At time t minutes, robot A is at $(t,2t+1)$ and robot B is at $(t^2,2t^2+1)$.

   (a) Where is each robot when $t=0$ and $t=1$?

   (b) Sketch the path each robot follows during the first minute.

   (c) Find the slope of the tangent line, $\dfrac{dy}{dx}$, to the path of each robot at $t=1$ minute.

   (d) Find the speed of each robot at $t=1$ minute.

   (e) Discuss the motion of a robot that follows the path $(\sin(t),2\sin(t)+1)$ for 20 minutes.

44. Let $x(t)=t+1$ and $y(t)=t^2$.

   (a) Graph $(x(t),y(t))$ for $-1\le t\le 4$.

   (b) Find $\dfrac{dx}{dt},\dfrac{dy}{dt}$, the tangent slope $\dfrac{dy}{dx}$, and speed when $t=1$ and $t=4$.

45. For the parametric graph shown below, determine whether $\frac{dx}{dt}, \frac{dy}{dt}$ and $\frac{dy}{dx}$ are positive, negative or 0 when $t = 1$ and $t = 3$.

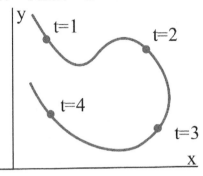

46. For the parametric graph shown below, determine whether $\frac{dx}{dt}, \frac{dy}{dt}$ and $\frac{dy}{dx}$ are positive, negative or 0 when $t = 1$ and $t = 3$.

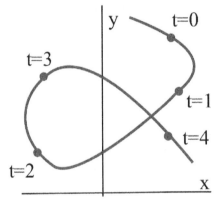

47. The parametric graph $(x(t), y(t))$ defined by $x(t) = R \cdot (t - \sin(t))$ and $y(t) = R \cdot (1 - \cos(t))$ is called a **cycloid**, the path of a light attached to the edge of a rolling wheel with radius R.

    (a) Graph $(x(t), y(t))$ for $0 \leq t \leq 4\pi$.

    (b) Find $\frac{dx}{dt}, \frac{dy}{dt}$, the tangent slope $\frac{dy}{dx}$, and speed when $t = \frac{\pi}{2}$ and $t = \pi$.

48. Describe the motion of particles whose locations at time t are $(\cos(t), \sin(t))$ and $(\cos(t), -\sin(t))$.

49. (a) Describe the path of a robot whose location at time t is $(3 \cdot \cos(t), 5 \cdot \sin(t))$.

    (b) Describe the path of a robot whose location at time t is $(A \cdot \cos(t), B \cdot \sin(t))$.

    (c) Give parametric equations so the robot will move along the same path as in part (a) but in the opposite direction.

50. After t seconds, a projectile hurled with initial velocity v and angle θ will be at $x(t) = v \cdot \cos(\theta) \cdot t$ feet and $y(t) = v \cdot \sin(\theta) \cdot t - 16t^2$ feet (see figure below). (This formula neglects air resistance.)

    (a) For an initial velocity of 80 feet/second and an angle of $\frac{\pi}{4}$, find $T > 0$ so that $y(T) = 0$. What does this value for t represent physically? Evaluate $x(T)$.

    (b) For v and θ in part (a), calculate $\frac{dy}{dx}$. Find T so that $\frac{dy}{dx} = 0$ at $t = T$, and evaluate $x(T)$. What does $x(T)$ represent physically?

    (c) What initial velocity is needed so a ball hit at an angle of $\frac{\pi}{4} \approx 0.7854$ will go over a 40-foot-high fence 350 feet away?

    (d) What initial velocity is needed so a ball hit at an angle of 0.7 radians will go over a 40-foot-high fence 350 feet away?

initial speed = v

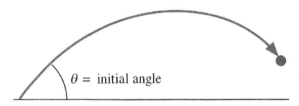

θ = initial angle

51. Use the method from the proof that $\mathbf{D}(\ln(x)) = \frac{1}{x}$ to compute the derivative $\mathbf{D}(\arctan(x))$:

    (a) Rewrite $y = \arctan(x)$ as $\tan(y) = x$.

    (b) Differentiate both sides using the Chain Rule and solve for y'.

    (c) Use the identity $1 + \tan^2(\theta) = \sec^2(\theta)$ and the fact that $\tan(y) = x$ to show that $y' = \frac{1}{1 + x^2}$.

52. Use the method from the proof that $\mathbf{D}(\ln(x)) = \frac{1}{x}$ to compute the derivative $\mathbf{D}(\arcsin(x))$:

    (a) Rewrite $y = \arcsin(x)$ as $\sin(y) = x$.

    (b) Differentiate both sides using the Chain Rule and solve for y'.

    (c) Use the identity $\cos^2(\theta) + \sin^2(\theta) = 1$ and the fact that $\sin(y) = x$ to show that $y' = \frac{1}{\sqrt{1 - x^2}}$.

2.5 Practice Answers

1. $\log_9(20) = \dfrac{\log(20)}{\log(9)} \approx 1.3634165 \approx \dfrac{\ln(20)}{\ln(9)}$

$\log_3(20) = \dfrac{\log(20)}{\log(3)} \approx 2.726833 \approx \dfrac{\ln(20)}{\ln(3)}$

$\log_\pi(e) = \dfrac{\log(e)}{\log(\pi)} \approx 0.8735685 \approx \dfrac{\ln(e)}{\ln(\pi)} = \dfrac{1}{\ln(\pi)}$

2. $\mathbf{D}\left(\log_{10}(\sin(x))\right) = \dfrac{1}{\sin(x) \cdot \ln(10)} \, \mathbf{D}(\sin(x)) = \dfrac{\cos(x)}{\sin(x) \cdot \ln(10)}$

$\mathbf{D}\left(\log_\pi(e^x)\right) = \dfrac{1}{e^x \cdot \ln(\pi)} \, \mathbf{D}(e^x) = \dfrac{e^x}{e^x \cdot \ln(\pi)} = \dfrac{1}{\ln(\pi)}$

3. $\mathbf{D}\left(\sin(2^x)\right) = \cos(2^x) \, \mathbf{D}\left(2^x\right) = \cos(2^x) \cdot 2^x \cdot \ln(2)$

$\dfrac{d}{dt}\left(3^{t^2}\right) = 3^{t^2} \ln(3) \, \mathbf{D}(t^2) = 3^{t^2} \ln(3) \cdot 2t$

4. $T = \dfrac{72}{1+h} = \dfrac{72}{1+t+\sin(t)} \Rightarrow$

$\dfrac{dT}{dt} = \dfrac{(1+t+\sin(t)) \cdot 0 - 72 \cdot \mathbf{D}(1+t+\sin(t))}{(1+t+\sin(t))^2} = \dfrac{-72(1+\cos(t))}{(1+t+\sin(t))^2}$

When $t = 5$, $\dfrac{dT}{dt} = \dfrac{-72(1+\cos(5))}{(1+5+\sin(5))^2} \approx -3.63695$.

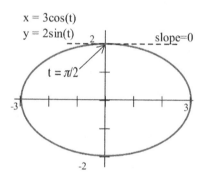

x = 3cos(t)

y = 2sin(t)

slope=0

t = π/2

5. $x(t) = 3\cos(t) \Rightarrow \dfrac{dx}{dt} = -3\sin(t)$, $y(t) = 2\sin(t) \Rightarrow \dfrac{dy}{dt} = 2\cos(t)$:

$\dfrac{dy}{dx} = \dfrac{\frac{dy}{dt}}{\frac{dx}{dt}} = \dfrac{2\cos(t)}{-3\sin(t)} \Rightarrow \dfrac{dy}{dx}\bigg|_{t=\frac{\pi}{2}} = \dfrac{2\cos(\frac{\pi}{2})}{-3\sin(\frac{\pi}{2})} = \dfrac{2 \cdot 0}{-3 \cdot 1} = 0$

(See margin for graph.)

6. $x = 1$: positive, positive, positive. $x = 3$: positive, negative, negative.

7. $x(t) = 3\sin(t) \Rightarrow \dfrac{dx}{dt} = 3\cos(t)$ and $y(t) = 3\cos(t) \Rightarrow \dfrac{dy}{dt} = -3\sin(t)$. So:

$$\text{speed} = \sqrt{\left(\dfrac{dx}{dt}\right)^2 + \left(\dfrac{dy}{dt}\right)^2} = \sqrt{(3\cos(t))^2 + (-3\sin(t))^2}$$

$$= \sqrt{9 \cdot \cos^2(t) + 9 \cdot \sin^2(t)} = \sqrt{9} = 3 \quad \text{(a constant)}$$

8. $x(t) = 3\cos(t) \Rightarrow \dfrac{dx}{dt} = -3\sin(t)$ and $y(t) = 2\sin(t) \Rightarrow \dfrac{dy}{dt} = 2\cos(t)$ so:

$$\text{speed} = \sqrt{\left(\dfrac{dx}{dt}\right)^2 + \left(\dfrac{dy}{dt}\right)^2} = \sqrt{(-3\sin(t))^2 + (2\cos(t))^2}$$

$$= \sqrt{9 \cdot \sin^2(t) + 4 \cdot \cos^2(t)}$$

When $t = 0$, the speed is $\sqrt{9 \cdot 0^2 + 4 \cdot 1^2} = 2$.

When $t = \frac{\pi}{2}$, the speed is $\sqrt{9 \cdot 1^2 + 4 \cdot 0^2} = 3$ (faster).

2.6 Related Rates

Throughout the next several sections we'll look at a variety of applications of derivatives. Probably no single application will be of interest or use to everyone, but at least some of them should be useful to you. Applications also reinforce what you have been practicing: they require that you recall what a derivative means and require you to use the differentiation techniques covered in the last several sections. Most people gain a deeper understanding and appreciation of a tool as they use it, and differentiation is both a powerful concept and a useful tool.

The Derivative as a Rate of Change

In Section 2.1, we discussed several interpretations of the derivative of a function. Here we will examine the "rate of change of a function" interpretation. If several variables or quantities are related to each other and some of the variables are changing at a known rate, then we can use derivatives to determine how rapidly the other variables must be changing.

Example 1. The radius of a circle is increasing at a rate of 10 feet each second (see margin figure) and we want to know how fast the **area** of the circle is increasing when the radius is 5 feet. What can we do?

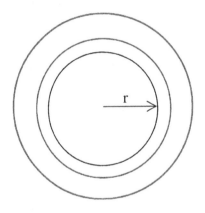

Solution. We could get an *approximate* answer by calculating the area of the circle when the radius is 5 feet:

$$A = \pi r^2 = \pi(5 \text{ feet})^2 \approx 78.6 \text{ feet}^2$$

and the area 1 second later when the radius is 10 feet larger than before:

$$A = \pi r^2 = \pi(15 \text{ feet})^2 \approx 706.9 \text{ feet}^2$$

and then computing:

$$\frac{\Delta \text{area}}{\Delta \text{time}} = \frac{706.9 \text{ feet}^2 - 78.6 \text{ feet}^2}{1 \text{ second}} = 628.3 \frac{\text{ft}^2}{\text{sec}}$$

This approximate answer represents the average change in area during the 1-second period when the radius increased from 5 feet to 15 feet. It is also the slope of the secant line through the points P and Q in the margin figure, and it is clearly not a very good approximation of the instantaneous rate of change of the area, the slope of the tangent line at the point P.

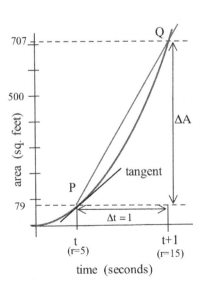

We could get a *better approximation* by calculating $\dfrac{\Delta A}{\Delta t}$ over a shorter time interval, say $\Delta t = 0.1$ seconds. In this scenario, the original area is still 78.6 ft² but the new area (after $t = 0.1$ seconds has passed) is $A = \pi(6 \text{ feet})^2 \approx 113.1 \text{ ft}^2$ (why is the new radius 6 feet?) so:

$$\frac{\Delta A}{\Delta t} = \frac{113.1 \text{ feet}^2 - 78.6 \text{ feet}^2}{0.1 \text{ second}} = 345\frac{\text{ft}^2}{\text{sec}}$$

This is the slope of the secant line through the points P and Q in the margin figure, which represents a much better approximation of the slope of the tangent line at P—but it is still only an approximation. Using derivatives, we can get an **exact** answer without doing very much work at all.

We know that the two variables in this problem, the radius r and the area A, are related to each other by the formula $A = \pi r^2$. We also know that both r and A are changing over time, so each of them is a function of an additional variable t (time, in seconds): $r(t)$ and $A(t)$.

We want to know the rate of change of the area "when the radius is 5 feet" so if $t = 0$ corresponds to the particular moment in time when the radius is 5 feet, we can write $r(0) = 5$.

The statement that "the radius is increasing at a rate of 10 feet each second" can be translated into a mathematical statement about the rate of change, the derivative of r (radius) with respect to t (time): if $t = 0$ corresponds to the moment when the radius is 5 feet, then $r'(0) = \dfrac{dr}{dt} = 10$ ft/sec.

The question about the rate of change of the area is a question about $A'(t) = \dfrac{dA}{dt}$.

Collecting all of this information...

- **variables**: $r(t)$ = radius at time t, $A(t)$ = area at time t

- **we know**: $r(0) = 5$ feet and $r'(0) = 10$ ft/sec

- **we want to know**: $A'(0)$, the rate of change of area with respect to time at the moment when $r = 5$ feet

- **connecting equation**: $A = \pi r^2$ or $A(t) = \pi\left[r(t)\right]^2$

To find $A'(0)$ we must first find $A'(t)$ and then evaluate this derivative at $t = 0$. Differentiating both sides of the connecting equation, we get:

$$A(t) = \pi\left[r(t)\right]^2 \Rightarrow A'(t) = 2\pi\left[r(t)\right]^1 \cdot r'(t) \Rightarrow A'(t) = 2\pi \cdot r(t) \cdot r'(t)$$

Now we can plug in $t = 0$ and use the information we know:

$$A'(0) = 2\pi \cdot r(0) \cdot r'(0) = 2\pi \cdot 5 \cdot 10 = 100\pi$$

Notice that we have used the Power Rule for Functions (or, more generally, the Chain Rule) because the area is a function of the radius, which is a function of time.

When the radius is 5 feet, the area is increasing at 100π ft²/sec ≈ 314.2 square feet per second. ◄

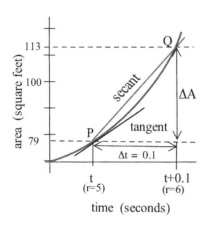

Before considering other examples, let's review the solution to the previous example. The statement "the radius is increasing at a rate of 10 feet each second" implies that this rate of change is the same at $t = 0$ (the moment in time we were interested in) as at any other time during this process, say $t = 1.5$ or $t = 98$: $r'(0) = r'(1.5) = r'(98) = 10$. But we only used the fact that $r'(0) = 10$ in our solution.

Next, notice that we let $t = 0$ correspond to the particular moment in time the question asked about (the moment when $r = 5$). But this choice was arbitrary: we could have let this moment correspond to $t = 2.8$ or $t = 7\pi$ and the eventual answer would have been the same.

Finally, notice that we explicitly wrote each variable (and their derivatives) as a function of the time variable, t: $A(t)$, $r(t)$, $A'(t)$ and $r'(t)$. Consequently, we used the composition form of the Chain Rule:

$$(A \circ r)'(t) = A'(r(t)) \cdot r'(t)$$

Let's redo the previous example using the Leibniz form of the Chain Rule, keeping the above observations in mind.

Solution. We know that the two variables in this problem, the radius r and the area A, are related to each other by the formula $A = \pi r^2$. We also know that both r and A are changing over time, so each of them is a function of an additional variable t (time, in seconds).

We want to know the rate of change of the area "when the radius is 5 feet," which translates to evaluating $\dfrac{dA}{dt}$ at the moment when $r = 5$. We write this in Leibniz notation as:

$$\left.\frac{dA}{dt}\right|_{r=5}$$

The statement that "the radius is increasing at a rate of 10 feet each second" translates into $\dfrac{dr}{dt} = 10$. From the connecting equation $A = \pi r^2$ we know that $\dfrac{dA}{dr} = 2\pi r$. Furthermore, the Chain Rule tells us that:

$$\frac{dA}{dt} = \frac{dA}{dr} \cdot \frac{dr}{dt}$$

We know that $\dfrac{dA}{dr} = 2\pi r$ and $\dfrac{dr}{dt} = 10$ are *always* true, so we can rewrite the Chain Rule statement above as:

$$\frac{dA}{dt} = 2\pi r \cdot 10 = 20\pi r$$

Finally, we evaluate both sides at the moment in time we are interested in (the moment when $r = 5$):

$$\left.\frac{dA}{dt}\right|_{r=5} = \left.20\pi r\right|_{r=5} = 20\pi \cdot 5 = 100\pi \approx 314.2$$

which is the same answer we found in the original solution.   ◄

We should take care in future problems to consider whether the information we are given about rates of change holds true all the time or just at a particular moment in time. That didn't matter in our first example, but it might in other situations.

The key steps in finding the rate of change of the area of the circle were:

- write the known information in a mathematical form, expressing rates of change as derivatives: $\dfrac{dr}{dt} = 10$ ft/sec

- write the question in a mathematical form: $\dfrac{dA}{dt} = ?$

- find an equation connecting or relating the variables: $A = \pi r^2$

- differentiate both sides of the connecting equation using the Chain Rule (and other differentiation patterns as necessary): $\dfrac{dA}{dt} = \dfrac{dA}{dr}\dfrac{dr}{dt}$

- substitute all of the known values that are *always* true into the equation resulting from the previous step and (if necessary) solve for the desired quantity in the resulting equation: $\dfrac{dA}{dt} = 2\pi r \cdot 10$

- substitute all of the known values that are true at the particular moment in time the question asks about into the equation resulting from the previous step: $\dfrac{dA}{dt}\bigg|_{r=5} = 2\pi r \cdot 10 \bigg|_{r=5} = 100\pi$

Example 2. Divers' lives depend on understanding situations involving related rates. In water, the pressure at a depth of x feet is approximately $P(x) = 15\left(1 + \dfrac{x}{33}\right)$ pounds per square inch (compared to approximately $P(0) = 15$ pounds per square inch at sea level). Volume is inversely proportional to the pressure, $V = \dfrac{k}{P}$, so doubling the pressure will result in half the original volume. Remember that volume is a function of the pressure: $V = V(P)$.

(a) Suppose a diver's lungs, at a depth of 66 feet, contained 1 cubic foot of air and the diver ascended to the surface without releasing any air. What would happen?

(b) If a diver started at a depth of 66 feet and ascended at a rate of 2 feet per second, how fast would the pressure be changing?

(Dives deeper than 50 feet also involve a risk of the "bends," or decompression sickness, if the ascent is too rapid. Tables are available that show the safe rates of ascent from different depths.)

Solution. (a) The diver would risk rupturing his or her lungs. The 1 cubic foot of air at a depth of 66 feet would be at a pressure of $P(66) = 15\left(1 + \frac{66}{33}\right) = 45$ pounds per square inch (psi). Because the pressure at sea level, $P(0) = 15$ psi, is only $\frac{1}{3}$ as great, each cubic foot of air would expand to 3 cubic feet, and the diver's lungs would be in danger. Divers are taught to release air as they ascend to avoid this danger. (b) The diver is ascending at a rate of 2 feet/second so the rate of change of the diver's depth with respect to time is $\frac{dx}{dt} = -2$ ft/s. (Why is this rate of change negative?) The pressure is $P = 15\left(1 + \frac{x}{33}\right) = 15 + \frac{15}{33}x$, a function of x, so using the Chain Rule:

$$\frac{dP}{dt} = \frac{dP}{dx} \cdot \frac{dx}{dt} = \frac{15}{33}\frac{\text{psi}}{\text{ft}} \cdot \left(-2\frac{\text{ft}}{\text{sec}}\right) = -\frac{30}{33}\frac{\text{psi}}{\text{sec}} \approx -0.91\frac{\text{psi}}{\text{sec}}$$

The rates of change in this problem are constant (they hold true at any moment in time during the ascent) so we are done.   ◀

Example 3. The height of a cylinder is increasing at 7 meters per second and the radius is increasing at 3 meters per second. How fast is the volume changing when the cylinder is 5 meters high and has a radius of 6 meters? (See margin.)

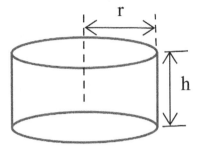

Solution. First we need to translate our known information into a mathematical format. The height and radius are given: at the particular moment in time the question asks about, $h = $ height $= 5$ m and $r = $ radius $= 6$ m. We are also told how fast h and r are changing at this moment in time: $\frac{dh}{dt} = 7$ m/sec and $\frac{dr}{dt} = 3$ m/sec. Finally, we are asked to find $\frac{dV}{dt}$, and we should expect the units of $\frac{dV}{dt}$ to be the same as $\frac{\Delta V}{\Delta t}$, which are m^3/sec.

- **variables:** $h(t) = $ height at time t seconds, $r(t) = $ radius at time t, $V(t) = $ volume at time t.

- **we know:** at a particular moment in time, $h = 5$ m, $\frac{dh}{dt} = 7$ m/sec, $r = 6$ m and $\frac{dr}{dt} = 3$ m/sec

- **we want to know:** $\frac{dV}{dt}$ at this particular moment in time

We also need an equation that relates the variables h, r and V (all of which are functions of time t) to each other:

- **connecting equation:** $V = \pi r^2 h$

Differentiating each side of this equation with respect to t (remembering that h, r and V are functions of t), we have:

$$\frac{dV}{dt} = \frac{d}{dt}\left(\pi r^2 h\right) = \pi r^2 \cdot \frac{dh}{dt} + h \cdot \frac{d}{dt}\left(\pi r^2\right)$$
$$= \pi r^2 \cdot \frac{dh}{dt} + h \cdot 2\pi r \cdot \frac{dr}{dt}$$

using the Product Rule (on the product $\pi r^2 \cdot h$) and the Power Rule for Functions (on πr^2, remembering that r is actually a function of t).

The rest of the solution just involves substituting values and doing some arithmetic. At the particular moment in time we're interested in:

$$\frac{dV}{dt} = \pi \cdot 6^2 \, \text{m}^2 \cdot 7\frac{\text{m}}{\text{sec}} + 5\,\text{m} \cdot 2\pi \cdot 6\,\text{m} \cdot 3\frac{\text{m}}{\text{sec}}$$
$$= 432\pi\frac{\text{m}^3}{\text{sec}} \approx 1357.2\frac{\text{m}^3}{\text{sec}}$$

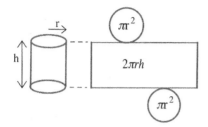

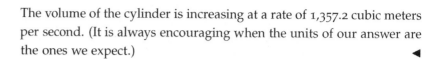

The volume of the cylinder is increasing at a rate of 1,357.2 cubic meters per second. (It is always encouraging when the units of our answer are the ones we expect.) ◀

Practice 1. How fast is the **surface area** of the cylinder changing in the previous example? (Assume that h, r, $\frac{dh}{dt}$ and $\frac{dr}{dt}$ have the same values as in the example and use the figure in the margin to help you determine an equation relating the surface area of the cylinder to the variables h and r. The cylinder includes a top and bottom.)

Practice 2. How fast is the **volume** of the cylinder in the previous example changing if the radius is decreasing at a rate of 3 meters per second? (The height, radius and rate of change of the height are the same as in the previous example: 5 m, 6 m and 7 m/sec respectively.)

Usually, the most difficult part of Related Rates problems is to find an equation that relates or connects all of the variables. In the previous problems, the relating equations required a knowledge of geometry and formulas for areas and volumes (or knowing where to look them up). Other Related Rates problems may require information about similar triangles, the Pythagorean Theorem or trigonometric identities: the information required varies from problem to problem.

It is a good idea—a very good idea—to draw a picture of the physical situation whenever possible. It is also a good idea, particularly if the problem is very important (your next raise depends on getting the right answer), to calculate at least one *approximate* answer as a check of your exact answer.

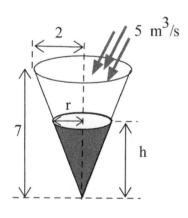

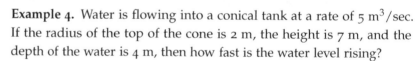

Example 4. Water is flowing into a conical tank at a rate of 5 m^3/sec. If the radius of the top of the cone is 2 m, the height is 7 m, and the depth of the water is 4 m, then how fast is the water level rising?

Solution. Let's define our variables to be $h =$ height (or depth) of the water in the cone and $V =$ the volume of the water in the cone. Both h and V are changing, and both of them are functions of time t. We are told in the problem that $h = 4$ m and $\frac{dV}{dt} = 5$ m^3/sec, and we are asked to find $\frac{dh}{dt}$. We expect that the units of $\frac{dh}{dt}$ will be the same as $\frac{\Delta h}{\Delta t}$, which are meters/second.

- **variables**: $h(t) =$ height at time t seconds, $r(t) =$ radius of the top surface of the water at time t, $V(t) =$ volume of water at time t

- **we know**: $\frac{dV}{dt} = 5$ m^3/sec (always true) and $h = 4$ m (at a particular moment)

- **we want to know**: $\frac{dh}{dt}$ at this particular moment

Unfortunately, the equation for the volume of a cone, $V = \frac{1}{3}\pi r^2 h$, also involves an additional variable r, the radius of the cone at the top of the water. This is a situation in which a picture can be a great help by suggesting that we have a pair of similar triangles:

$$\frac{r}{h} = \frac{\text{top radius}}{\text{total height}} = \frac{2\text{ m}}{7\text{ m}} = \frac{2}{7} \quad \Rightarrow \quad r = \frac{2}{7}h$$

Knowing this, we can rewrite the volume of the water contained in the cone, $V = \frac{1}{3}\pi r^2 h$, as a function of the single variable h:

- **connecting equation**: $V = \dfrac{1}{3}\pi r^2 h = \dfrac{1}{3}\pi \left(\dfrac{2}{7}h\right)^2 h = \dfrac{4}{147}\pi h^3$

The rest of the solution is reasonably straightforward.

$$\frac{dV}{dt} = \frac{dV}{dh} \cdot \frac{dh}{dt} = \frac{d}{dh}\left(\frac{4}{147}\pi h^3\right) \cdot \frac{dh}{dt}$$

We know $\dfrac{dV}{dt} = 5$ always holds, and the derivative is easy to compute:

$$5 = \frac{4}{49}\pi h^2 \cdot \frac{dh}{dt}$$

At the particular moment in time we want to know about (when $h = 4$):

$$5 = \frac{4}{49}\pi h^2 \Big|_{h=4} \cdot \frac{dh}{dt}\Big|_{h=4} \quad \Rightarrow \quad 5 = \frac{64\pi}{49} \cdot \frac{dh}{dt}\Big|_{h=4}$$

and we can now solve for the quantity of interest:

$$\frac{dh}{dt}\Big|_{h=4} = \frac{5}{\frac{64\pi}{49}} = \frac{245}{64\pi} \approx 1.22 \,\frac{\text{m}}{\text{sec}}$$

This example was a bit more challenging because we needed to use similar triangles to get an equation relating V to h and because we eventually needed to do some arithmetic to solve for $\dfrac{dh}{dt}$. ◀

Practice 3. A rainbow trout has taken the fly at the end of a 60-foot fishing line, and the line is being reeled in at a rate of 30 feet per minute. If the tip of the rod is 10 feet above the water and the trout is at the surface of the water, how fast is the trout being pulled toward the angler? (Hint: Draw a picture and use the Pythagorean Theorem.)

Example 5. When rain is falling vertically, the amount (volume) of rain collected in a cylinder is proportional to the area of the **opening** of the cylinder. If you place a narrow cylindrical glass and a wide cylindrical glass out in the rain:

(a) which glass will collect water faster?

(b) in which glass will the water level rise faster?

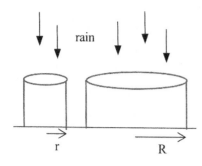

Solution. Let's assume that the smaller glass has a radius of r and the larger glass has a radius of R, so that $R > r$. The areas of their openings are πr^2 and πR^2, respectively. Call the volume of water collected in each glass v (for the smaller glass) and V (for the larger glass).

(a) The smaller glass will collect water at the rate $\dfrac{dv}{dt} = K \cdot \pi r^2$ and the larger at the rate $\dfrac{dV}{dt} = K \cdot \pi R^2$ so $\dfrac{dV}{dt} > \dfrac{dv}{dt}$ and the larger glass will collect water faster than the smaller glass.

(b) The volume of water in each glass is a function of the radius of the glass and the height of the water in the glass: $v = \pi r^2 h$ and $V = \pi R^2 H$ where h and H are the heights of the water levels in the smaller and larger glasses, respectively. The heights h and H vary with t (in other words, they are each functions of t) while the radii (r and R) remain constant, so:

$$\frac{dv}{dt} = \frac{d}{dt}\left(\pi r^2 h\right) = \pi r^2 \frac{dh}{dt} \quad \Rightarrow \quad \frac{dh}{dt} = \frac{\frac{dv}{dt}}{\pi r^2} = \frac{K\pi r^2}{\pi r^2} = K$$

Similarly:

$$\frac{dV}{dt} = \frac{d}{dt}\left(\pi R^2 H\right) = \pi R^2 \frac{dH}{dt} \quad \Rightarrow \quad \frac{dH}{dt} = \frac{\frac{dV}{dt}}{\pi R^2} = \frac{K\pi R^2}{\pi R^2} = K$$

So $\dfrac{dh}{dt} = K = \dfrac{dH}{dt}$, which tells us the water level in each glass is rising at the same rate. In a one-minute period, the larger glass will collect more rain, but the larger glass also requires more rain to raise its water level by a fixed amount. How do you think the volumes and water levels would change if we placed a small glass and a large plastic (rectangular) box side by side in the rain? ◀

2.6 Problems

1. An expandable sphere is being filled with liquid at a constant rate from a tap (imagine a water balloon connected to a faucet). When the radius of the sphere is 3 inches, the radius is increasing at 2 inches per minute. How fast is the liquid coming out of the tap? ($V = \frac{4}{3}\pi r^3$)

2. The 12-inch base of a right triangle is growing at 3 inches per hour, and the 16-inch height of the triangle is shrinking at 3 inches per hour (see figure in the margin).

    (a) Is the area increasing or decreasing?

    (b) Is the perimeter increasing or decreasing?

    (c) Is the hypotenuse increasing or decreasing?

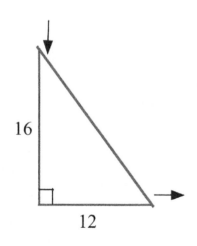

3. One hour later the right triangle in the previous problem is 15 inches long and 13 inches high (see figure below) and the base and height are changing at the same rate as in Problem 2.

   (a) Is the area increasing or decreasing now?

   (b) Is the hypotenuse increasing or decreasing?

   (c) Is the perimeter increasing or decreasing?

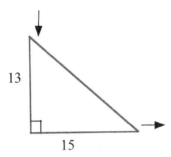

4. A young woman and her boyfriend plan to elope, but she must rescue him from his mother, who has locked him in his room. The young woman has placed a 20-foot long ladder against his house and is knocking on his window when his mother begins pulling the bottom of the ladder away from the house at a rate of 3 feet per second (see figure below). How fast is the top of the ladder (and the young couple) falling when the bottom of the ladder is:

   (a) 12 feet from the bottom of the wall?

   (b) 16 feet from the bottom of the wall?

   (c) 19 feet from the bottom of the wall?

5. The length of a 12-foot by 8-foot rectangle is increasing at a rate of 3 feet per second and the width is decreasing at 2 feet per second (see figure below).

   (a) How fast is the perimeter changing?

   (b) How fast is the area changing?

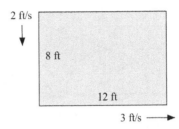

6. A circle of radius 3 inches is inside a square with 12-inch sides (see figure below). How fast is the area between the circle and square changing if the radius is increasing at 4 inches per minute and the sides are increasing at 2 inches per minute?

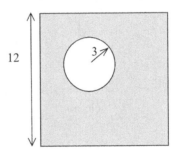

7. An oil tanker in Puget Sound has sprung a leak, and a circular oil slick is forming. The oil slick is 4 inches thick everywhere, is 100 feet in diameter, and the diameter is increasing at 12 feet per hour. Your job, as the Coast Guard commander or the tanker's captain, is to determine how fast the oil is leaking from the tanker.

oil spill

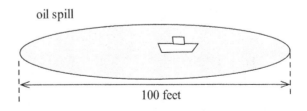

100 feet

8. A mathematical species of slug has a semicircular cross section and is always 5 times as long as it is high (see figure below). When the slug is 5 inches long, it is growing at 0.2 inches per week.

   (a) How fast is its volume increasing?

   (b) How fast is the area of its "foot" (the part of the slug in contact with the ground) increasing?

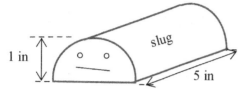

9. Lava flowing from a hole at the top of a hill is forming a conical mountain whose height is always the same as the width of its base (see figure below). If the mountain is increasing in height at 2 feet per hour when it is 500 feet high, how fast is the lava flowing (that is, how fast is the volume of the mountain increasing)? ($V = \frac{1}{3}\pi r^2 h$)

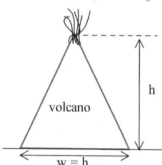

10. A 6-foot-tall person is walking away from a 14-foot lamp post at 3 feet per second. When the person is 10 feet away from the lamp post:

    (a) how fast is the length of the shadow changing?

    (b) how fast is the tip of the shadow moving away from the lamp post?

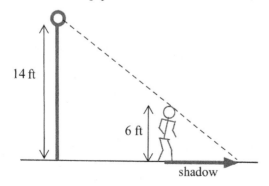

11. Redo the previous problem if the person is 20 feet from the lamp post.

12. Water is being poured at a rate of 15 cubic feet per minute into a conical reservoir that is 20 feet deep and has a top radius of 10 feet (see below).

    (a) How long will it take to fill the empty reservoir?

    (b) How fast is the water level rising when the water is 4 feet deep?

    (c) How fast is the water level rising when the water is 16 feet deep?

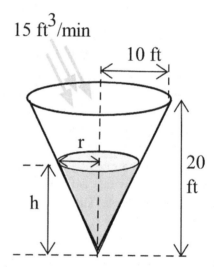

13. The string of a kite is perfectly taut and always makes an angle of 35° above horizontal.

    (a) If the kite flyer has let out 500 feet of string, how high is the kite?

    (b) If the string is let out at a rate of 10 feet per second, how fast is the kite's height increasing?

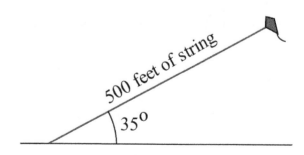

14. A small tracking telescope is viewing a hot-air balloon rise from a point 1,000 meters away from a point directly under the balloon.

    (a) When the viewing angle is $20°$, it is increasing at a rate of $3°$ per minute. How high is the balloon, and how fast is it rising?

    (b) When the viewing angle is $80°$, it is increasing at a rate of $2°$ per minute. How high is the balloon, and how fast is it rising?

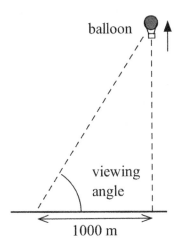

15. The 8-foot diameter of a spherical gas bubble is increasing at 2 feet per hour, and the 12-foot-long edges of a cube containing the bubble are increasing at 3 feet per hour. Is the volume contained between the spherical bubble and the cube increasing or decreasing? At what rate?

16. In general, the strength S of an animal is proportional to the cross-sectional area of its muscles, and this area is proportional to the square of its height H, so the strength $S = aH^2$. Similarly, the weight W of the animal is proportional to the cube of its height, so $W = bH^3$. Finally, the relative strength R of an animal is the ratio of its strength to its weight. As the animal grows, show that its strength and weight increase, but that the relative strength decreases.

17. The snow in a hemispherical pile melts at a rate proportional to its exposed surface area (the surface area of the hemisphere). Show that the height of the snow pile is decreasing at a constant rate.

18. If the rate at which water vapor condenses onto a spherical raindrop is proportional to the surface area of the raindrop, show that the radius of the raindrop will increase at a constant rate.

19. Define $A(x)$ to be the area bounded by the t- and y-axes, and the lines $y = 5$ and $t = x$.

    (a) Find a formula for A as a function of x.

    (b) Determine $A'(x)$ when $x = 1, 2, 4$ and 9.

    (c) If x is a function of time, $x(t) = t^2$, find a formula for A as a function of t.

    (d) Determine $A'(t)$ when $t = 1, 2$ and 3.

    (e) Suppose instead $x(t) = 2 + \sin(t)$. Find a formula for $A(t)$ and determine $A'(t)$.

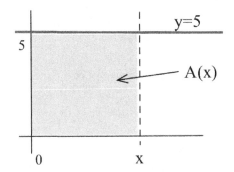

20. The point P is going around the circle $x^2 + y^2 = 1$ twice a minute. How fast is the distance between the point P and the point $(4, 3)$ changing:

    (a) when $P = (1, 0)$?

    (b) when $P = (0, 1)$?

    (c) when $P = (0.8, 0.6)$?

    (Suggestion: Write x and y as parametric functions of time t.)

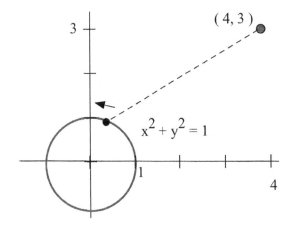

21. You are walking along a sidewalk toward a 40-foot-wide sign adjacent to the sidewalk and perpendicular to it. If your viewing angle θ is $10°$:

    (a) how far are you from the corner of the sign?

    (b) how fast is your viewing angle changing if you are walking at 25 feet per minute?

    (c) how fast are you walking if the angle is increasing at $2°$ per minute?

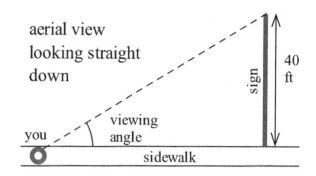

2.6 Practice Answers

1. The surface area is $S = 2\pi rh + 2\pi r^2$. From the Example, we know that $\frac{dh}{dt} = 7$ m/sec and $\frac{dr}{dt} = 3$ m/sec, and we want to know how fast the surface area is changing when $h = 5$ m and $r = 6$ m.

$$\frac{dS}{dt} = 2\pi r \cdot \frac{dh}{dt} + 2\pi \frac{dr}{dt} \cdot h + 2\pi \cdot 2r \cdot \frac{dr}{dt}$$
$$= 2\pi(6\,\text{m})\left(7\,\frac{\text{m}}{\text{sec}}\right) + 2\pi\left(3\,\frac{\text{m}}{\text{sec}}\right)(5m) + 2\pi\left(2 \cdot 6\,\text{m}\right)\left(3\,\frac{\text{m}}{\text{sec}}\right)$$

> Note that the units represent a rate of change of area.

$$= 186\pi\,\frac{\text{m}^2}{\text{sec}} \approx 584.34\,\frac{\text{m}^2}{\text{sec}}$$

2. The volume is $V = \pi r^2 h$. We know that $\frac{dr}{dt} = -3$ m/sec and that $h = 5$ m, $r = 6$ m and $\frac{dh}{dt} = 7$ m/sec.

$$\frac{dV}{dt} = \pi r^2 \cdot \frac{dh}{dt} + \pi \cdot 2r \cdot \frac{dr}{dt}$$
$$= h\pi(6\,\text{m})^2\left(7\,\frac{\text{m}}{\text{sec}}\right) + \pi(2 \cdot 6\,\text{m})\left(-3\,\frac{\text{m}}{\text{sec}}\right)(5\,\text{m})$$

> Note that the units represent a rate of change of volume.

$$= 72\pi\,\frac{\text{m}^3}{\text{sec}} \approx 226.19\,\frac{\text{m}^3}{\text{sec}}$$

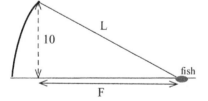

3. See margin figure. We know $\frac{dL}{dt} = -30\,\frac{\text{ft}}{\text{min}}$ (always true); F represents the distance from the fish to a point directly below the tip of the rod, and the distance from that point to the angler remains constant, so $\frac{dF}{dt}$ will equal the rate at which the fish is moving toward the angler. We want to know $\frac{dF}{dt}\Big|_{L=60}$. The Pythagorean Theorem connects F and L: $F^2 + 10^2 = L^2$. Differentiating with respect to t and using the Power Rule for Functions:

$$2F \cdot \frac{dF}{dt} + 0 = 2L \cdot \frac{dL}{dt} \quad \Rightarrow \quad \frac{dF}{dt} = \frac{L}{F} \cdot \frac{dL}{dt}$$

At a particular moment in time, $L = 60 \Rightarrow F^2 + 10^2 = 60^2 \Rightarrow F = \sqrt{3600 - 100} = \sqrt{3500} = 10\sqrt{35}$ so:

$$\frac{dF}{dt}\Big|_{L=60} = -30 \cdot \frac{60}{10\sqrt{35}} = -\frac{180}{\sqrt{35}} \approx -30.43\,\frac{\text{ft}}{\text{min}}$$

2.7 Newton's Method

Newton's method is a process that can find roots of functions whose graphs cross or just "kiss" the x-axis. Although this method is a bit harder to apply than the Bisection Algorithm, it often finds roots that the Bisection Algorithm misses, and it usually finds them faster.

Off on a Tangent

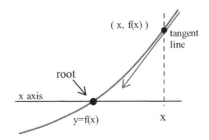

The basic idea of Newton's Method is remarkably simple and graphical: at a point $(x, f(x))$ on the graph of f, the tangent line to the graph "points toward" a root of f, a place where the graph touches the x-axis.

To find a root of f, we just pick a starting value x_0, go to the point $(x_0, f(x_0))$ on the graph of f, build a tangent line there, and follow the tangent line to where it crosses the x-axis, say at x_1.

If x_1 is a root of f, we are done. If x_1 is not a root of f, then x_1 is usually closer to the root than x_0 was, and we can repeat the process, using x_1 as our new starting point. Newton's method is an **iterative** procedure—that is, the output from one application of the method becomes the starting point for the next application.

Let's begin with the function $f(x) = x^2 - 5$, whose roots we already know ($x = \pm\sqrt{5} \approx \pm 2.236067977$), to illustrate Newton's method. First, pick some value for x_0, say $x_0 = 4$, and move to the point $(x_0, f(x_0)) = (4, 11)$ on the graph of f. The tangent line to the graph of f at $(4, 11)$ "points to" a location on the x-axis that is closer to the root of f than the point we started with. We calculate this location on the x-axis by finding an equation of the line tangent to the graph of f at $(4, 11)$ and then finding where this line intersects the x-axis.

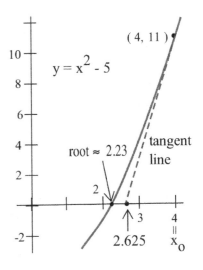

At $(4, 11)$, the line tangent to f has slope $f'(4) = 2(4) = 8$, so an equation of the tangent line is $y - 11 = 8(x - 4)$. Setting $y = 0$, we can find where this line crosses the x-axis:

$$0 - 11 = 8(x - 4) \Rightarrow x = 4 - \frac{11}{8} = \frac{21}{8} = 2.625$$

Call this new value x_1: The point $x_1 = 2.625$ is closer to the actual root $\sqrt{5}$, but it certainly does not equal the actual root. So we can use this new x-value, $x_1 = 2.625$, to repeat the procedure:

- move to the point $(x_1, f(x_1)) = (2.625, 1.890625)$

- find an equation of the tangent line at $(x_1, f(x_1))$:

$$y - 1.890625 = 5.25(x - 2.625)$$

- find x_2, the x-value where this new line intersects the x-axis:

$$y - 1.890625 = 5.25(x - 2.625) \Rightarrow 0 - 1.890625 = 5.25(x_2 - 2.625)$$
$$\Rightarrow x_2 = 2.264880952$$

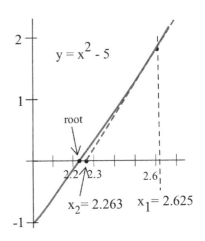

Repeating this process, each new estimate for the root of $f(x) = x^2 - 5$ becomes the starting point to calculate the next estimate. We get:

$x_0 = 4$	(0 correct digits)
$x_1 = 2.625$	(1 correct digit)
$x_2 = 2.262880952$	(2 correct digits)
$x_3 = 2.236251252$	(4 correct digits)
$x_4 = 2.236067985$	(8 correct digits)

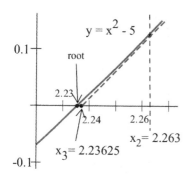

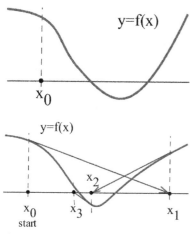

The process for Newton's Method, starting with x_0 and graphically finding the locations on the x-axis of x_1, x_2 and x_3.

It only took 4 iterations to get an approximation within 0.000000008 of the exact value of $\sqrt{5}$. One more iteration gives an approximation x_5 that has 16 correct digits. If we start with $x_0 = -2$ (or any negative number), then the values of x_n approach $-\sqrt{5} \approx -2.23606$.

Practice 1. Find where the tangent line to $f(x) = x^3 + 3x - 1$ at $(1, 3)$ intersects the x-axis.

Practice 2. A starting point and a graph of f appear in the margin. Label the approximate locations of the next two points on the x-axis that will be found by Newton's method.

The Algorithm for Newton's Method

Rather than deal with each particular function and starting point, let's find a pattern for a general function f.

For the starting point x_0, the slope of the tangent line at the point $(x_0, f(x_0))$ is $f'(x_0)$ so the equation of the tangent line is $y - f(x_0) = f'(x_0) \cdot (x - x_0)$. This line intersects the x-axis at a point $(x_1, 0)$, so:

$$0 - f(x_0) = f'(x_0) \cdot (x_1 - x_0) \Rightarrow x_1 = x_0 - \frac{f(x_0)}{f'(x_0)}$$

Starting with x_1 and repeating this process we get $x_2 = x_1 - \frac{f(x_1)}{f'(x_1)}$, $x_3 = x_2 - \frac{f(x_2)}{f'(x_2)}$ and so on. In general, starting with x_n, the line tangent to the graph of f at $(x_n, f(x_n))$ intersects the x-axis at $(x_{n+1}, 0)$ with $x_{n+1} = x_n - \frac{f(x_n)}{f'(x_n)}$, our new estimate for the root of f.

Algorithm for Newton's Method:

1. Pick a starting value x_0 (preferably close to a root of $f(x)$).

2. For each x_n, calculate a new estimate $x_{n+1} = x_n - \frac{f(x_n)}{f'(x_n)}$

3. Repeat step 2 until the estimates are "close enough" to a root or until the method "fails."

When we use Newton's method with $f(x) = x^2 - 5$, the function in our first example, we have $f'(x) = 2x$ so

$$x_{n+1} = x_n - \frac{f(x_n)}{f'(x_n)} = x_n - \frac{x_n^2 - 5}{2x_n} = \frac{2x_n^2 - x_n^2 + 5}{2x_n}$$

$$= \frac{x_n^2 + 5}{2x_n} = \frac{1}{2}\left(x_n + \frac{5}{x_n}\right)$$

The new approximation, x_{n+1}, is the average of the previous approximation, x_n, and 5 divided by the previous approximation, $\frac{5}{x_n}$.

Example 1. Use Newton's method to approximate the root(s) of $f(x) = 2x + x \cdot \sin(x + 3) - 5$.

Solution. $f'(x) = 2 + x \cos(x+3) + \sin(x+3)$ so:

$$x_{n+1} = x_n - \frac{f(x_n)}{f'(x_n)} = x_n - \frac{2x_n + x_n \cdot \sin(x_n + 3) - 5}{2 + x_n \cdot \cos(x_n + 3) + \sin(x_n + 3)}$$

The graph of $f(x)$ (see margin) indicates only one root of f, which is near $x = 3$, so pick $x_0 = 3$. Then Newton's method yields the values $x_0 = 3$, $x_1 = \underline{2.96}484457$, $x_2 = \underline{2.96446}277$, $x_3 = \underline{2.96446273}$ (the underlined digits agree with the exact answer). ◀

If we had picked $x_0 = 4$ in the previous example, Newton's method would have required 4 iterations to get 9 digits of accuracy. For $x_0 = 5$, 7 iterations are needed to get 9 digits of accuracy. If we pick $x_0 = 5.1$, then the values of x_n are not close to the actual root after even 100 iterations: $x_{100} \approx -49.183$. Picking a "good" value for x_0 can result in values of x_n that get close to the root quickly. Picking a "poor" value for x_0 can result in x_n values that take many more iterations to get close to the root—or that don't approach the root at all.

> **The graph of the function can help you pick a "good" x_0.**

Practice 3. Put $x_0 = 3$ and use Newton's method to find the first two iterates, x_1 and x_2, for the function $f(x) = x^3 - 3x^2 + x - 1$.

Example 2. The function graphed in the margin has roots at $x = 3$ and $x = 7$. If we pick $x_0 = 1$ and apply Newton's method, which root do the iterates (the values of x_n) approach?

Solution. The iterates of $x_0 = 1$ are labeled in the margin graph. They are approaching the root at 7. ◀

Practice 4. For the function graphed in the margin, which root do the iterates of Newton's method approach if:

(a) $x_0 = 2$?   (b) $x_0 = 3$?   (c) $x_0 = 5$?

Problem 16 helps you show this pattern—called Heron's method—approximates the square root of any positive number: just replace 5 with the number whose square root you want to find.

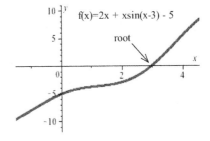

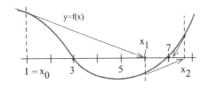

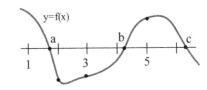

Iteration

We have been emphasizing the geometric nature of Newton's method, but Newton's method is also an example of **iterating a function**. If $N(x) = x - \dfrac{f(x)}{f'(x)}$, the "pattern" in the algorithm, then:

$$x_1 = x_0 - \frac{f(x_0)}{f'(x_0)} = N(x_0)$$

$$x_2 = x_1 - \frac{f(x_1)}{f'(x_1)} = N(x_1) = N(N(x_0)) = N \circ N(x_0)$$

$$x_3 = x_2 - \frac{f(x_2)}{f'(x_2)} = N(x_2) = N(N(N(x_0))) = N \circ N \circ N(x_0)$$

and, in general:

$$x_n = N(x_{n-1}) = n\text{th iteration of } N \text{ starting with } x_0$$

At each step, we use the output from N as the next input into N.

What Can Go Wrong?

When Newton's method works, it usually works very well and the values of x_n approach a root of f very quickly, often doubling the number of correct digits with each iteration. There are, however, several things that can go wrong.

An obvious problem with Newton's method is that $f'(x_n)$ can be 0. Then the algorithm tells us to divide by 0 and x_{n+1} is undefined. Geometrically, if $f'(x_n) = 0$, the tangent line to the graph of f at x_n is horizontal and does not intersect the x-axis at any point. If $f'(x_n) = 0$, just pick another starting value x_0 and begin again. In practice, a second or third choice of x_0 usually succeeds.

There are two other less obvious difficulties that are not as easy to overcome — the values of the iterates x_n may become locked into an infinitely repeating loop (see margin), or they may actually move farther away from a root (see lower margin figure).

Example 3. Put $x_0 = 1$ and use Newton's method to find the first two iterates, x_1 and x_2, for the function $f(x) = x^3 - 3x^2 + x - 1$.

Solution. This is the function from the previous Practice Problem, but with a different starting value for x_0: $f'(x) = 3x^2 - 6x + 1$ so,

$$x_1 = x_0 - \frac{f(x_0)}{f'(x_0)} = 1 - \frac{f(1)}{f'(1)} = 1 - \frac{-2}{-2} = 0$$

$$\text{and } x_2 = x_1 - \frac{f(x_1)}{f'(x_1)} = 0 - \frac{f(0)}{f'(0)} = 0 - \frac{-1}{1} = 1$$

which is the same as x_0, so $x_3 = x_1 = 0$ and $x_4 = x_2 = 1$. The values of x_n alternate between 1 and 0 and do not approach a root. ◄

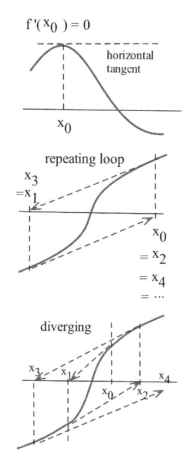

$f'(x_0) = 0$

horizontal tangent

x_0

repeating loop

$x_3 = x_1$

$x_0 = x_2 = x_4 = \cdots$

diverging

Newton's method behaves badly at only a few starting points for this particular function — for most starting points, Newton's method converges to the root of this function. There are some functions, however, that defeat Newton's method for **almost every** starting point.

Practice 5. For $f(x) = \sqrt[3]{x} = x^{\frac{1}{3}}$ and $x_0 = 1$, verify that $x_1 = -2$, $x_2 = 4$ and $x_3 = -8$. Also try $x_0 = -3$ and verify that the same pattern holds: $x_{n+1} = -2x_n$. Graph f and explain why the Newton's method iterates get farther and farther away from the root at 0.

Newton's method is powerful and quick and very easy to program on a calculator or computer. It usually works so well that many people routinely use it as the first method they apply. If Newton's method fails for their particular function, they simply try some other method.

Chaotic Behavior and Newton's Method

An algorithm leads to **chaotic behavior** if two starting points that are close together generate iterates that are sometimes far apart and sometimes close together: $|a_0 - b_0|$ is small but $|a_n - b_n|$ is large for lots (infinitely many) of values of n and $|a_n - b_n|$ is small for lots of values of n. The iterates of the next simple algorithm exhibit chaotic behavior.

A Simple Chaotic Algorithm: Starting with any number between 0 and 1, double the number and keep the fractional part of the result: x_1 is the fractional part of $2x_0$, x_2 is the fractional part of $2x_1$, and in general, $x_{n+1} = 2x_n - \lfloor 2x_n \rfloor$.

If $x_0 = 0.33$, then the iterates of this algorithm are 0.66, 0.32 = fractional part of $2 \cdot 0.66$, 0.64, 0.28, 0.56, ... The iterates for two other starting values close to 0.33 are given below as well as the iterates of 0.470 and 0.471:

x_0	0.32	0.33	0.34	0.470	0.471
x_1	0.64	0.66	0.68	0.940	0.942
x_2	0.28	0.32	0.36	0.880	0.884
x_3	0.56	0.64	0.72	0.760	0.768
x_4	0.12	0.28	0.44	0.520	0.536
x_5	0.24	0.56	0.88	0.040	0.072
x_6	0.48	0.12	0.76	0.080	0.144
x_7	0.96	0.24	0.56	0.160	0.288
x_8	0.92	0.48	0.12	0.320	0.576
x_9	0.84	0.96	0.24	0.640	0.152

There are starting values as close together as we want whose iterates are far apart infinitely often.

Many physical, biological and financial phenomena exhibit chaotic behavior. Atoms can start out within inches of each other and several weeks later be hundreds of miles apart. The idea that small initial differences can lead to dramatically diverse outcomes is sometimes called the "butterfly effect" from the title of a talk ("Predictability: Does the Flap of a Butterfly's Wings in Brazil Set Off a Tornado in Texas?") given by Edward Lorenz, one of the first people to investigate chaos. The "butterfly effect" has important implications about the possibility— or rather the impossibility—of accurate long-range weather forecasting. Chaotic behavior is also an important aspect of studying turbulent air and water flows, the incidence and spread of diseases, and even the fluctuating behavior of the stock market.

Newton's method often exhibits chaotic behavior and—because it is relatively easy to study—is often used as a model to investigate the properties of chaotic behavior. If we use Newton's method to approximate the roots of $f(x) = x^3 - x$ (with roots 0, +1 and −1), then starting points that are very close together can have iterates that converge to different roots. The iterates of 0.4472 and 0.4473 converge to the roots 0 and +1, respectively. The iterates of the median value 0.44725 converge to the root −1, and the iterates of another nearby point, $\dfrac{1}{\sqrt{5}} \approx 0.44721$, simply cycle between $-\dfrac{1}{\sqrt{5}}$ and $+\dfrac{1}{\sqrt{5}}$ and do not converge at all.

Practice 6. Find the first four Newton's method iterates of $x_0 = 0.997$ and $x_0 = 1.02$ for $f(x) = x^2 + 1$. Try two other starting values very close to 1 (but not equal to 1) and find their first four iterates. Use the graph of $f(x) = x^2 + 1$ to explain how starting points so close together can quickly have iterates so far apart.

2.7 Problems

1. The graph of $y = f(x)$ appears below. Estimate the locations of x_1 and x_2 when you apply Newton's method with the given starting value x_0.

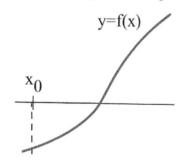

2. The graph of $y = g(x)$ appears below. Estimate the locations of x_1 and x_2 when you apply Newton's method starting value with the value x_0 shown in the graph.

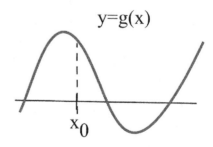

3. The function graphed below has several roots. Which root do the iterates of Newton's method converge to if we start with $x_0 = 1$? With $x_0 = 5$?

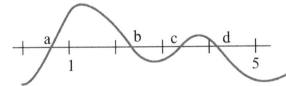

4. The function graphed below has several roots. Which root do the iterates of Newton's method converge to if we start with $x_0 = 2$? With $x_0 = 6$?

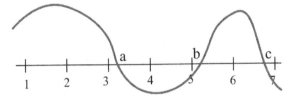

5. What happens to the iterates if we apply Newton's method to the function graphed below and start with $x_0 = 1$? With $x_0 = 5$?

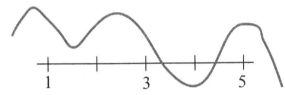

6. What happens if we apply Newton's method to a function f and start with $x_0 = $ a root of f?

7. What happens if we apply Newton's method to a function f and start with $x_0 = $ a maximum of f?

In Problems 8–9, a function and a value for x_0 are given. Apply Newton's method to find x_1 and x_2.

8. $f(x) = x^3 + x - 1$ and $x_0 = 1$

9. $f(x) = x^4 - x^3 - 5$ and $x_0 = 2$

In Problems 10–11, use Newton's method to find a root, accurate to 2 decimal places, of the given functions using the given starting points.

10. $f(x) = x^3 - 7$ and $x_0 = 2$

11. $f(x) = x - \cos(x)$ and $x_0 = 0.7$

In Problems 12–15, use Newton's method to find all roots or solutions, accurate to 2 decimal places, of the given equation. It is helpful to examine a graph to determine a "good" starting value x_0.

12. $2 + x = e^x$

13. $\dfrac{x}{x + 3} = x^2 - 2$

14. $x = \sin(x)$

15. $x = \sqrt[5]{3}$

16. Show that if you apply Newton's method to $f(x) = x^2 - A$ to approximate $\sqrt{A}$, then

$$x_{n+1} = \frac{1}{2}\left(x_n + \frac{A}{x_n}\right)$$

so the new estimate of the square root is the average of the previous estimate and A divided by the previous estimate. This method of approximating square roots is called Heron's method.

17. Use Newton's method to devise an algorithm for approximating the cube root of a number A.

18. Use Newton's method to devise an algorithm for approximating the n-th root of a number A.

Problems 19–22 involve chaotic behavior.

19. The iterates of numbers using the Simple Chaotic Algorithm have some interesting properties.

   (a) Verify that the iterates starting with $x_0 = 0$ are all equal to 0.

   (b) Verify that if $x_0 = \frac{1}{2}, \frac{1}{4}, \frac{1}{8}$ and, in general, $\frac{1}{2^n}$, then the n-th iterate of x_0 is 0 (and so are all iterates beyond the n-th iterate.)

20. When Newton's method is applied to the function $f(x) = x^2 + 1$, most starting values for x_0 lead to chaotic behavior for x_n. Find a value for x_0 so that the iterates alternate: $x_1 = -x_0$ and $x_2 = -x_1 = x_0$.

21. The function $f(x)$ defined as:

$$f(x) = \begin{cases} 2x & \text{if } 0 \le x < \frac{1}{2} \\ 2 - 2x & \text{if } \frac{1}{2} \le x \le 1 \end{cases}$$

is called a "stretch and fold" function.

(a) Describe what f does to the points in the interval $[0, 1]$.

(b) Examine and describe the behavior of the iterates of $\frac{2}{3}$, $\frac{2}{5}$, $\frac{2}{7}$ and $\frac{2}{9}$.

(c) Examine and describe the behavior of the iterates of 0.10, 0.105 and 0.11.

(d) Do the iterates of f lead to chaotic behavior?

22. (a) After many iterations (50 is fine) what happens when you apply Newton's method starting with $x_0 = 0.5$ to:

    i. $f(x) = 2x(1 - x)$

    ii. $f(x) = 3.3x(1 - x)$

    iii. $f(x) = 3.83x(1 - x)$

(b) What do you think happens to the iterates of $f(x) = 3.7x(1 - x)$? What actually happens?

(c) Repeat parts (a)–(b) with some other starting values x_0 between 0 and 1 ($0 < x_0 < 1$). Does the starting value seem to effect the eventual behavior of the iterates?

(The behavior of the iterates of f depends in a strange way on the numerical value of the leading coefficient. The behavior exhibited in part (b) is an example of "chaos.")

2.7 Practice Answers

1. $f'(x) = 3x^2 + 3$, so the slope of the tangent line at $(1, 3)$ is $f'(1) = 6$ and an equation of the tangent line is $y - 3 = 6(x - 1)$ or $y = 6x - 3$. The y-coordinate of a point on the x-axis is 0 so putting $y = 0$ in this equation: $0 = 6x - 3 \Rightarrow x = \frac{1}{2}$. The line tangent to the graph of $f(x) = x^3 + 3x + 1$ at the point $(1, 3)$ intersects the x-axis at the point $(\frac{1}{2}, 0)$.

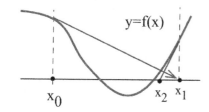

y=f(x)

2. The approximate locations of x_1 and x_2 appear in the margin.

3. Using $f'(x) = 3x^2 + 3$ and $x_0 = 3$:

$$x_1 = x_0 - \frac{f(x_0)}{f'(x_0)} = 3 - \frac{f(3)}{f'(3)} = 3 - \frac{2}{10} = 2.8$$

$$x_2 = x_1 - \frac{f(x_1)}{f'(x_1)} = 2.8 - \frac{f(2.8)}{f'(2.8)} = 2.8 - \frac{0.232}{7.72} \approx 2.769948187$$

$$x_3 = x_2 - \frac{f(x_2)}{f'(x_2)} \approx 2.769292663$$

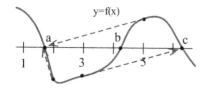

y=f(x)

4. The margin figure shows the first iteration of Newton's Method for $x_0 = 2$, 3 and 5: If $x_0 = 2$, the iterates approach the root at a; if $x_0 = 3$, they approach the root at c; and if $x_0 = 5$, they approach the root at a.

5. $f(x) = x^{\frac{1}{3}} \Rightarrow f'(x) = \frac{1}{3}x^{-\frac{2}{3}}$. If $x_0 = 1$, then:

$$x_1 = 1 - \frac{f(1)}{f'(1)} = 1 - \frac{1}{\frac{1}{3}} = 1 - 3 = -2$$

$$x_2 = -2 - \frac{f(-2)}{f'(-2)} = -2 - \frac{(-2)^{\frac{1}{3}}}{\frac{1}{3}(-2)^{-\frac{2}{3}}} = -2 - \frac{-2}{\frac{1}{3}} = 4$$

$$x_3 = 4 - \frac{f(4)}{f'(4)} = 4 - \frac{(4)^{\frac{1}{3}}}{\frac{1}{3}(4)^{-\frac{2}{3}}} = 4 - \frac{4}{\frac{1}{3}} = -8$$

and so on. If $x_0 = -3$, then:

$$x_1 = -3 - \frac{f(-3)}{f'(-3)} = -3 - \frac{(-3)^{\frac{1}{3}}}{\frac{1}{3}(-3)^{-\frac{2}{3}}} = -3 + 9 = 6$$

$$x_2 = 6 - \frac{f(6)}{f'(6)} = 6 - \frac{(6)^{\frac{1}{3}}}{\frac{1}{3}(6)^{-\frac{2}{3}}} = 6 - \frac{6}{\frac{1}{3}} = -12$$

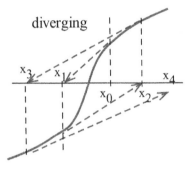
diverging

The graph of $f(x) = \sqrt[3]{x}$ has a shape similar to the margin figure and the behavior of the iterates is similar to the pattern shown in that figure. Unless $x_0 = 0$ (the only root of f) the iterates alternate in sign and double in magnitude with each iteration: they get progressively farther from the root with each iteration.

6. If $x_0 = 0.997$, then $x_1 \approx -0.003$, $x_2 \approx 166.4$, $x_3 \approx 83.2$, $x_4 \approx 41.6$. If $x_0 = 1.02$, then $x_1 \approx 0.0198$, $x_2 \approx -25.2376$, $x_3 \approx -12.6$ and $x_4 \approx -6.26$.

2.8   Linear Approximation and Differentials

Newton's method used tangent lines to "point toward" a root of a function. In this section we examine and use another geometric characteristic of tangent lines:

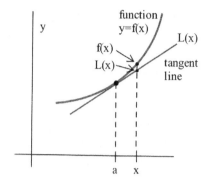

> If f is differentiable at a, c is close to a
>   and $y = L(x)$ is the line tangent to $f(x)$ at $x = a$
> then $L(c)$ is close to $f(c)$.

We can use this idea to approximate the values of some commonly used functions and to predict the "error" or uncertainty in a computation if we know the "error" or uncertainty in our original data. At the end of this section, we will define a related concept called the **differential** of a function.

Linear Approximation

Because this section uses tangent lines extensively, it is worthwhile to recall how we find the equation of the line tangent to $f(x)$ where $x = a$: the tangent line goes through the point $(a, f(a))$ and has slope $f'(a)$ so, using the point-slope form $y - y_0 = m(x - x_0)$ for linear equations, we have $y - f(a) = f'(a) \cdot (x - a) \Rightarrow y = f(a) + f'(a) \cdot (x - a)$.

> If f is differentiable at $x = a$
> then   an equation of the line L tangent to f at $x = a$ is:
> $$L(x) = f(a) + f'(a) \cdot (x - a)$$

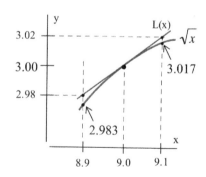

Example 1. Find a formula for $L(x)$, the linear function tangent to the graph of $f(x) = \sqrt{x}$ at the point $(9, 3)$. Evaluate $L(9.1)$ and $L(8.88)$ to approximate $\sqrt{9.1}$ and $\sqrt{8.88}$.

Solution. $f(x) = \sqrt{x} = x^{\frac{1}{2}} \Rightarrow f'(x) = \frac{1}{2}x^{-\frac{1}{2}} = \frac{1}{2\sqrt{x}}$ so $f(9) = 3$ and $f'(9) = \frac{1}{2\sqrt{9}} = \frac{1}{6}$. Thus:

$$L(x) = f(9) + f'(9) \cdot (x - 9) = 3 + \frac{1}{6}(x - 9)$$

If x is close to 9, then the value of $L(x)$ should be a good approximation of the value of x. The number 9.1 is close to 9 so $\sqrt{9.1} = f(9.1) \approx L(9.1) = 3 + \frac{1}{6}(9.1 - 9) \approx 3.016666$. Similarly, $\sqrt{8.88} = f(8.88) \approx L(8.88) = 3 + \frac{1}{6}(8.88 - 9) = 2.98$. In fact, $\sqrt{9.1} \approx 3.016621$, so our estimate using $L(9.1)$ is within 0.000045 of the exact answer; $\sqrt{8.88} \approx 2.979933$ (accurate to 6 decimal places) and our estimate is within 0.00007 of the exact answer.   ◀

In each case in the previous example, we got a good estimate of a square root with very little work. The graph in the margin indicates the graph of the tangent line $y = L(x)$ lies slightly above the graph of $y = f(x)$; consequently (as we observed), each estimate is slightly larger than the exact value.

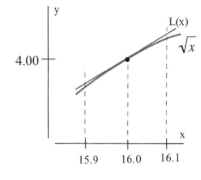

Practice 1. Find a formula for $L(x)$, the linear function tangent to the graph of $f(x) = \sqrt{x}$ at the point $(16,4)$. Evaluate $L(16.1)$ and $L(15.92)$ to approximate $\sqrt{16.1}$ and $\sqrt{15.92}$. Are your approximations using L larger or smaller than the exact values of the square roots?

Practice 2. Find a formula for $L(x)$, the linear function tangent to the graph of $f(x) = x^3$ at the point $(1,1)$ and use $L(x)$ to approximate $(1.02)^3$ and $(0.97)^3$. Do you think your approximations using L are larger or smaller than the exact values?

The process we have used to approximate square roots and cubics can be used to approximate values of any differentiable function, and the main result about the linear approximation follows from the two statements in the boxes. Putting these two statements together, we have the process for Linear Approximation.

Linear Approximation Process:

If f is differentiable at a and $L(x) = f(a) + f'(a) \cdot (x - a)$

then (geometrically) the graph of $L(x)$ is close to the graph of $f(x)$ when x is close to a

and (algebraically) the values of the $L(x)$ approximate the values of $f(x)$ when x is close to a:
$$f(x) \approx L(x) = f(a) + f'(a) \cdot (x - a)$$

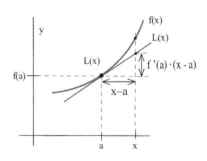

Sometimes we replace "$x - a$" with "Δx" in the last equation, and the statement becomes $f(x) \approx f(a) + f'(a) \cdot \Delta x$.

Example 2. Use the linear approximation process to approximate the value of $e^{0.1}$.

Solution. $f(x) = e^x \Rightarrow f'(x) = e^x$ so we need to pick a value a near $x = 0.1$ for which we know the exact value of $f(a) = e^a$ and $f'(a) = e^a$: $a = 0$ is an obvious choice. Then:

$$e^{0.1} = f(0.1) \approx L(0.1) = f(0) + f'(0) \cdot (0.1 - 0)$$
$$= e^0 + e^0 \cdot (0.1) = 1 + 1 \cdot (0.1) = 1.1$$

You can use your calculator to verify that this approximation is within 0.0052 of the exact value of $e^{0.1}$.   ◄

Practice 3. Approximate the value of $(1.06)^4$, the amount $1 becomes after 4 years in a bank account paying 6% interest compounded annually. (Take $f(x) = x^4$ and $a = 1$.)

Practice 4. Use the linear approximation process and the values in the table below to estimate the value of f when $x = 1.1$, 1.23 and 1.38.

x	$f(x)$	$f'(x)$
1.0	0.7854	0.5
1.2	0.8761	0.4098
1.4	0.9505	0.3378

We can approximate **functions** as well as numbers (specific values of those functions).

Example 3. Find a linear approximation formula $L(x)$ for $\sqrt{1+x}$ when x is small. Use your result to approximate $\sqrt{1.1}$ and $\sqrt{0.96}$.

Solution. $f(x) = \sqrt{1+x} = (1+x)^{\frac{1}{2}} \Rightarrow f'(x) = \frac{1}{2}(1+x)^{-\frac{1}{2}} = \frac{1}{2\sqrt{1+x}}$, so because "$x$ is small," we know that x is close to 0 and we can pick $a = 0$. Then $f(a) = f(0) = 1$ and $f'(a) = f'(0) = \frac{1}{2}$ so

$$\sqrt{1+x} \approx L(x) = f(0) + f'(0) \cdot (x - 0) = 1 + \frac{1}{2}x = 1 + \frac{x}{2}$$

Taking $x = 0.1$, $\sqrt{1.1} = \sqrt{1+0.1} \approx 1 + \frac{0.1}{2} = 1.05$; taking $x = -0.04$, $\sqrt{0.96} = \sqrt{1+(-.04)} \approx 1 + \frac{-0.04}{2} = 0.98$. Use your calculator to determine by how much each estimate differs from the true value. ◀

Applications of Linear Approximation to Measurement "Error"

Most scientific experiments use instruments to take measurements, but these instruments are not perfect, and the measurements we get from them are only accurate up to a certain level of precision. If we know this level of accuracy of our instruments and measurements, we can use the idea of linear approximation to estimate the level of accuracy of results we calculate from our measurements.

If we measure the side x of a square to be 8 inches, then we would of course calculate its area to be $8^2 = 64$ square inches. Suppose, as would reasonable with a real measurement, that our measuring instrument could only measure or be read to the nearest 0.05 inches. Then our measurement of 8 inches would really mean some number between $8 - 0.05 = 7.95$ inches and $8 + 0.05 = 8.05$ inches, so the true area of the square would be between $7.95^2 = 63.2025$ and $8.05^2 = 64.8025$ square inches. Our possible "error" or "uncertainty," because of the limitations of the instrument, could be as much as $64.8025 - 64 = 0.8025$ square inches, so we could report the area of the square to be 64 ± 0.8025

square inches. We can also use the linear approximation method to estimate the "error" or uncertainty of the area.

For a square with side x, the area is $A(x) = x^2$ and $A'(x) = 2x$. If Δx represents the "error" or uncertainty of our measurement of the side then, using the linear approximation technique for $A(x)$, $A(x) \approx A(a) + A'(a) \cdot \Delta x$ so the uncertainty of our calculated area is $A(x) - A(a) \approx A'(a) \cdot \Delta x$. In this example, $a = 8$ inches and $\Delta x = 0.05$ inches, so $A(8.05) \approx A(8) + A'(8) \cdot (0.05) = 64 + 2(8) \cdot (0.05) = 64.8$ square inches, and the uncertainty in our calculated area is approximately $A(8 + 0.05) - A(8) \approx A'(8) \cdot \Delta x = 2(8 \text{ inches})(0.05 \text{ inches}) = 0.8$ square inches. (Compare this approximation of the biggest possible error with the exact answer of 0.8025 square inches computed previously.) This process can be summarized as:

For a function as simple as the area of a square, this linear approximation method really isn't needed, but it serves as a useful and easily understood illustration of the technique.

Linear Approximation Error:

If   the value of the x-variable is measured to be $x = a$ with
        a maximum "error" of Δx units
then Δf, the "error" in estimating $f(x)$, is:
$$\Delta f = f(x) - f(a) \approx f'(a) \cdot \Delta x.$$

Practice 5. If we measure the side of a cube to be 4 cm with an uncertainty of 0.1 cm, what is the volume of the cube and the uncertainty of our calculation of the volume? (Use linear approximation.)

Example 4. We are using a tracking telescope to follow a small rocket. Suppose we are 3,000 meters from the launch point of the rocket, and, 2 seconds after the launch we measure the angle of the inclination of the rocket to be 64° with a possible "error" of 2°. How high is the rocket and what is the possible error in this calculated height?

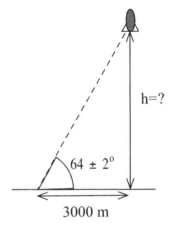

Solution. Our measured angle is $x = 1.1170$ radians with $\Delta x = 0.0349$ radians (all trigonometric work should be in radians), and the height of the rocket at an angle x is $f(x) = 3000 \tan(x)$ so $f(1.1170) \approx 6151$ m. Our uncertainty in the height is $\Delta f \approx f'(x) \cdot \Delta x \approx 3000 \cdot \sec^2(x) \cdot \Delta x = 3000 \sec^2(1.1170) \cdot 0.0349 \approx 545$ m. If our measured angle of 64° can be in error by as much as 2°, then our calculated height of 6,151 m can be in error by as much as 545 m. The height is 6151 ± 545 meters.   ◄

Practice 6. Suppose we measured the angle of inclination in the previous Example to be $43° \pm 1°$. Estimate the height of the rocket in the form "height ± error."

In some scientific and engineering applications, the calculated **result** must be within some given specification. You might need to determine how accurate the initial measurements must be in order to guarantee the final calculation is within that specification. Added precision usually costs time and money, so it is important to choose a measuring instrument good enough for the job but which is not too expensive.

Example 5. Your company produces ball bearings (small metal spheres) with a volume of 10 cm^3 and the volume must be accurate to within 0.1 cm^3. What radius should the bearings have—and what error can you tolerate in the radius measurement to meet the accuracy specification for the volume?

Solution. We want $V = 10$ and we know that the volume of a sphere is $V = \frac{4}{3}\pi r^3$, so solve $10 = \frac{4}{3}\pi r^3$ for r to get $r = 1.3365$ cm. $V(r) = \frac{4}{3}\pi r^3 \Rightarrow V'(r) = 4\pi r^2$ so $\Delta V \approx V'(r) \cdot \Delta r$. In this case we know that $\Delta V = 0.1$ cm^3 and we have calculated $r = 1.3365$ cm, so 0.1 cm$^3 = V'(1.3365 \text{ cm}) \cdot \Delta r = (22.45 \text{ cm}^2) \cdot \Delta r$. Solving for Δr, we get $\Delta r \approx 0.0045$ cm. To meet the specification for allowable error in volume, we must allow the radius to vary no more than 0.0045 cm. If we instead measure the diameter of the sphere, then we want the diameter to be $d = 2r = 2(1.3365 \pm 0.0045) = 2.673 \pm 0.009$ cm. ◀

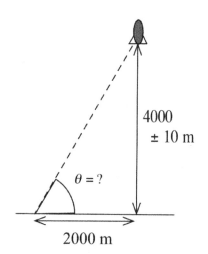

4000
± 10 m

$\theta = ?$

2000 m

Practice 7. You want to determine the height of a rocket to within 10 meters when it is 4,000 meters high (see margin figure). How accurate must your angle of measurement be? (Do your calculations in radians.)

Relative Error and Percentage Error

The "error" we've been examining is called the **absolute error** to distinguish it from two other terms, the **relative error** and the **percentage error**, which compare the absolute error with the magnitude of the number being measured. An "error" of 6 inches in measuring the Earth's circumference would be extremely small, but a 6-inch error in measuring your head for a hat would result in a very bad fit.

Definitions:

The **Relative** Error of f is $\dfrac{\text{error of } f}{\text{value of } f} = \dfrac{\Delta f}{f}$

The **Percentage** Error of f is $\dfrac{\Delta f}{f} \cdot 100\%$.

Example 6. If the relative error in the calculation of the area of a circle must be less than 0.4, then what relative error can we tolerate in the measurement of the radius?

Solution. $A(r) = \pi r^2 \Rightarrow A'(r) = 2\pi r$ and $\Delta A \approx A'(r) \cdot \Delta r = 2\pi r \Delta r$. The Relative Error of A is:

$$\frac{\Delta A}{A} \approx \frac{2\pi r \Delta r}{\pi r^2} = 2\frac{\Delta r}{r}$$

We can guarantee that the Relative Error of A, $\dfrac{\Delta A}{A}$, is less than 0.4 if the Relative Error of r, $\dfrac{\Delta r}{r} = \dfrac{1}{2}\dfrac{\Delta A}{A}$, is less than $\frac{1}{2}(0.4) = 0.2$.   ◀

Practice 8. If you can measure the side of a cube with a percentage error less than 3%, then what will the percentage error for your calculation of the surface area of the cube be?

The Differential of f

As shown in the margin, the change in value of the function f near the point $(x, f(x))$ is $\Delta f = f(x + \Delta x) - f(x)$ and the change along the tangent line is $f'(x) \cdot \Delta x$. If Δx is small, then we have used the approximation that $\Delta f \approx f'(x) \cdot \Delta x$. This leads to the definition of a new quantity, df, called the **differential** of f.

Definition:

The **differential** of f is $df = f'(x) \cdot dx$ where dx is any real number.

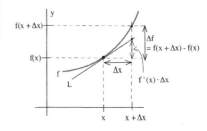

The differential of f represents the change in f, as x changes from x to $x + dx$, along the tangent line to the graph of f at the point $(x, f(x))$. If we take dx to be the number Δx, then the differential is an approximation of Δf: $\Delta f \approx f'(x) \cdot \Delta x = f'(x) \cdot dx = df$.

Example 7. Determine the differential for the functions $f(x) = x^3 - 7x$, $g(x) = \sin(x)$ and $h(r) = \pi r^2$.

Solution. $df = f'(x) \cdot dx = (3x^2 - 7)\,dx$, $dg = g'(x) \cdot dx = \cos(x)\,dx$, and $dh = h'(r)\,dr = 2\pi r\,dr$.   ◀

While we will do very little with differentials for a while, we will use them extensively in integral calculus.

Practice 9. Determine the differentials of $f(x) = \ln(x)$, $u = \sqrt{1 - 3x}$ and $r = 3\cos(\theta)$.

The Linear Approximation "Error" $|f(x) - L(x)|$

An approximation is most valuable if we also have have some measure of the size of the "error," the distance between the approximate value and the value being approximated. Typically, we will not know the exact value of the error (why not?), but it is useful to know an upper bound for the error. For example, if one scale gives the weight of a gold pendant as 10.64 grams with an error less than 0.3 grams (10.64 ± 0.3 grams) and another scale gives the weight of the same pendant as 10.53 grams with an error less than 0.02 grams (10.53 ± 0.02 grams), then we can have more faith in the second approximate weight because of the smaller "error" guarantee. Before finding a guarantee on the size of the error of the linear approximation process, we will check how well the linear approximation process approximates values of some functions we can compute exactly. Then we will prove one bound on the possible error and state a somewhat stronger bound.

Example 8. Given the function $f(x) = x^2$, evaluate the expressions $f(2 + \Delta x)$, $L(2 + \Delta x)$ and $|f(2 + \Delta x) - L(2 + \Delta x)|$ for $\Delta x = 0.1, 0.05, 0.01, 0.001$ and for a general value of Δx.

Solution. $f(2 + \Delta x) = (2 + \Delta x)^2 = 2^2 + 4\Delta x + (\Delta x)^2$ and $L(2 + \Delta x) = f(2) + f'(2) \cdot \Delta x = 2^2 + 4 \cdot \Delta x$. Then:

Δx	$f(2 + \Delta x)$	$L(2 + \Delta x)$	$|f(2 + \Delta x) - L(2 + \Delta x)|$
0.1	4.41	4.4	0.01
0.05	4.2025	4.2	0.0025
0.01	4.0401	4.04	0.0001
0.001	4.004001	4.004	0.000001

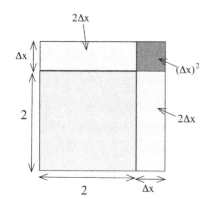

Cutting the value of Δx in half makes the error one fourth as large. Cutting Δx to $\frac{1}{10}$ as large makes the error $\frac{1}{100}$ as large. In general:

$$|f(2 + \Delta x) - L(2 + \Delta x)| = \left| \left(2^2 + 4 \cdot \Delta x + (\Delta x)^2 \right) - \left(2^2 + 4 \cdot \Delta x \right) \right|$$
$$= (\Delta x)^2$$

This function and error also have a nice geometric interpretation (see margin): $f(x) = x^2$ is the area of a square of side x so $f(2 + \Delta x)$ is the area of a square of side $2 + \Delta x$, and that area is the sum of the pieces with areas 2^2, $2 \cdot \Delta x$, $2 \cdot \Delta x$ and $(\Delta x)^2$. The linear approximation $L(2 + \Delta x) = 2^2 + 4 \cdot \Delta x$ to the area of the square includes the three largest pieces, 2^2, $2 \cdot \Delta x$ and $2 \cdot \Delta x$, but omits the small square with area $(\Delta x)^2$ so the approximation is in error by the amount $(\Delta x)^2$.   ◄

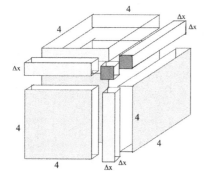

Practice 10. Given $f(x) = x^3$, evaluate $f(4 + \Delta x)$, $L(4 + \Delta x)$ and $|f(4 + \Delta x) - L(4 + \Delta x)|$ for $\Delta x = 0.1, 0.05, 0.01, 0.001$ and for a general value of Δx. Use the margin figure to give a geometric interpretation of $f(4 + \Delta x)$, $L(4 + \Delta x)$ and $|f(4 + \Delta x) - L(4 + \Delta x)|$.

In the previous Example and previous Practice problem, the error $|f(a + \Delta x) - L(a + \Delta x)|$ was very small, proportional to $(\Delta x)^2$, when Δx was small. In general, this error approaches 0 as $\Delta x \to 0$.

Theorem:

If $f(x)$ is differentiable at a

and $L(a + \Delta x) = f(a) + f'(a) \cdot \Delta x$

then $\displaystyle\lim_{\Delta x \to 0} |f(a + \Delta x) - L(a + \Delta x)| = 0$

and $\displaystyle\lim_{\Delta x \to 0} \frac{|f(a + \Delta x) - L(a + \Delta x)|}{\Delta x} = 0.$

Proof. First rewrite the quantity inside the absolute value as:

$$f(a + \Delta x) - L(a + \Delta x) = f(a + \Delta x) - f(a) - f'(a) \cdot \Delta x$$

$$= \left[\frac{f(a + \Delta x) - f(a)}{\Delta x} - f'(a) \right] \cdot \Delta x$$

But f is differentiable at $x = a$ so $\displaystyle\lim_{\Delta x \to 0} \frac{f(a + \Delta x) - f(a)}{\Delta x} = f'(a),$

which we can rewrite as $\displaystyle\lim_{\Delta x \to 0} \left[\frac{f(a + \Delta x) - f(a)}{\Delta x} - f'(a) \right] = 0.$ Thus:

$$\lim_{\Delta x \to 0} [f(a + \Delta x) - L(a + \Delta x)] = \lim_{\Delta x \to 0} \left[\frac{f(a + \Delta x) - f(a)}{\Delta x} - f'(a) \right] \cdot \lim_{\Delta x \to 0} \Delta x = 0 \cdot 0 = 0$$

Not only does the difference $f(a + \Delta x) - L(a + \Delta x)$ approach 0, but this difference approaches 0 so fast that we can divide it by Δx, another quantity approaching 0, and the quotient still approaches 0.   □

In the next chapter we will be able to prove that the error of the linear approximation process is in fact proportional to $(\Delta x)^2$. For now, we just state the result.

Theorem:

If f is differentiable at a

and $|f''(x)| \leq M$ for all x between a and $a + \Delta x$

then $|\text{"error"}| = |f(a + \Delta x) - L(a + \Delta x)| \leq \frac{1}{2} M \cdot (\Delta x)^2.$

2.8 Problems

1. The figure in the margin shows the tangent line to a function g at the point $(2, 2)$ and a line segment Δx units long.

   (a) On the figure, label the locations of

      i. $2 + \Delta x$ on the x-axis

      ii. the point $(2 + \Delta x, g(2 + \Delta x))$

      iii. the point $(2 + \Delta x, g(2) + g'(2) \cdot \Delta x)$

   (b) How large is the "error," $(g(2) + g'(2) \cdot \Delta x) - (g(2 + \Delta x))$?

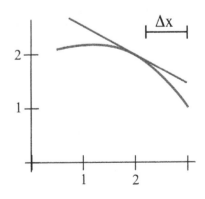

2. In the figure below, is the linear approximation $L(a + \Delta x)$ larger or smaller than the value of $f(a + \Delta x)$ when:

(a) $a = 1$ and $\Delta x = 0.2$?

(b) $a = 2$ and $\Delta x = -0.1$?

(c) $a = 3$ and $\Delta x = 0.1$?

(d) $a = 4$ and $\Delta x = 0.2$?

(e) $a = 4$ and $\Delta x = -0.2$?

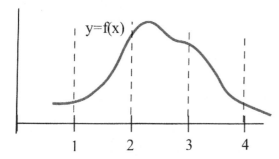

In Problems 3–4, find a formula for the linear function $L(x)$ tangent to the given function f at the given point $(a, f(a))$. Use the value $L(a + \Delta x)$ to approximate the value of $f(a + \Delta x)$.

3. (a) $f(x) = \sqrt{x}$, $a = 4$, $\Delta x = 0.2$

(b) $f(x) = \sqrt{x}$, $a = 81$, $\Delta x = -1$

(c) $f(x) = \sin(x)$, $a = 0$, $\Delta x = 0.3$

4. (a) $f(x) = \ln(x)$, $a = 1$, $\Delta x = 0.3$

(b) $f(x) = e^x$, $a = 0$, $\Delta x = 0.1$

(c) $f(x) = x^5$, $a = 1$, $\Delta x = 0.03$

5. Show that $(1 + x)^n \approx 1 + nx$ if x is "close to" 0. (Suggestion: Put $f(x) = (1 + x)^n$ and $a = 0$ and then replace Δx with x.)

In 6–7, use the linear approximation process to obtain each formula for x "close to" 0.

6. (a) $(1 - x)^n \approx 1 - nx$

(b) $\sin(x) \approx x$

(c) $e^x \approx 1 + x$

7. (a) $\ln(1 + x) \approx x$

(b) $\cos(x) \approx 1$

(c) $\tan(x) \approx x$

(d) $\sin\left(\frac{\pi}{2} + x\right) \approx 1$

8. The height of a triangle is exactly 4 inches, and the base is measured to be 7 ± 0.5 inches (see figure below). Shade a part of the figure that represents the "error" in the calculation of the area of the triangle.

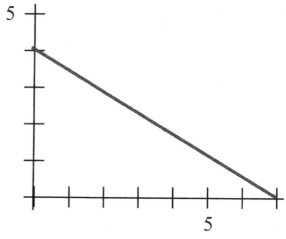

9. A rectangle has one side on the x-axis, one side on the y-axis and a corner on the graph of $y = x^2 + 1$ (see figure below).

(a) Use Linear Approximation of the area formula to estimate the increase in the area of the rectangle if the base grows from 2 to 2.3 inches.

(b) Calculate exactly the increase in the area of the rectangle as the base grows from 2 to 2.3 inches.

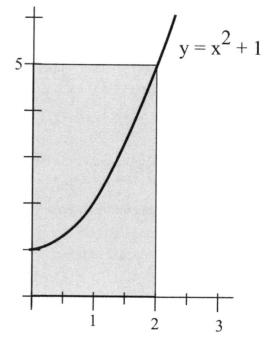

10. You know that you can measure the diameter of a circle to within 0.3 cm of the exact value.

    (a) How large is the "error" in the calculated area of a circle with a measured diameter of 7.4 cm?

    (b) How large is the "error" with a measured diameter of 13.6 cm?

    (c) How large is the percentage error in the calculated area with a measured diameter of d?

11. You are minting gold coins that must have a volume of 47.3 ± 0.1 cm^3. If you can manufacture the coins to be exactly 2 cm high, how much variation can you allow for the radius?

12. If F is the fraction of carbon-14 remaining in a plant sample Y years after it died, then $Y = 5700\ln(0.5) \cdot \ln(F)$.

    (a) Estimate the age of a plant sample in which $83\pm2\%$ (0.83 ± 0.02) of the carbon-14 remains.

    (b) Estimate the age of a plant sample in which $13\pm2\%$ (0.13 ± 0.02) of the carbon-14 remains.

13. Your company is making dice (cubes) and specifications require that their volume be 87 ± 2 cm^3. How long should each side be and how much variation can be allowed?

14. If the specifications require a cube with a surface area of 43 ± 0.2 cm^2, how long should each side be and how much variation can be allowed in order to meet the specifications?

15. The period P, in seconds, for a pendulum to make one complete swing and return to the release point is $P = 2\pi\sqrt{\dfrac{L}{g}}$ where L is the length of the pendulum in feet and g is 32 feet/sec^2.

    (a) If $L = 2$ feet, what is the period?

    (b) If $P = 1$ second, how long is the pendulum?

    (c) Estimate the change in P if L increases from 2 feet to 2.1 feet.

    (d) The length of a 24-foot pendulum is increasing 2 inches per hour. Is the period getting longer or shorter? How fast is the period changing?

16. A ball thrown at an angle θ (with the horizontal) with an initial velocity v will land $\dfrac{v^2}{g} \cdot \sin(2\theta)$ feet from the thrower.

    (a) How far away will the ball land if $\theta = \frac{\pi}{4}$ and $v = 80$ feet/second?

    (b) Which will result in a greater change in the distance: a 5% error in the angle θ or a 5% error in the initial velocity v?

17. For the function graphed below, estimate the value of df when

    (a) $x = 2$ and $dx = 1$

    (b) $x = 4$ and $dx = -1$

    (c) $x = 3$ and $dx = 2$

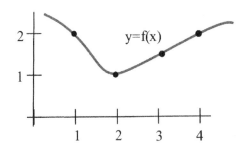

18. For the function graphed below, estimate the value of df when

    (a) $x = 1$ and $dx = 2$

    (b) $x = 2$ and $dx = -1$

    (c) $x = 3$ and $dx = 1$

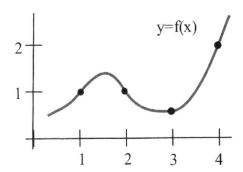

19. Calculate the differentials df for the following functions:

    (a) $f(x) = x^2 - 3x$

    (b) $f(x) = e^x$

    (c) $f(x) = \sin(5x)$

    (d) $f(x) = x^3 + 2x$ with $x = 1$ and $dx = 0.2$

    (e) $f(x) = \ln(x)$ with $x = e$ and $dx = -0.1$

    (f) $f(x) = \sqrt{2x + 5}$ with $x = 22$ and $dx = 3$.

2.8 Practice Answers

1. $f(x) = x^{\frac{1}{2}} \Rightarrow f'(x) = \frac{1}{2\sqrt{x}}$. At the point $(16, 4)$ on the graph of f, the slope of the tangent line is $f'(16) = \frac{1}{2\sqrt{16}} = \frac{1}{8}$. An equation of the tangent line is $y - 4 = \frac{1}{8}(x - 16)$ or $y = \frac{1}{8}x + 2$: $L(x) = \frac{1}{8}x + 2$. So:

$$\sqrt{16.1} \approx L(16.1) = \frac{1}{8}(16.1) + 2 = 4.0125$$
$$\sqrt{15.92} \approx L(15.92) = \frac{1}{8}(15.92) + 2 = 3.99$$

2. $f(x) = x^3 \Rightarrow f'(x) = 3x^2$. At $(1, 1)$, the slope of the tangent line is $f'(1) = 3$. An equation of the tangent line is $y - 1 = 3(x - 1)$ or $y = 3x - 2$: $L(x) = 3x - 2$. So:

$$(1.02)^3 \approx L(1.02) = 3(1.02) - 2 = 1.06$$
$$(0.97)^3 \approx L(0.97) = 3(0.97) - 2 = 0.91$$

3. $f(x) = x^4 \Rightarrow f'(x) = 4x^3$. Taking $a = 1$ and $\Delta x = 0.06$:

$$(1.06)^4 = f(1.06) \approx L(1.06) = f(1) + f'(1) \cdot (0.06)$$
$$= 1^4 + 4(1^3)(0.06) = 1.24$$

4. Using values given in the table:

$$f(1.1) \approx f(1) + f'(1) \cdot (0.1)$$
$$= 0.7854 + (0.5)(0.1) = 0.8354$$
$$f(1.23) \approx f(1.2) + f'(1.2) \cdot (0.03)$$
$$= 0.8761 + (0.4098)(0.03) = 0.888394$$
$$f(1.38) \approx f(1.4) + f'(1.4) \cdot (-0.02)$$
$$= 0.9505 + (0.3378)(-0.02) = 0.943744$$

5. $f(x) = x^3 \Rightarrow f'(x) = 3x^2$ so $f(4) = 4^3 = 64$ cm^3 and the "error" is:

$$\Delta f \approx f'(x) \cdot \Delta x = 3x^2 \cdot \Delta x$$

When $x = 4$ and $\Delta x = 0.1$, $\Delta f \approx 3(4)^2(0.1) = 4.8$ cm^3.

6. $43° \pm 1°$ is equivalent to 0.75049 ± 0.01745 radians, so with $f(x) = 3000 \tan(x)$ we have $f(0.75049) = 3000 \tan(0.75049) \approx 2797.5$ m and $f'(x) = 3000 \sec^2(x)$. So:

$$\Delta f \approx f'(x) \cdot \Delta x = 3000 \sec^2(x) \cdot \Delta x$$
$$= 3000 \sec^2(0.75049) \cdot (0.01745) = 97.9 \text{ m}$$

The height of the rocket is 2797.5 ± 97.9 m.

7. $f(\theta) = 2000\tan(\theta) \Rightarrow f'(\theta) = 2000\sec^2(\theta)$ and we know $f(\theta) = 4000$, so:

$$4000 = 2000\tan(\theta) \Rightarrow \tan(\theta) = 2 \Rightarrow \theta \approx 1.10715 \text{ (radians)}$$

and thus $f'(1.10715) = 2000\sec^2(1.10715) \approx 10000$. Finally, the "error" is given by $\Delta f \approx f'(\theta)\cdot\Delta\theta$ so:

$$10 \approx 10000\cdot\Delta\theta \Rightarrow \Delta\theta \approx \frac{10}{10000} = 0.001 \text{ (radians)} \approx 0.057°$$

8. $A(r) = 6r^2 \Rightarrow A'(r) = 12r \Rightarrow \Delta A \approx A'(r)\cdot\Delta r = 12r\cdot\Delta r$ and we also know that $\frac{\Delta r}{r} < 0.03$, so the percentage error is:

$$\frac{\Delta A}{A}\cdot 100\% = \frac{12r\cdot\Delta r}{6r^2}\cdot 100\% = \frac{2\Delta r}{r}\cdot 100\% < 200(0.03)\% = 6\%$$

9. Computing differentials:

$$df = f'(x)\cdot dx = \frac{1}{x}\,dx$$
$$du = \frac{du}{dx}\cdot dx = \frac{-3}{2\sqrt{1-3x}}\,dx$$
$$dr = \frac{dr}{d\theta}\,d\theta = -3\sin(\theta)\,d\theta$$

10. $f(x) = x^3 \Rightarrow f'(x) = 3x^2$ so:

$$L(4+\Delta x) = f(4) + f'(4)\Delta x = 4^3 + 3(4)^2\Delta x = 64 + 48\Delta x$$

Evaluating the various quantities at the indicated points:

Δx	$f(4+\Delta x)$	$L(4+\Delta x)$	$|f(4+\Delta x) - L(4+\Delta x)|$
0.1	68.921	68.8	0.121
0.05	66.430125	66.4	0.030125
0.01	64.481201	64.48	0.001201
0.001	64.048012	64.048	0.000012

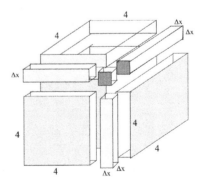

$f(4+\Delta x)$ is the actual volume of the cube with side length $4+\Delta x$. $L(4+\Delta x)$ is the volume of the cube with side length 4 ($V = 64$) plus the volume of the three "slabs" ($V = 3\cdot 4^2\cdot\Delta x$).

$|f(4+\Delta x) - L(4+\Delta x)|$ is the volume of the "leftover" pieces from L: the three "rods" ($V = 3\cdot 4\cdot(\Delta x)^2$) and the tiny cube ($V = (\Delta x)^3$).

2.9 Implicit and Logarithmic Differentiation

This short section presents two more differentiation techniques, both more specialized than the ones we have already seen — and consequently used on a smaller class of functions. For some functions, however, one of these techniques may be the only method that works. The idea of each method is straightforward, but actually using each of them requires that you proceed carefully and practice.

Implicit Differentiation

In our work up until now, the functions we needed to differentiate were either given **explicitly** as a function of x, such as $y = f(x) = x^2 + \sin(x)$, or it was fairly straightforward to find an explicit formula, such as solving $y^3 - 3x^2 = 5$ to get $y = \sqrt[3]{5 + 3x^2}$. Sometimes, however, we will have an equation relating x and y that is either difficult or impossible to solve explicitly for y, such as $y^2 + 2y = \sin(x) + 4$ (difficult) or $y + \sin(y) = x^3 - x$ (impossible). In each case, we can still find $y' = f'(x)$ by using **implicit differentiation**.

The key idea behind implicit differentiation is to *assume* that y is a function of x even if we cannot explicitly solve for y. This assumption does not require any work, but we need to be very careful to treat y as a function when we differentiate and to use the Chain Rule or the Power Rule for Functions.

Example 1. Assume y is a function of x and compute each derivative:

$$\text{(a) } \mathbf{D}(y^3) \qquad \text{(b) } \frac{d}{dx}\left(x^3 y^2\right) \qquad \text{(c) } (\sin(y))'$$

Solution. (a) We need the Power Rule for Functions because y is a function of x:

$$\mathbf{D}(y^3) = 3y^2 \cdot \mathbf{D}(y) = 3y^2 \cdot y'$$

(b) We need to use the Product Rule and the Chain Rule:

$$\frac{d}{dx}\left(x^3 y^2\right) = x^3 \cdot \frac{d}{dx}\left(y^2\right) + y^2 \cdot \frac{d}{dx}\left(x^3\right) = x^3 \cdot 2y \cdot \frac{dy}{dx} + y^2 \cdot 3x^2$$

(c) We just need to remember that $\mathbf{D}(\sin(u)) = \cos(u)$ and then use the Chain Rule: $(\sin(y))' = \cos(y) \cdot y'$. ◄

Practice 1. Assume that y is a function of x. Calculate:

$$\text{(a) } \mathbf{D}\left(x^2 + y^2\right) \qquad \text{(b) } \frac{d}{dx}\left(\sin(2 + 3y)\right).$$

Implicit Differentiation:

To determine y', differentiate each side of the defining equation, treating y as a function of x, and then algebraically solve for y'.

Example 2. Find the slope of the tangent line to the circle $x^2 + y^2 = 25$ at the point $(3, 4)$ with and without implicit differentiation.

Solution. **Explicitly**: We can solve $x^2 + y^2 = 25$ for y: $y = \pm\sqrt{25 - x^2}$ but because the point $(3, 4)$ is on the top half of the circle, we just need $y = \sqrt{25 - x^2}$ so:

$$\mathbf{D}(y) = \mathbf{D}\left(\sqrt{25 - x^2}\right) = \frac{1}{2}\left(25 - x^2\right)^{-\frac{1}{2}} \cdot (-2x) = \frac{-x}{\sqrt{25 - x^2}}$$

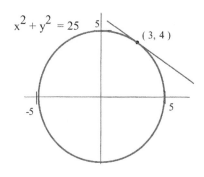

Replacing x with 3, we have $y' = \frac{-3}{\sqrt{25 - 3^2}} = -\frac{3}{4}$.

 Implicitly: We differentiate each side of the equation $x^2 + y^2 = 25$ treating y as a function of x and then solve for y':

$$\mathbf{D}\left(x^2 + y^2\right) = \mathbf{D}(25) \;\Rightarrow\; 2x + 2y \cdot y' = 0 \;\Rightarrow\; y' = \frac{-2x}{2y} = -\frac{x}{y}$$

so at the point $(3, 4)$, $y' = -\frac{3}{4}$, the same answer we found explicitly. ◀

Practice 2. Find the slope of the tangent line to $y^3 - 3x^2 = 15$ at the point $(2, 3)$ with and without implicit differentiation.

 In the previous Example and Practice problem, it was easy to explicitly solve for y, and then we could differentiate y to get y'. Because we could explicitly solve for y, we had a choice of methods for calculating y'. Sometimes, however, we cannot explicitly solve for y and the only way to determine y' is with implicit differentiation.

Example 3. Determine y' at $(0, 2)$ for $y^2 + 2y = \sin(x) + 8$.

Solution. Assuming that y is a function of x and differentiating each side of the equation, we get:

$$\mathbf{D}\left(y^2 + 2y\right) = \mathbf{D}\left(\sin(x) + 8\right) \;\Rightarrow\; 2y \cdot y' + 2y' = \cos(x)$$

$$\Rightarrow\; (2y + 2)y' = \cos(x) \;\Rightarrow\; y' = \frac{\cos(x)}{2y + 2}$$

so, at the point $(0, 2)$, $y' = \frac{\cos(0)}{2(2) + 2} = \frac{1}{6}$. ◀

We could have first solved the equation explicitly for y using the quadratic formula. Do you see how? Would that make the problem easier or harder than using implicit differentiation?

Practice 3. Determine y' at $(1, 0)$ for $y + \sin(y) = x^3 - x$.

 In practice, the equations may be rather complicated, but if you proceed carefully and step by step, implicit differentiation is not difficult. Just remember that y **must be treated as a function** so every time you differentiate a term containing a y you should use the Chain Rule and get something that has a y'. The algebra needed to solve for y' is always easy — if you differentiated correctly, the resulting equation will be a linear equation in the variable y'.

Example 4. Find an equation of the tangent line L to the "tilted" parabola graphed below at the point $(1,2)$.

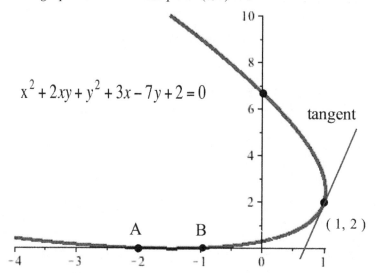

$$x^2 + 2xy + y^2 + 3x - 7y + 2 = 0$$

Solution. The line goes through the point $(1,2)$ so we need to find the slope there. Differentiating each side of the equation, we get:

$$\mathbf{D}\left(x^2 + 2xy + y^2 + 3x - 7y + 2\right) = \mathbf{D}(0)$$

which yields:

$$2x + 2x \cdot y' + 2y + 2y \cdot y' + 3 - 7y' = 0$$
$$\Rightarrow (2x + 2y - 7)y' = -2x - 2y - 3$$
$$\Rightarrow y' = \frac{-2x - 2y - 3}{2x + 2y - 7}$$

so the slope at $(1,2)$ is $m = y' = \dfrac{-2 - 4 - 3}{2 + 4 - 7} = 9$. Finally, an equation for the line is $y - 2 = 9(x - 1)$ so $y = 9x - 7$.   ◀

Practice 4. Find the points where the parabola graphed above crosses the y-axis, and find the slopes of the tangent lines at those points.

Implicit differentiation provides an alternate method for differentiating equations that can be solved explicitly for the function we want, and it is the *only* method for finding the derivative of a function that we cannot describe explicitly.

Logarithmic Differentiation

In Section 2.5 we saw that $D\left(\ln(f(x))\right) = \dfrac{f'(x)}{f(x)}$. If we simply multiply each side by $f(x)$, we have: $f'(x) = f(x) \cdot \mathbf{D}\left(\ln(f(x))\right)$. When the logarithm of a function is simpler than the function itself, it is often easier to differentiate the logarithm of f than to differentiate f itself.

Logarithmic Differentiation:

$$f'(x) = f(x) \cdot \mathbf{D}\left(\ln(f(x))\right)$$

In words: The derivative of f is f times the derivative of the natural logarithm of f. Usually it is easiest to proceed in three steps:

- Calculate $\ln\left(f(x)\right)$ and simplify.

- Calculate $\mathbf{D}\left(\ln(f(x))\right)$ and simplify

- Multiply the result in the previous step by $f(x)$.

Let's examine what happens when we use this process on an "easy" function, $f(x) = x^2$, and a "hard" one, $f(x) = 2^x$. Certainly we don't need to use logarithmic differentiation to find the derivative of $f(x) = x^2$, but sometimes it is instructive to try a new algorithm on a familiar function. Logarithmic differentiation is the easiest way to find the derivative of $f(x) = 2^x$ (if we don't remember the pattern for differentiating a^x from Section 2.5).

$f(x) = x^2$
$\ln\left(f(x)\right) = \ln(x^2) = 2 \cdot \ln(x)$
$\mathbf{D}\left(\ln\left(f(x)\right)\right) = \mathbf{D}\left(2 \cdot \ln(x)\right) = \frac{2}{x}$
$f'(x) = f(x) \cdot \mathbf{D}\left(\ln\left(f(x)\right)\right) = x^2 \cdot \frac{2}{x} = 2x$

$f(x) = 2^x$
$\ln\left(f(x)\right) = \ln(2^x) = x \cdot \ln(2)$
$\mathbf{D}\left(\ln\left(f(x)\right)\right) = \mathbf{D}\left(x \cdot \ln(2)\right) = \ln(2)$
$f'(x) = f(x) \cdot \mathbf{D}\left(\ln\left(f(x)\right)\right) = 2^x \cdot \ln(2)$

Example 5. Use the pattern $f'(x) = f(x) \cdot \mathbf{D}\left(\ln(f(x))\right)$ to find the derivative of $f(x) = (3x+7)^5 \sin(2x)$.

Solution. Apply the natural logarithm to both sides and rewrite:

$$\ln\left(f(x)\right) = \ln\left((3x+7)^5 \cdot \sin(2x)\right) = 5\ln(3x+7) + \ln\left(\sin(2x)\right)$$

so:

$$\mathbf{D}\left(\ln(f(x))\right) = \mathbf{D}\left(5\ln(3x+7) + \ln\left(\sin(2x)\right)\right)$$
$$= 5 \cdot \frac{3}{3x+7} + 2 \cdot \frac{\cos(2x)}{\sin(2x)}$$

Then:

$$f'(x) = f(x) \cdot \mathbf{D}\left(\ln(f(x))\right)$$
$$= (3x+7)^5 \sin(2x)\left(\frac{15}{3x+7} + 2 \cdot \frac{\cos(2x)}{\sin(2x)}\right)$$
$$= 15(3x+7)^4 \sin(2x) + 2(3x+7)^5 \cos(2x)$$

the same result we would obtain using the Product Rule.   ◀

Practice 5. Use logarithmic differentiation to find the derivative of $f(x) = (2x + 1)^3 (3x^2 - 4)^7 (x + 7)^4$.

We could have differentiated the functions in the previous Example and Practice problem without logarithmic differentiation. There are, however, functions for which logarithmic differentiation is the **only** method we can use. We know how to differentiate x raised to a constant power, $\mathbf{D}(x^p) = p \cdot x^{p-1}$, and a constant to a variable power, $\mathbf{D}(b^x) = b^x \ln(b)$, but the function $f(x) = x^x$ has both a variable base and a variable power, so neither differentiation rule applies. We need to use logarithmic differentiation.

Example 6. Find $\mathbf{D}(x^x)$, assuming that $x > 0$.

Solution. Apply the natural logarithm to both sides and rewrite:

$$\ln(f(x)) = \ln(x^x) = x \cdot \ln(x)$$

so:

$$\mathbf{D}(\ln(f(x))) = \mathbf{D}(x \cdot \ln(x)) = x \cdot \mathbf{D}(\ln(x)) + \ln(x) \cdot \mathbf{D}(x)$$
$$= x \cdot \frac{1}{x} + \ln(x) \cdot 1 = 1 + \ln(x)$$

Then $\mathbf{D}(x^x) = f'(x) = f(x)\,\mathbf{D}(\ln(f(x))) = x^x(1 + \ln(x))$. ◀

Practice 6. Find $\mathbf{D}\left(x^{\sin(x)}\right)$ assuming that $x > 0$.

Logarithmic differentiation is an alternate method for differentiating some functions such as products and quotients, and it is the only method we've seen for differentiating some other functions such as variable bases to variable exponents.

2.9 Problems

In Problems 1–10 find $\frac{dy}{dx}$ in two ways: (a) by differentiating implicitly and (b) by explicitly solving for y and then differentiating. Then find the value of $\frac{dy}{dx}$ at the given point using your results from both the implicit and the explicit differentiation.

1. $x^2 + y^2 = 100$, point: $(6, 8)$

2. $x^2 + 5y^2 = 45$, point: $(5, 2)$

3. $x^2 - 3xy + 7y = 5$, point: $(2, 1)$

4. $\sqrt{x} + \sqrt{y} = 5$, point: $(4, 9)$

5. $\dfrac{x^2}{9} + \dfrac{y^2}{16} = 1$, point: $(0, 4)$

6. $\dfrac{x^2}{9} + \dfrac{y^2}{16} = 1$, point: $(3, 0)$

7. $\ln(y) + 3x - 7 = 0$, point: $(2, e)$

8. $x^2 - y^2 = 16$, point: $(5, 3)$

9. $x^2 - y^2 = 16$, point: $(5, -3)$

10. $y^2 + 7x^3 - 3x = 8$, point: $(1, 2)$

11. Find the slopes of the lines tangent to the graph below at the points $(3,1)$, $(3,3)$ and $(4,2)$.

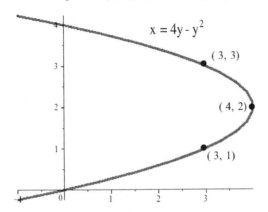

12. Find the slopes of the lines tangent to the graph in the figure above at the points where the graph crosses the y-axis.

13. Find the slopes of the lines tangent to the graph below at the points $(5,0)$, $(5,6)$ and $(-4,3)$.

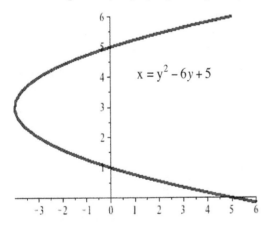

14. Find the slopes of the lines tangent to the graph in the figure above at the points where the graph crosses the y-axis.

In Problems 15–22 , find $\frac{dy}{dx}$ using implicit differentiation and then find the slope of the line tangent to the graph of the equation at the given point.

15. $y^3 - 5y = 5x^2 + 7$, point: $(1,3)$

16. $y^2 - 5xy + x^2 + 21 = 0$, point: $(2,5)$

17. $y^2 + \sin(y) = 2x - 6$, point: $(3,0)$

18. $y + 2x^2y^3 = 4x + 7$, point: $(3,1)$

19. $e^y + \sin(y) = x^2 - 3$, point: $(2,0)$

20. $\left(x^2 + y^2 + 1\right)^2 - 4x^2 = 81$, point: $(0, 2\sqrt{2})$

21. $x^{\frac{2}{3}} + y^{\frac{2}{3}} = 5$, point: $(8,1)$

22. $x + \cos(xy) = y + 3$, point: $(2,0)$

23. Find the slope of the line tangent to the ellipse shown in the figure below at the point $(1,2)$.

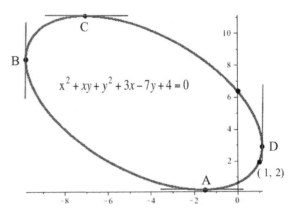

24. Find the slopes of the tangent lines at the points where the ellipse shown above crosses the y-axis.

25. Find y' for $y = Ax^2 + Bx + C$ and for the equation $x = Ay^2 + By + C$.

26. Find y' for $y = Ax^3 + B$ and for $x = Ay^3 + B$.

27. Find y' for $Ax^2 + Bxy + Cy^2 + Dx + Ey + F = 0$.

28. In Chapter 1 we assumed that the tangent line to a circle at a point was perpendicular to the radial line passing through that point and the center of the circle. Use implicit differentiation to prove that the line tangent to the circle $x^2 + y^2 = r^2$ (see below) at an arbitrary point (x,y) is perpendicular to the line passing through $(0,0)$ and (x,y).

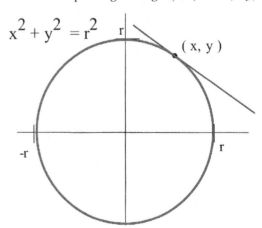

Problems 29–31 use the figure from Problems 23–24.

29. Find the coordinates of point A where the tangent line to the ellipse is horizontal.

30. Find the coordinates of point B where the tangent line to the ellipse is vertical.

31. Find the coordinates of points C and D.

In 32–40, find $\frac{dy}{dx}$ in two ways: (a) by using the "usual" differentiation patterns and (b) by using logarithmic differentiation.

32. $y = x \cdot \sin(3x)$

33. $y = (x^2 + 5)^7 (x^3 - 1)^4$

34. $y = \dfrac{\sin(3x - 1)}{x + 7}$

35. $y = x^5 \cdot (3x + 2)^4$

36. $y = 7^x$

37. $y = e^{\sin(x)}$

38. $y = \cos^7(2x + 5)$

39. $y = \sqrt{25 - x^2}$

40. $y = \dfrac{x \cdot \cos(x)}{x^2 + 1}$

In 41–46, use logarithmic differentiation to find $\frac{dy}{dx}$.

41. $y = x^{\cos(x)}$

42. $y = (\cos(x))^x$

43. $y = x^4 \cdot (x - 2)^7 \cdot \sin(3x)$

44. $y = \dfrac{\sqrt{x + 10}}{(2x + 3)^3 \cdot (5x - 1)^7}$

45. $y = (3 + \sin(x))^x$

46. $y = \sqrt{\dfrac{x^2 + 1}{x^2 - 1}}$

In 47–50, use the values in each table to calculate the values of the derivative in the last column.

47.

x	$f(x)$	$\ln(f(x))$	$\mathbf{D}(\ln(f(x)))$	$f'(x)$
1	1	0.0	1.2	
2	9	2.2	1.8	
3	64	4.2	2.1	

48.

x	$g(x)$	$\ln(g(x))$	$\mathbf{D}(\ln(g(x)))$	$g'(x)$
1	5	1.6	0.6	
2	10	2.3	0.7	
3	20	3.0	0.8	

49.

x	$f(x)$	$\ln(f(x))$	$\mathbf{D}(\ln(f(x)))$	$f'(x)$
1	5	1.6	-1	
2	2	0.7	0	
3	7	1.9	2	

50.

x	$g(x)$	$\ln(g(x))$	$\mathbf{D}(\ln(g(x)))$	$g'(x)$
2	1.4	0.3	1.2	
3	3.3	1.2	0.6	
7	13.6	2.6	0.2	

Problems 51–55 illustrate how logarithmic differentiation can be used to verify some differentiation patterns we already know (51–52, 54) and to derive some new patterns (53, 55). Assume that all of the functions are differentiable and that the function combinations are defined.

51. Use logarithmic differentiation on $f \cdot g$ to re-derive the Product Rule: $\mathbf{D}(f \cdot g) = f \cdot g' + g \cdot f'$.

52. Use logarithmic differentiation on $\dfrac{f}{g}$ to re-derive the quotient rule: $\mathbf{D}\left(\dfrac{f}{g}\right) = \dfrac{g \cdot f' - f \cdot g'}{g^2}$.

53. Use logarithmic differentiation to obtain a product rule for three functions: $\mathbf{D}(f \cdot g \cdot h) = ?$.

54. Use logarithmic differentiation on the exponential function a^x (with $a > 0$) to show that its derivative is $a^x \ln(a)$.

55. Use logarithmic differentiation to determine a pattern for the derivative of f^g: $\mathbf{D}(f^g) = ?$.

56. In Section 2.1 we proved the Power Rule $\mathbf{D}(x^n) = n \cdot x^{n-1}$ for any positive integer n.

    (a) Why does this formula hold for $n = 0$?

    (b) Use the Quotient Rule to prove that $\mathbf{D}(x^{-m}) = -m \cdot x^{-m-1}$ for any positive integer m and conclude that the Power Rule holds for all integers.

    (c) Now let $y = x^{\frac{p}{q}}$ where p and q are integers so that $y^q = x^p$. Use implicit differentiation to show that the Power Rule holds for all rational exponents. (We still have not considered the case where $y = x^a$ with a an irrational number, because we haven't actually *defined* what x^a means for a irrational. We will take care of that — and the extension of the Power Rule to all real exponents — in Chapter 7.)

2.9 Practice Answers

1. $\mathbf{D}(x^2 + y^2) = 2x + 2y \cdot y'$

   $\frac{d}{dx}(\sin(2 + 3y)) = \cos(2 + 3y) \cdot \mathbf{D}(2 + 3y) = \cos(2 + 3y) \cdot 3y'$

2. Explicitly: $y' = \frac{1}{3}\left(3x^2 + 15\right)^{-\frac{2}{3}} \mathbf{D}\left(3x^2 + 15\right) = \frac{1}{3}\left(3x^2 + 15\right)^{-\frac{2}{3}}(6x)$.

   When $(x, y) = (2, 3)$, $y' = \frac{1}{3}\left(3(2)^2 + 15\right)^{\frac{2}{3}}(6 \cdot 2) = 4(27)^{-\frac{2}{3}} = \frac{4}{9}$.

   Implicitly: $\mathbf{D}\left(y^3 - 3x^2\right) = \mathbf{D}(15) \Rightarrow 3y^2 \cdot y' - 6x = 0$ so $y' = \frac{2x}{y^2}$.

   When $(x, y) = (2, 3)$, $y' = \frac{2 \cdot 2}{3^2} = \frac{4}{9}$.

3. $\mathbf{D}(y + \sin(y)) = \mathbf{D}(x^3 - x) \Rightarrow y' + \cos(y) \cdot y' = 3x^2 - 1 \Rightarrow y' \cdot$
   $(1 + \cos(y)) = 3x^2 - 1$, so we have $y' = \frac{3x^2 - 1}{1 + \cos(y)}$. When $(x, y) =$
   $(1, 0)$, $y' = \frac{3(1)^2 - 1}{1 + \cos(0)} = 1$.

4. To find where the parabola crosses the y-axis, we can set $x = 0$ and
   solve for the values of y: $y^2 - 7y + 2 = 0 \Rightarrow y = \frac{7 \pm \sqrt{(-7)^2 - 4(1)(2)}}{2(1)} =$
   $\frac{7 \pm \sqrt{41}}{2} \approx 0.3$ and 6.7. The parabola crosses the y-axis (approximately)
   at the points $(0, 0.3)$ and $(0, 6.7)$. From Example 4, we know that
   $y' = \frac{-2x - 2y - 3}{2x + 2y - 7}$, so at the point $(0, 0.3)$, the slope is approximately $\frac{0 - 0.6 - 3}{0 + 0.6 - 7} \approx 0.56$, and at the point $(0, 6.7)$, the slope is
   approximately $\frac{0 - 13.4 - 3}{0 + 13.4 - 7} \approx -2.56$.

5. Applying the formula $f'(x) = f(x) \cdot \mathbf{D}(\ln(f(x)))$ to the function
   $f(x) = (2x + 1)^3(3x^2 - 4)^7(x + 7)^4$, we have:

   $$\ln(f(x)) = 3 \cdot \ln(2x + 1) + 7 \cdot \ln(3x^2 - 4) + 4 \cdot \ln(x + 7)$$

   so:

   $$\mathbf{D}(\ln(f(x))) = \frac{3}{2x + 1}(2) + \frac{7}{3x^2 - 4}(6x) + \frac{4}{x + 7}(1)$$

   and thus:

   $$f'(x) = f(x) \cdot \mathbf{D}(\ln(f(x))) = (2x + 1)^3(3x^2 - 4)^7(x + 7)^4 \cdot \left[\frac{6}{2x + 1} + \frac{42x}{3x^2 - 4} + \frac{4}{x + 7}\right]$$

6. Using $f'(x) = f(x) \cdot \mathbf{D}(\ln(f(x)))$ with $f(x) = x^{\sin(x)}$:

   $$\ln(f(x)) = \ln\left(x^{\sin(x)}\right) = \sin(x) \cdot \ln(x)$$

   so:

   $$\mathbf{D}(\ln(f(x))) = \mathbf{D}(\sin(x) \cdot \ln(x)) = \sin(x) \cdot \mathbf{D}(\ln(x)) + \ln(x) \cdot \mathbf{D}(\sin(x)) = \sin(x) \cdot \frac{1}{x} + \ln(x) \cdot \cos(x)$$

   and thus:

   $$f'(x) = f(x) \cdot \mathbf{D}(\ln(f(x))) = x^{\sin(x)} \cdot \left[\frac{\sin(x)}{x} + \ln(x) \cdot \cos(x)\right]$$

3

Derivatives and Graphs

In this chapter, we explore what the first and second derivatives of a function tell us about the graph of that function and apply this graphical knowledge to locate the extreme values of a function.

3.1 Finding Maximums and Minimums

In theory and applications, we often want to maximize or minimize some quantity. An engineer may want to maximize the speed of a new computer or minimize the heat produced by an appliance. A manufacturer may want to maximize profits and market share or minimize waste. A student may want to maximize a grade in calculus or minimize the hours of study needed to earn a particular grade.

Many natural objects follow minimum or maximum principles, so if we want to model natural phenomena we may need to maximize or minimize. A light ray travels along a "minimum time" path. The shape and surface texture of some animals tend to minimize or maximize heat loss. Systems reach equilibrium when their potential energy is minimized. A basic tenet of evolution is that a genetic characteristic that maximizes the reproductive success of an individual will become more common in a species.

Calculus provides tools for analyzing functions and their behavior and for finding maximums and minimums.

Methods for Finding Maximums and Minimums

We can try to find where a function f is largest or smallest by evaluating f at lots of values of x, a method that is not very efficient and may not find the exact place where f achieves its extreme value. If we try hundreds or thousands of values for x, however, then we can often find a value of f that is close to the maximum or minimum. In general, this type of exhaustive search is only practical if you have a computer do the work.

The graph of a function provides a visual way of examining lots of values of f, and it is a good method, particularly if you have a computer to do the work for you. It is still inefficient, however, as you (or a computer) still need to evaluate the function at hundreds or thousands of inputs in order to create the graph—and we still may not find the exact location of the maximum or minimum.

Calculus provides ways to drastically narrow the number of points we need to examine to find the exact locations of maximums and minimums. Instead of examining f at thousands of values of x, calculus can often guarantee that the maximum or minimum must occur at one of three or four values of x, a substantial improvement in efficiency.

A Little Terminology

Before we examine how calculus can help us find maximums and minimums, we need to carefully define these concepts.

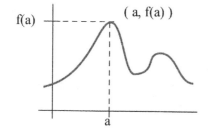

> **Definitions:**
>
> - f has a **maximum** or **global maximum** at $x = a$ if $f(a) \geq f(x)$ for all x in the domain of f.
>
> - The **maximum value** of f is then $f(a)$ and this maximum value of f **occurs at** a.
>
> - The **maximum point** on the graph of f is $(a, f(a))$.

The previous definition involves the overall biggest value a function attains on its entire domain. We are sometimes interested in how a function behaves locally rather than globally.

> **Definition:** f has a **local** or **relative maximum** at $x = a$ if $f(a) \geq f(x)$ for all x "near" a, (that is, in some open interval that contains a).

Global and local **minimums** are defined similarly by replacing the $\geq$ symbol with $\leq$ in the previous definitions.

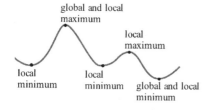

> **Definition:**
>
> f has a **global extreme** at $x = a$ if $f(a)$ is a global maximum or minimum.

See the margin figure for graphical examples of local and global extremes of a function.

You should notice that every global extreme is also a local extreme, but there are local extremes that are not global extremes. If $h(x)$ is the height of the earth above sea level at location x, then the global maximum of h is $h(\text{summit of Mt. Everest}) = 29{,}028$ feet. The local maximum of h for the United States is $h(\text{summit of Mt. McKinley}) = 20{,}320$ feet. The local minimum of h for the United States is $h(\text{Death Valley}) = -282$ feet.

Finding Maximums and Minimums of a Function

One way to narrow our search for a maximum value of a function f is to eliminate those values of x that, for some reason, cannot possibly make f maximum.

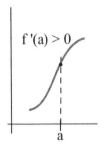

> **Theorem:**
>
> If $f'(a) > 0$ or $f'(a) < 0$
>
> then $f(a)$ is not a local maximum or minimum.

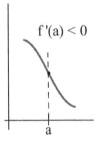

Proof. Assume that $f'(a) > 0$. By definition:

$$f'(a) = \lim_{\Delta x \to 0} \frac{f(a + \Delta x) - f(a)}{\Delta x}$$

so $f'(a) = \lim\limits_{\Delta x \to 0} \dfrac{f(a + \Delta x) - f(a)}{\Delta x} > 0$. This means that the right and left limits are both positive: $f'(a) = \lim\limits_{\Delta x \to 0^+} \dfrac{f(a + \Delta x) - f(a)}{\Delta x} > 0$ and $f'(a) = \lim\limits_{\Delta x \to 0^-} \dfrac{f(a + \Delta x) - f(a)}{\Delta x} > 0$.

Considering the right limit, we know that if we restrict $\Delta x > 0$ to be sufficiently small, we can guarantee that $\dfrac{f(a + \Delta x) - f(a)}{\Delta x} > 0$ so, multiplying each side of this last inequality by the positive number Δx, we have $f(a + \Delta x) - f(a) > 0 \Rightarrow f(a + \Delta x) > f(a)$ for all sufficiently small values of $\Delta x > 0$, so any open interval containing $x = a$ will also contain values of x with $f(x) > f(a)$. This tell us that $f(a)$ is not a maximum.

Considering the left limit, we know that if we restrict $\Delta x < 0$ to be sufficiently small, we can guarantee that $\dfrac{f(a + \Delta x) - f(a)}{\Delta x} > 0$ so, multiplying each side of this last inequality by the negative number Δx, we have $f(a + \Delta x) - f(a) < 0 \Rightarrow f(a + \Delta x) < f(a)$ for all sufficiently small values of $\Delta x < 0$, so any open interval containing $x = a$ will also contain values of x with $f(x) < f(a)$. This tell us that $f(a)$ is not a minimum.

The argument for the "$f'(a) < 0$" case is similar. $\square$

When we evaluate the derivative of a function f at a point $x = a$, there are only four possible outcomes: $f'(a) > 0$, $f'(a) < 0$, $f'(a) = 0$ or $f'(a)$ is undefined. If we are looking for extreme values of f, then we can eliminate those points at which f' is positive or negative, and only two possibilities remain: $f'(a) = 0$ or $f'(a)$ is undefined.

> **Theorem:**
>
> If f is defined on an open interval
>   and $f(a)$ is a local extreme of f
> then   either $f'(a) = 0$ or f is not differentiable at a.

Example 1. Find the local extremes of $f(x) = x^3 - 6x^2 + 9x + 2$.

Solution. An extreme value of f can occur only where $f'(x) = 0$ or where f is not differentiable; $f(x)$ is a polynomial, so it is differentiable for all values of x, and we can restrict our attention to points where $f'(x) = 0$.

$$f'(x) = 3x^2 - 12x + 9 = 3(x^2 - 4x + 3) = 3(x-1)(x-3)$$

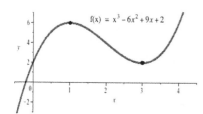

so $f'(x) = 0$ only at $x = 1$ and $x = 3$.

The only possible locations of local extremes of f are at $x = 1$ and $x = 3$. We don't know yet whether $f(1)$ or $f(3)$ is a local extreme of f, but we can be certain that no other point is a local extreme. The graph of f (see margin) shows that $(1, f(1)) = (1, 6)$ appears to be a local maximum and $(3, f(3)) = (3, 2)$ appears to be a local minimum. This function does not have a global maximum or minimum.   ◀

Practice 1. Find the local extremes of $f(x) = x^2 + 4x - 5$ and of $g(x) = 2x^3 - 12x^2 + 7$.

It is important to recognize that the two conditions "$f'(a) = 0$" or "f not differentiable at a" do not guarantee that $f(a)$ is a local maximum or minimum. They only say that $f(a)$ **might** be a local extreme or that $f(a)$ is a **candidate** for being a local extreme.

Example 2. Find all local extremes of $f(x) = x^3$.

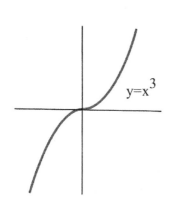

Solution. $f(x) = x^3$ is differentiable for all x, and $f'(x) = 3x^2$ equals 0 only at $x = 0$, so the only candidate is the point $(0, 0)$. But if $x > 0$ then $f(x) = x^3 > 0 = f(0)$, so $f(0)$ is not a local maximum. Similarly, if $x < 0$ then $f(x) = x^3 < 0 = f(0)$ so $f(0)$ is not a local minimum. The point $(0, 0)$ is the only candidate to be a local extreme of f, but this candidate did not turn out to be a local extreme of f. The function $f(x) = x^3$ does not have any local extremes.   ◀

> If $f'(a) = 0$ or f is not differentiable at a
>
> then the point $(a, f(a))$ is a candidate to be a local extreme but may not actually be a local extreme.

Practice 2. Sketch the graph of a differentiable function f that satisfies the conditions: $f(1) = 5$, $f(3) = 1$, $f(4) = 3$ and $f(6) = 7$; $f'(1) = 0$, $f'(3) = 0$, $f'(4) = 0$ and $f'(6) = 0$; the only local maximums of f are at $(1,5)$ and $(6,7)$; and the only local minimum is at $(3,1)$.

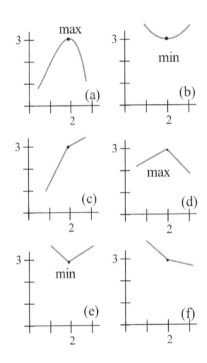

Is $f(a)$ a Maximum or Minimum or Neither?

Once we have found the candidates $(a, f(a))$ for extreme points of f, we still have the problem of determining whether the point is a maximum, a minimum or neither.

One method involves graphing (or letting a calculator graph) the function near a, and then drawing a conclusion from the graph. All of the graphs in the margin have $f(2) = 3$, and on each of the graphs $f'(2)$ either equals 0 or is undefined. It is clear from the graphs that the point $(2,3)$ is: a local maximum in (a) and (d); a local minimum in (b) and (e); and not a local extreme in (c) and (f).

In Sections 3.3 and 3.4, we will investigate how information about the first and **second** derivatives of f can help determine whether the candidate $(a, f(a))$ is a maximum, a minimum or neither.

Endpoint Extremes

So far we have discussed finding extreme values of functions over the entire real number line or on an open interval, but in practice we may need to find the extreme of a function over some closed interval $[c, d]$. If an extreme value of f occurs at $x = a$ between c and d ($c < a < d$) then the previous reasoning and results still apply: either $f'(a) = 0$ or f is not differentiable at a. On a closed interval, however, there is one more possibility: an extreme can occur at an **endpoint** of the closed interval (see margin): at $x = c$ or $x = d$.

We can extend our definition of a local extreme at $x = a$ (which requires $f(a) \geq f(x)$ [or $f(a) \leq f(x)$] for all x in some *open* interval containing a) to include $x = a$ being the endpoint of a closed interval: $f(a) \geq f(x)$ [or $f(a) \leq f(x)$] for all x in an interval of the form $[a, a + h)$ (for left endpoints) or $(a - h, a]$ (for right endpoints), where $h > 0$ is a number small enough to guarantee the "half-open" interval is in the domain of $f(x)$. Using this extended definition, the function in the margin has a local maximum (which is also a global maximum) at $x = c$ and a local minimum (also a global minimum) at $x = d$.

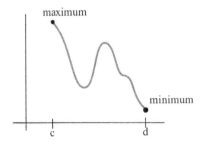

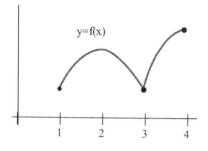

Practice 3. List all of the extremes $(a, f(a))$ of the function in the margin figure on the interval $[1, 4]$ and state whether $f'(a) = 0$, f is not differentiable at a, or a is an endpoint.

Example 3. Find the extreme values of $f(x) = x^3 - 3x^2 - 9x + 5$ for $-2 \leq x \leq 6$.

Solution. We need to find investigate points where $f'(x) = 0$, points where f is not differentiable, and the endpoints:

- $f'(x) = 3x^2 - 6x - 9 = 3(x + 1)(x - 3)$, so $f'(x) = 0$ only at $x = -1$ and $x = 3$.

- f is a polynomial, so it is differentiable everywhere.

- The endpoints of the interval are $x = -2$ and $x = 6$.

Altogether we have four points in the interval to examine, and any extreme values of f can only occur when x is one of those four points: $f(-2) = 3$, $f(-1) = 10$, $f(3) = -22$ and $f(6) = 59$. The (global) minimum of f on $[-2, 6]$ is -22 when $x = 3$, and the (global) maximum of f on $[-2, 6]$ is 59 when $x = 6$.   ◄

Sometimes the function we need to maximize or minimize is more complicated, but the same methods work.

Example 4. Find the extreme values of $f(x) = \frac{1}{3}\sqrt{64 + x^2} + \frac{1}{5}(10 - x)$ for $0 \leq x \leq 10$.

Solution. This function comes from an application we will examine in section 3.5. The only possible locations of extremes are where $f'(x) = 0$ or $f'(x)$ is undefined or where x is an endpoint of the interval $[0, 10]$.

$$f'(x) = \mathbf{D}\left(\frac{1}{3}\left(64 + x^2\right)^{\frac{1}{2}} + \frac{1}{5}(10 - x)\right)$$

$$= \frac{1}{3} \cdot \frac{1}{2}(64 + x^2)^{-\frac{1}{2}} \cdot 2x - \frac{1}{5}$$

$$= \frac{x}{3\sqrt{64 + x^2}} - \frac{1}{5}$$

To find where $f'(x) = 0$, set the derivative equal to 0 and solve for x:

$$\frac{x}{3\sqrt{64 + x^2}} - \frac{1}{5} = 0 \Rightarrow \frac{x}{3\sqrt{64 + x^2}} = \frac{1}{5} \Rightarrow \frac{x^2}{576 + 9x^2} = \frac{1}{25}$$

$$\Rightarrow 16x^2 = 576 \Rightarrow x = \pm 6$$

but only $x = 6$ is in the interval $[0, 10]$. Evaluating f at this point gives $f(6) \approx 4.13$.

We can evaluate the formula for $f'(x)$ for any value of x, so the derivative is always defined.

Finally, the interval $[0, 10]$ has two endpoints, $x = 0$ and $x = 10$, and $f(0) \approx 4.67$ while $f(10) \approx 4.27$.

The maximum of f on $[0, 10]$ must occur at one of the points $(0, 4.67)$, $(6, 4.13)$ and $(10, 4.27)$, and the minimum must occur at one of these three points as well.

The maximum value of f is 4.67 at $x = 0$, and the minimum value of f is 4.13 at $x = 6$. ◀

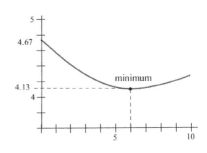

Practice 4. Rework the previous Example to find the extreme values of $f(x) = \frac{1}{3}\sqrt{64 + x^2} + \frac{1}{5}(10 - x)$ for $0 \le x \le 5$.

Critical Numbers

The points at which a function **might** have an extreme value are called **critical numbers**.

Definitions: A **critical number** for a function f is a value $x = a$ in the domain of f so that:

- $f'(a) = 0$ or

- f is not differentiable at a or

- a is an **endpoint** of a closed interval to which f is restricted.

If we are trying to find the extreme values of f on an **open** interval $c < x < d$ or on the entire number line, then the set of inputs to which f is restricted will not include any endpoints, so we will not need to worry about any endpoint critical numbers.

We can now give a very succinct description of where to look for extreme values of a function.

An extreme value of f can only occur at a critical number.

The critical numbers only give **possible** locations of extremes; some critical numbers are not locations of extremes. In other words, critical numbers are the **candidates** for the locations of maximums and minimums.

Section 3.5 is devoted entirely to translating and solving maximum and minimum problems.

each (a, f(a)) is a critical point but not an extreme point

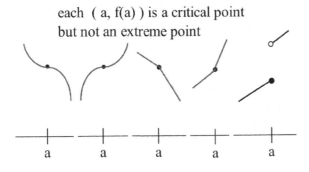

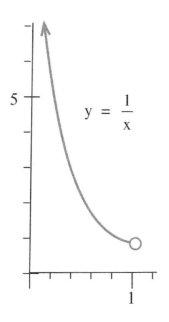

How would the situation change if we changed the interval in this example to $(0,1]$? To $[1,2]$?

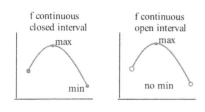

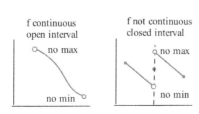

Which Functions Have Extremes?

Some functions don't have extreme values: Example 2 showed that $f(x) = x^3$ (defined on the entire number line) did not have a maximum or minimum.

Example 5. Find the extreme values of $f(x) = x$.

Solution. Because $f'(x) = 1 > 0$ for all x, the first theorem in this section guarantees that f has no extreme values. The function $f(x) = x$ does not have a maximum or minimum on the real number line.   ◀

With the previous function, the domain was so large that we could always make the function output larger or smaller than any given value by choosing an appropriate input x. The next example shows that we can encounter the same difficulty even on a "small" interval.

Example 6. Show that $f(x) = \dfrac{1}{x}$ does not have a maximum or minimum on the interval $(0,1)$.

Solution. f is continuous for all $x \neq 0$ so f is continuous on the interval $(0,1)$. For $0 < x < 1$, $f(x) = \dfrac{1}{x} > 0$ and for any number a strictly between 0 and 1, we can show that $f(a)$ is neither a maximum nor a minimum of f on $(0,1)$, as follows.

Pick b to be any number between 0 and a: $0 < b < a$. Then $f(b) = \dfrac{1}{b} > \dfrac{1}{a} = f(a)$, so $f(a)$ is not a maximum. Similarly, pick c to be any number between a and 1: $a < c < 1$. Then $f(a) = \dfrac{1}{a} > \dfrac{1}{c} = f(c)$, so $f(a)$ is not a minimum. The interval $(0,1)$ is not "large," yet f does not attain an extreme value anywhere in $(0,1)$.   ◀

The Extreme Value Theorem provides conditions that guarantee a function to have a maximum and a minimum.

Extreme Value Theorem:

If f is continuous on a **closed** interval $[a,b]$
then f attains both a maximum and minimum on $[a,b]$.

The proof of this theorem is difficult, so we omit it. The margin figure illustrates some of the possibilities for continuous and discontinuous functions on open and closed intervals.

The Extreme Value Theorem guarantees that certain functions (continuous ones) on certain intervals (closed ones) must have maximums and minimums. Other functions on other intervals **may** or **may not** have maximums and minimums.

3.1 Problems

1. Label all of the local maximums and minimums of the function in the figure below. Also label all of the critical points.

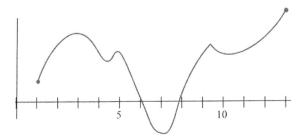

2. Label the local extremes and critical points of the function graphed below.

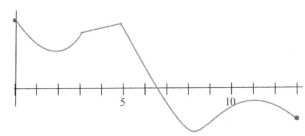

In Problems 3–22, find all of the critical points and local maximums and minimums of each function.

3. $f(x) = x^2 + 8x + 7$ 4. $f(x) = 2x^2 - 12x + 7$

5. $f(x) = \sin(x)$ 6. $f(x) = x^3 - 6x^2 + 5$

7. $f(x) = \sqrt[3]{x}$ 8. $f(x) = 5x - 2$

9. $f(x) = xe^{5x}$ 10. $f(x) = \sqrt[3]{1 + x^2}$

11. $f(x) = (x - 1)^2(x - 3)$

12. $f(x) = \ln(x^2 - 6x + 11)$

13. $f(x) = 2x^3 - 96x + 42$

14. $f(x) = 5x + \cos(2x + 1)$

15. $f(x) = e^{-(x-2)^2}$ 16. $f(x) = |x + 5|$

17. $f(x) = \dfrac{x}{1 + x^2}$ 18. $f(x) = \dfrac{x^3}{1 + x^4}$

19. $f(x) = (x - 2)^{\frac{2}{3}}$ 20. $f(x) = (x^2 - 1)^{\frac{2}{3}}$

21. $f(x) = \sqrt[3]{x^2 - 4}$ 22. $f(x) = \sqrt[3]{x - 2}$

23. Sketch the graph of a continuous function f with:

   (a) $f(1) = 3$, $f'(1) = 0$ and the point $(1, 3)$ a relative maximum of f.

   (b) $f(2) = 1$, $f'(2) = 0$ and the point $(2, 1)$ a relative minimum of f.

   (c) $f(3) = 5$, f is not differentiable at $x = 3$, and the point $(3, 5)$ a relative maximum of f.

   (d) $f(4) = 7$, f is not differentiable at $x = 4$, and the point $(4, 7)$ a relative minimum of f.

   (e) $f(5) = 4$, $f'(5) = 0$ and the point $(5, 4)$ not a relative minimum or maximum of f.

   (f) $f(6) = 3$, f not differentiable at 6, and $(6, 3)$ not a relative minimum or maximum of f.

In Problems 24–37, find all critical points and local extremes of each function on the given intervals.

24. $f(x) = x^2 - 6x + 5$ on the entire real number line

25. $f(x) = x^2 - 6x + 5$ on $[-2, 5]$

26. $f(x) = 2 - x^3$ on the entire real number line

27. $f(x) = 2 - x^3$ on $[-2, 1]$

28. $f(x) = x^3 - 3x + 5$ on the entire real number line

29. $f(x) = x^3 - 3x + 5$ on $[-2, 1]$

30. $f(x) = x^5 - 5x^4 + 5x^3 + 7$ on $(-\infty, \infty)$

31. $f(x) = x^5 - 5x^4 + 5x^3 + 7$ on $[0, 2]$

32. $f(x) = \dfrac{1}{x^2 + 1}$ on $(-\infty, \infty)$

33. $f(x) = \dfrac{1}{x^2 + 1}$ on $[1, 3]$

34. $f(x) = 3\sqrt{x^2 + 4} - x$ on $(-\infty, \infty)$

35. $f(x) = 3\sqrt{x^2 + 4} - x$ on $[0, 2]$

36. $f(x) = xe^{-5x}$ on $(-\infty, \infty)$

37. $f(x) = x^3 - \ln(x)$ on $\left[\frac{1}{2}, 2\right]$

38. (a) Find two numbers whose sum is 22 and whose product is as large as possible. (Suggestion: call the numbers x and $22 - x$).

   (b) Find two numbers whose sum is $A > 0$ and whose product is as large as possible.

39. Find the coordinates of the point in the first quadrant on the circle $x^2 + y^2 = 1$ so that the rectangle in the figure below has the largest possible area.

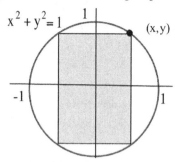

40. Find the coordinates of the point in the first quadrant on the ellipse $9x^2 + 16y^2 = 144$ so that the rectangle in the figure below has:

(a) the largest possible area.

(b) The smallest possible area.

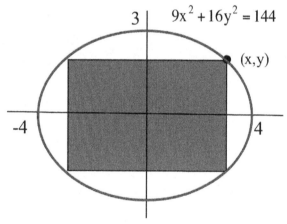

41. Find the value for x so the box shown below has:

(a) the largest possible volume.

(b) The smallest possible volume.

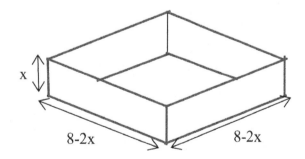

42. Find the radius and height of the cylinder that has the largest volume ($V = \pi r^2 h$) if the sum of the radius and height is 9.

43. Suppose you are working with a polynomial of degree 3 on a closed interval.

(a) What is the largest number of critical points the function can have on the interval?

(b) What is the smallest number of critical points it can have?

(c) What are the patterns for the most and fewest critical points a polynomial of degree n on a closed interval can have?

44. Suppose you have a polynomial of degree 3 divided by a polynomial of degree 2 on a closed interval.

(a) What is the largest number of critical points the function can have on the interval?

(b) What is the smallest number of critical points it can have?

45. Suppose $f(1) = 5$ and $f'(1) = 0$. What can we conclude about the point $(1, 5)$ if:

(a) $f'(x) < 0$ for $x < 1$ and $f'(x) > 0$ for $x > 1$?

(b) $f'(x) < 0$ for $x < 1$ and $f'(x) < 0$ for $x > 1$?

(c) $f'(x) > 0$ for $x < 1$ and $f'(x) < 0$ for $x > 1$?

(d) $f'(x) > 0$ for $x < 1$ and $f'(x) > 0$ for $x > 1$?

46. Suppose $f(2) = 3$ and f is continuous but not differentiable at $x = 2$. What can we conclude about the point $(2, 3)$ if:

(a) $f'(x) < 0$ for $x < 2$ and $f'(x) > 0$ for $x > 2$?

(b) $f'(x) < 0$ for $x < 2$ and $f'(x) < 0$ for $x > 2$?

(c) $f'(x) > 0$ for $x < 2$ and $f'(x) < 0$ for $x > 2$?

(d) $f'(x) > 0$ for $x < 2$ and $f'(x) > 0$ for $x > 2$?

47. The figure below shows the graph of $f'(x)$, which is continuous on $(0, 12)$ except at $x = 8$.

    (a) Which values of x are critical points of $f(x)$?
    (b) At which values of x does f attain a local maximum?
    (c) At which values of x does f attain a local minimum?

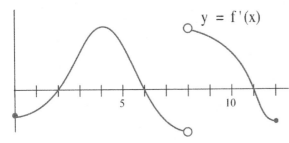

48. The figure below shows the graph of $f'(x)$, which is continuous on $(0, 13)$ except at $x = 7$.

    (a) Which values of x are critical points?
    (b) At which values of x does f attain a local maximum?
    (c) At which values of x does f attain a local minimum?

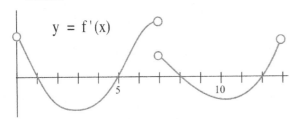

49. State the contrapositive form of the Extreme Value Theorem.

50. Imagine the graph of $f(x) = 1 - x$. Does f have a **maximum** value for x in the given interval?

    (a) $[0, 2]$ (b) $[0, 2)$ (c) $(0, 2]$

    (d) $(0, 2)$ (e) $(1, \pi]$

51. Imagine the graph of $f(x) = 1 - x$. Does f have a **minimum** value for x in the given interval?

    (a) $[0, 2]$ (b) $[0, 2)$ (c) $(0, 2]$

    (d) $(0, 2)$ (e) $(1, \pi]$

52. Imagine the graph of $f(x) = x^2$. Does f have a **maximum** value for x in the given interval?

    (a) $[-2, 3]$ (b) $[-2, 3)$ (c) $(-2, 3]$

    (d) $[-2, 1)$ (e) $(-2, 1]$

53. Imagine the graph of $f(x) = x^2$. Does f have a **minimum** value for x in the interval I?

    (a) $[-2, 3]$ (b) $[-2, 3)$ (c) $(-2, 3]$

    (d) $[-2, 1)$ (e) $(-2, 1]$

54. Define $A(x)$ to be the **area** bounded between the t-axis, the graph of $y = f(t)$ and a vertical line at $t = x$ (see figure below).

    (a) At what value of x is $A(x)$ minimum?
    (b) At what value of x is $A(x)$ maximum?

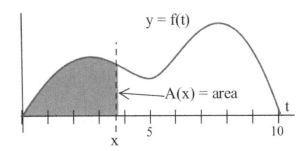

55. Define $S(x)$ to be the **slope** of the line through the points $(0, 0)$ and $(x, f(x))$ in the figure below.

    (a) At what value of x is $S(x)$ minimum?
    (b) At what value of x is $S(x)$ maximum?

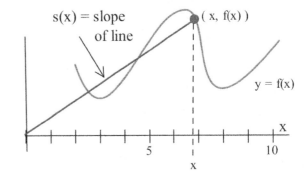

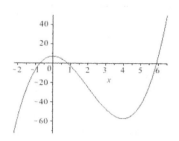

3.1 Practice Answers

1. $f(x) = x^2 + 4x - 5$ is a polynomial so f is differentiable for all x and $f'(x) = 2x + 4$; $f'(x) = 0$ when $x = -2$ so the only candidate for a local extreme is $x = -2$. Because the graph of f is a parabola opening up, the point $(-2, f(-2)) = (-2, -9)$ is a local minimum.

   $g(x) = 2x^3 - 12x^2 + 7$ is a polynomial so g is differentiable for all x and $g'(x) = 6x^2 - 24x = 6x(x - 4)$ so $g'(x) = 0$ when $x = 0$ or 4, so the only candidates for a local extreme are $x = 0$ and $x = 4$. The graph of g (see margin) indicates that g has a local maximum at $(0, 7)$ and a local minimum at $(4, -57)$.

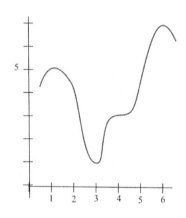

2. See the margin figure.

x	$f(x)$	$f'(x)$	max/min
1	5	0	local max
3	1	0	local min
4	3	0	neither
6	7	0	local max

3. $(1, f(1))$ is a global minimum; $x = 1$ is an endpoint
   $(2, f(2))$ is a local maximum; $f'(2) = 0$
   $(3, f(3))$ is a local/global minimum; f is not differentiable at $x = 3$
   $(4, f(4))$ is a global maximum; $x = 4$ is an endpoint

4. This is the same function used in Example 4, but now the interval is $[0, 5]$ instead of $[0, 10]$. See the Example for the calculations.

   Critical points:

   - endpoints: $x = 0$ and $x = 5$

   - f is differentiable for all $0 < x < 5$: none

   - $f'(x) = 0$: none in $[0, 5]$

   $f(0) \approx 4.67$ is the maximum of f on $[0, 5]$;
   $f(5) \approx 4.14$ is the minimum of f on $[0, 5]$.

3.2 Mean Value Theorem

If you averaged 30 miles per hour during a trip, then at some instant during the trip you were traveling exactly 30 miles per hour.

That relatively obvious statement is the Mean Value Theorem as it applies to a particular trip. It may seem strange that such a simple statement would be important or useful to anyone, but the Mean Value Theorem is important and some of its consequences are very useful in a variety of areas. Many of the results in the rest of this chapter depend on the Mean Value Theorem, and one of the corollaries of the Mean Value Theorem will be used every time we calculate an "integral" in later chapters. A truly delightful aspect of mathematics is that an idea as simple and obvious as the Mean Value Theorem can be so powerful.

Before we state and prove the Mean Value Theorem and examine some of its consequences, we will consider a simplified version called Rolle's Theorem.

Rolle's Theorem

Pick any two points on the x-axis and think about all of the differentiable functions that pass through those two points. Because our functions are differentiable, they must be continuous and their graphs cannot have any holes or breaks. Also, since these functions are differentiable, their derivatives are defined everywhere between our two points and their graphs can not have any "corners" or vertical tangents.

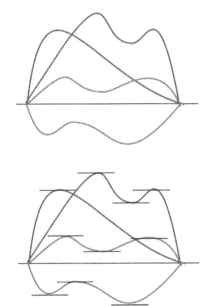

The graphs of the functions in the margin figure can still have all sorts of shapes, and it may seem unlikely that they have any common properties other than the ones we have stated, but Michel Rolle (1652–1719) found one. He noticed that every one of these functions has one or more points where the tangent line is horizontal (see margin), and this result is named after him.

Rolle's Theorem:

If $\quad f(a) = f(b)$
and $f(x)$ is continuous for $a \leq x \leq b$
and differentiable for $a < x < b$

then there is at least one number c between a and b so that
$f'(c) = 0$.

Proof. We consider three cases: when $f(x) = f(a)$ for all x in (a,b), when $f(x) > f(a)$ for some x in (a,b), and when $f(x) < f(a)$ for some x in (a,b).

Case I: If $f(x) = f(a)$ for all x between a and b, then the graph of f is a horizontal line segment and $f'(c) = 0$ for *all* values of c strictly between a and b.

Case II: Suppose $f(x) > f(a)$ for some x in (a,b). Because f is continuous on the closed interval $[a,b]$, we know from the Extreme Value Theorem that f must attain a maximum value on the closed interval $[a,b]$. Because $f(x) > f(a)$ for some value of x in $[a,b]$, then the maximum of f must occur at some value c strictly between a and b: $a < c < b$. (Why can't the maximum be at a or b?) Because $f(c)$ is a local maximum of f, c is a critical number of f, meaning $f'(c) = 0$ or $f'(c)$ is undefined. But f is differentiable at all x between a and b, so the only possibility is that $f'(c) = 0$.

Case III: Suppose $f(x) < f(a)$ for some x in (a,b). Then, arguing as we did in Case II, f attains a minimum at some value $x = c$ strictly between a and b, and so $f'(c) = 0$.

In each case, there is *at least one* value of c between a and b so that $f'(c) = 0$. ☐

Notice that Rolle's Theorem tells us that (at least one) number c with the required properties exists, but does not tell us how to find c.

Example 1. Show that $f(x) = x^3 - 6x^2 + 9x + 2$ satisfies the hypotheses of Rolle's Theorem on the interval $[0,3]$ and find a value of c that the theorem tells you must exist.

Solution. Because f is a polynomial, it is continuous and differentiable everywhere. Furthermore, $f(0) = 2 = f(3)$, so Rolle's Theorem applies. Differentiating:

$$f'(x) = 3x^2 - 12x + 9 = 3(x - 1)(x - 3)$$

so $f'(x) = 0$ when $x = 1$ and when $x = 3$. The value $c = 1$ is between 0 and 3. The margin figure shows a graph of f. ◀

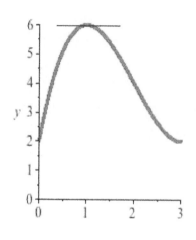

Practice 1. Find the value(s) of c for Rolle's Theorem for the functions graphed below.

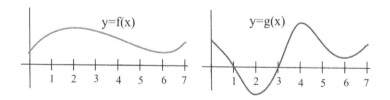

The Mean Value Theorem

Geometrically, the Mean Value Theorem is a "tilted" version of Rolle's Theorem (see margin). In each theorem we conclude that there is a number c so that the slope of the tangent line to f at $x = c$ is the same as the slope of the line connecting the two ends of the graph of f on the interval $[a,b]$. In Rolle's Theorem, the two ends of the graph of f are at the same height, $f(a) = f(b)$, so the slope of the line connecting the ends is zero. In the Mean Value Theorem, the two ends of the graph

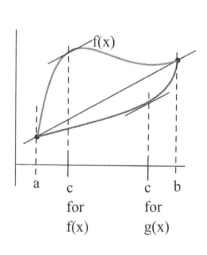

of f do not have to be at the same height, so the line through the two ends does not have to have a slope of zero.

Mean Value Theorem:

    If $f(x)$ is continuous for $a \leq x \leq b$

              and differentiable for $a < x < b$

    then there is at least one number c between a and b so the line tangent to the graph of f at $x = c$ is parallel to the secant line through $(a, f(a))$ and $(b, f(b))$:

$$f'(c) = \frac{f(b) - f(a)}{b - a}$$

Proof. The proof of the Mean Value Theorem uses a tactic common in mathematics: introduce a new function that satisfies the hypotheses of some theorem we already know and then use the conclusion of that previously proven theorem. For the Mean Value Theorem we introduce a new function, $h(x)$, which satisfies the hypotheses of Rolle's Theorem. Then we can be certain that the conclusion of Rolle's Theorem is true for $h(x)$ and the Mean Value Theorem for f will follow from the conclusion of Rolle's Theorem for h.

First, let $g(x)$ be the linear function passing through the points $(a, f(a))$ and $(b, f(b))$ of the graph of f. The function g goes through the point $(a, f(a))$ so $g(a) = f(a)$. Similarly, $g(b) = f(b)$. The slope of the linear function $g(x)$ is $\frac{f(b) - f(a)}{b - a}$ so $g'(x) = \frac{f(b) - f(a)}{b - a}$ for all x between a and b, and g is continuous and differentiable. (The formula for g is $g(x) = f(a) + m(x - a)$ with $m = \frac{f(b) - f(a)}{b - a}$.)

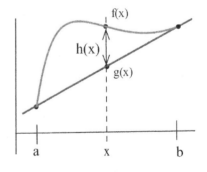

Define $h(x) = f(x) - g(x)$ for $a \leq x \leq b$ (see margin). The function h satisfies the hypotheses of Rolle's theorem:

- $h(a) = f(a) - g(a) = 0$ and $h(b) = f(b) - g(b) = 0$

- $h(x)$ is continuous for $a \leq x \leq b$ because both f and g are continuous there

- $h(x)$ is differentiable for $a < x < b$ because both f and g are differentiable there

so the conclusion of Rolle's Theorem applies to h: there is a c between a and b so that $h'(c) = 0$.

The derivative of $h(x) = f(x) - g(x)$ is $h'(x) = f'(x) - g'(x)$ so we know that there is a number c between a and b with $h'(c) = 0$. But:

$$0 = h'(c) = f'(c) - g'(c) \Rightarrow f'(c) = g'(c) = \frac{f(b) - f(a)}{b - a}$$

which is exactly what we needed to prove.   □

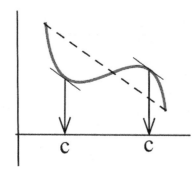

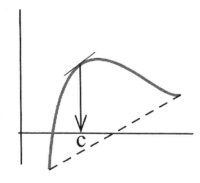

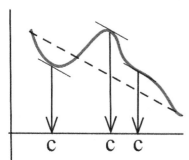

Graphically, the Mean Value Theorem says that there is at least one point c where the slope of the tangent line, $f'(c)$, equals the slope of the line through the end points of the graph segment, $(a, f(a))$ and $(b, f(b))$. The margin figure shows the locations of the parallel tangent lines for several functions and intervals.

The Mean Value Theorem also has a very natural interpretation if $f(x)$ represents the position of an object at time x: $f'(x)$ represents the velocity of the object at the **instant** x and $\dfrac{f(b) - f(a)}{b - a} = \dfrac{\text{change in position}}{\text{change in time}}$ represents the **average** (mean) velocity of the object during the time interval from time a to time b. The Mean Value Theorem says that there is a time c (between a and b) when the **instantaneous** velocity, $f'(c)$, is equal to the **average** velocity for the entire trip, $\dfrac{f(b) - f(a)}{b - a}$. If your average velocity during a trip is 30 miles per hour, then at some instant during the trip you were traveling exactly 30 miles per hour.

Practice 2. For $f(x) = 5x^2 - 4x + 3$ on the interval $[1, 3]$, calculate $m = \dfrac{f(b) - f(a)}{b - a}$ and find the value(s) of c so that $f'(c) = m$.

Some Consequences of the Mean Value Theorem

If the Mean Value Theorem was just an isolated result about the existence of a particular point c, it would not be very important or useful. However, the Mean Value Theorem is the basis of several results about the behavior of functions over entire intervals, and it is these consequences that give it an important place in calculus for both theoretical and applied uses.

The next two corollaries are just the first of many results that follow from the Mean Value Theorem.

We already know, from the Main Differentiation Theorem, that the derivative of a constant function $f(x) = K$ is always 0, but can a nonconstant function have a derivative that is always 0? The first corollary says no.

Corollary 1:

    If $f'(x) = 0$ for all x in an interval I

    then $f(x) = K$, a constant, for all x in I.

Proof. Assume $f'(x) = 0$ for all x in an interval I. Pick any two points a and b (with $a \neq b$) in the interval. Then, by the Mean Value Theorem, there is a number c between a and b so that $f'(c) = \dfrac{f(b) - f(a)}{b - a}$. By our assumption, $f'(x) = 0$ for all x in I, so we know that $0 = f'(c) =$

$\dfrac{f(b) - f(a)}{b - a}$ and thus $f(b) - f(a) = 0 \Rightarrow f(b) = f(a)$. But a and b were two arbitrary points in I, so the value of $f(x)$ is the same for any two values of x in I, and f is a constant function on the interval I. $\square$

We already know that if two functions are "parallel" (differ by a constant), then their derivatives are equal, but can two non-parallel functions have the same derivative? The second corollary says no.

Corollary 2:

 If $f'(x) = g'(x)$ for all x in an interval I

 then $f(x) - g(x) = K$, a constant, for all x in I, so the
     graphs of f and g are "parallel" on the interval I.

Proof. This corollary involves two functions instead of just one, but we can imitate the proof of the Mean Value Theorem and introduce a new function $h(x) = f(x) - g(x)$. The function h is differentiable and $h'(x) = f'(x) - g'(x) = 0$ for all x in I so, by Corollary 1, $h(x)$ is a constant function and $K = h(x) = f(x) - g(x)$ for all x in the interval. Thus $f(x) = g(x) + K$. $\square$

We will use Corollary 2 hundreds of times in Chapters 4 and 5 when we work with "integrals." Typically you will be given the derivative of a function, $f'(x)$, and be asked to find **all** functions f that have that derivative. Corollary 2 tells us that if we can find **one** function f that has the derivative we want, then the only other functions that have the same derivative are of the form $f(x) + K$ where K is a constant: once you find one function with the right derivative, you have essentially found all of them.

Example 2. (a) Find **all** functions whose derivatives equal $2x$. (b) Find a function $g(x)$ with $g'(x) = 2x$ and $g(3) = 5$.

Solution. (a) Observe that $f(x) = x^2 \Rightarrow f'(x) = 2x$, so one function with the derivative we want is $f(x) = x^2$. Corollary 2 guarantees that every function g whose derivative is $2x$ has the form $g(x) = f(x) + K = x^2 + K$. (b) Because $g'(x) = 2x$, we know that g must have the form $g(x) = x^2 + K$, but this gives a whole "family" of functions (see margin) and we want to find one member of that family. We also know that $g(3) = 5$ so we want to find the member of the family that passes through the point $(3, 5)$. Replacing $g(x)$ with 5 and x with 3 in the formula $g(x) = x^2 + K$, we can solve for the value of K: $5 = g(3) = (3)^2 + K \Rightarrow K = -4$. The function we want is $g(x) = x^2 - 4$.   ◄

Practice 3. Restate Corollary 2 as a statement about the positions and velocities of two cars.

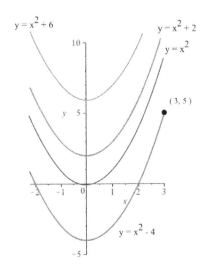

3.2 *Problems*

1. In the figure below, find the number(s) "c" that Rolle's Theorem promises (guarantees).

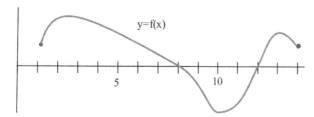

For Problems 2–4, verify that the hypotheses of Rolle's Theorem are satisfied for each of the functions on the given intervals, and find the value of the number(s) "c" that Rolle's Theorem promises.

2. (a) $f(x) = x^2$ on $[-2, 2]$
   (b) $f(x) = x^2 - 5x + 8$ on $[0, 5]$

3. (a) $f(x) = \sin(x)$ on $[0, \pi]$
   (b) $f(x) = \sin(x)$ on $[\pi, 5\pi]$

4. (a) $f(x) = x^3 - x + 3$ on $[-1, 1]$
   (b) $f(x) = x \cdot \cos(x)$ on $[0, \frac{\pi}{2}]$

5. Suppose you toss a ball straight up and catch it when it comes down. If $h(t)$ is the height of the ball t seconds after you toss it, what does Rolle's Theorem say about the velocity of the ball? Why is it easier to catch a ball that someone on the ground tosses up to you on a balcony, than for you to be on the ground and catch a ball that someone on a balcony tosses down to you?

6. If $f(x) = \dfrac{1}{x^2}$, then $f(-1) = 1$ and $f(1) = 1$ but $f'(x) = -\dfrac{2}{x^3}$ is never equal to 0. Why doesn't this function violate Rolle's Theorem?

7. If $f(x) = |x|$, then $f(-1) = 1$ and $f(1) = 1$ but $f'(x)$ is never equal to 0. Why doesn't this function violate Rolle's Theorem?

8. If $f(x) = x^2$, then $f'(x) = 2x$ is never 0 on the interval $[1, 3]$. Why doesn't this function violate Rolle's Theorem?

9. If I take off in an airplane, fly around for a while and land at the same place I took off from, then my starting and stopping heights are the same but the airplane is always moving. Why doesn't this violate Rolle's theorem, which says there is an instant when my velocity is 0?

10. Prove the following corollary of Rolle's Theorem: If $P(x)$ is a polynomial, then between any two roots of P there is a root of P'.

11. Use the corollary in Problem 10 to justify the conclusion that the only root of $f(x) = x^3 + 5x - 18$ is 2. (Suggestion: What could you conclude about f' if f had another root?)

12. In the figure below, find the location(s) of the "c" that the Mean Value Theorem promises.

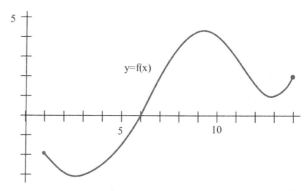

In Problems 13–15, verify that the hypotheses of the Mean Value Theorem are satisfied for each of the functions on the given intervals, and find the number(s) "c" that the Mean Value Theorem guarantees.

13. (a) $f(x) = x^2$ on $[0, 2]$
    (b) $f(x) = x^2 - 5x + 8$ on $[1, 5]$

14. (a) $f(x) = \sin(x)$ on $[0, \frac{\pi}{2}]$
    (b) $f(x) = x^3$ on $[-1, 3]$

15. (a) $f(x) = 5 - \sqrt{x}$ on $[1, 9]$
    (b) $f(x) = 2x + 1$ on $[1, 7]$

16. For the quadratic functions in parts (a) and (b) of Problem 13, the number c turned out to be the midpoint of the interval: $c = \frac{a+b}{2}$.

    (a) For $f(x) = 3x^2 + x - 7$ on $[1,3]$, show that $f'(2) = \frac{f(3) - f(1)}{3 - 1}$.

    (b) For $f(x) = x^2 - 5x + 3$ on $[2,5]$, show that $f'\left(\frac{7}{2}\right) = \frac{f(5) - f(2)}{5 - 2}$.

    (c) For $f(x) = Ax^2 + Bx + C$ on $[a,b]$, show that $f'\left(\frac{a+b}{2}\right) = \frac{f(b) - f(a)}{b - a}$.

17. If $f(x) = |x|$, then $f(-1) = 1$ and $f(3) = 3$ but $f'(x)$ is never equal to $\frac{f(3) - f(-1)}{3 - (-1)} = \frac{1}{2}$. Why doesn't this violate the Mean Value Theorem?

In Problems 18–19, you are a traffic-court judge. In each case, a driver has challenged a speeding ticket and you need to decide if the ticket is appropriate.

18. A tolltaker says, "Your Honor, based on the elapsed time from when the car entered the toll road until the car stopped at my booth, I know the average speed of the car was 83 miles per hour. I did not actually see the car speeding, but I know it was and I gave the driver a ticket."

19. The driver in the next case says, "Your Honor, my average velocity on that portion of the toll road was only 17 miles per hour, so I could not have been speeding."

20. Find three different functions (f, g and h) so that $f'(x) = g'(x) = h'(x) = \cos(x)$.

21. Find a function f so that $f'(x) = 3x^2 + 2x + 5$ and $f(1) = 10$.

22. Find $g(x)$ so that $g'(x) = x^2 + 3$ and $g(0) = 2$.

23. Find values for A and B so that the graph of the parabola $f(x) = Ax^2 + B$ is:

    (a) tangent to $y = 4x + 5$ at the point $(1,9)$.

    (b) tangent to $y = 7 - 2x$ at the point $(2,3)$.

    (c) tangent to $y = x^2 + 3x - 2$ at the point $(0,2)$.

24. Sketch the graphs of several members of the "family" of functions whose derivatives always equal 3. Give a formula that defines every function in this family.

25. Sketch the graphs of several members of the "family" of functions whose derivatives always equal $3x^2$. Give a formula that defines every function in this family.

26. At t seconds after takeoff, the upward velocity of a helicopter was $v(t) = 3t^2 + 2t$ feet/second. Two seconds after takeoff, the helicopter was 80 feet above sea level. Find a formula for the height of the helicopter at every time t.

27. Assume that a rocket is fired from the ground and has the upward velocity shown in the figure below. Estimate the height of the rocket when $t = 1, 2$ and 5 seconds.

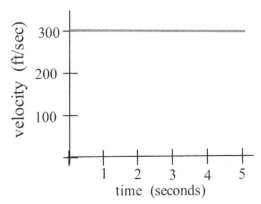

28. The figure below shows the upward velocity of a rocket. Use the information in the graph to estimate the height of the rocket when $t = 1, 2$ and 5 seconds.

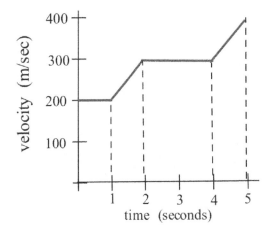

29. Determine a formula for $f(x)$ if you know: $f''(x) = 6$, $f'(0) = 4$ and $f(0) = -5$.

30. Determine a formula for $g(x)$ if you know: $g''(x) = 12x$, $g'(1) = 9$ and $g(2) = 30$.

31. Define $A(x)$ to be the **area** bounded by the t-axis, the line $y = 3$ and a vertical line at $t = x$.

    (a) Find a formula for $A(x)$.

    (b) Determine $A'(x)$.

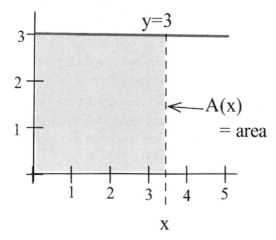

32. Define $A(x)$ to be the **area** bounded by the t-axis, the line $y = 2t$ and a vertical line at $t = x$.

    (a) Find a formula for $A(x)$.

    (b) Determine $A'(x)$.

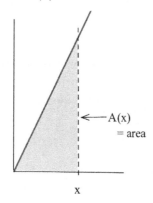

33. Define $A(x)$ to be the **area** bounded by the t-axis, the line $y = 2t + 1$ and a vertical line at $t = x$.

    (a) Find a formula for $A(x)$.

    (b) Determine $A'(x)$.

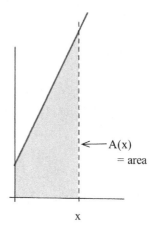

In Problems 34–36, given a list of numbers $a_1, a_2, a_3, a_4, \ldots$, the **consecutive differences** between numbers in the list are: $a_2 - a_1, a_3 - a_2, a_4 - a_3, \ldots$

34. If $a_1 = 5$ and the consecutive difference is always 0, what can you conclude about the numbers in the list?

35. If $a_1 = 5$ and the consecutive difference is always 3, find a formula for a_n.

36. Suppose the "a" list starts with 3, 4, 7, 8, 6, 10, 13,..., and there is a "b" list that has the same consecutive differences as the "a" list.

    (a) If $b_1 = 5$, find the next six numbers in the "b" list. How is b_n related to a_n?

    (b) If $b_1 = 2$, find the next six numbers in the "b" list. How is b_n related to a_n?

    (c) If $b_1 = B$, find the next six numbers in the "b" list. How is b_n related to a_n?

3.2 Practice Answers

1. $f'(x) = 0$ when $x = 2$ and 6, so $c = 2$ and $c = 6$.
   $g'(x) = 0$ when $x = 2, 4$ and 6, so $c = 2$, $c = 4$ and $c = 6$.

2. With $f(x) = 5x^2 - 4x + 3$ on $[1, 3]$, $f(1) = 4$ and $f(3) = 36$ so:

$$m = \frac{f(b) - f(a)}{b - a} = \frac{36 - 4}{3 - 1} = 16$$

$f'(x) = 10x - 4$ so $f'(c) = 10c - 4 = 16 \Rightarrow 10c = 20 \Rightarrow c = 2$. The graph of f showing the location of c appears below.

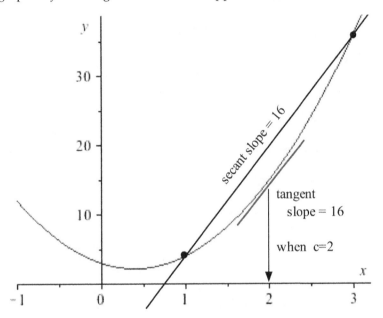

3. If two cars have the same velocities during an interval of time (so that $f'(t) = g'(t)$ for t in I) then the cars are always a constant distance apart during that time interval. (Note: "Same velocity" means **same speed** and **same direction**. If two cars are traveling at the same speed but in different directions, then the distance between them changes and is not constant.)

3.3   The First Derivative and the Shape of f

This section examines some of the interplay between the shape of the graph of a function f and the behavior of its derivative, f'. If we have a graph of f, we will investigate what we can conclude about the values of f'. And if we know values of f', we will investigate what we can conclude about the graph of f.

In this definition, I can be of the form (a, b), $[a, b)$, $(a, b]$, $[a, b]$, $(-\infty, b)$, $(-\infty, b]$, (a, ∞), $[a, \infty)$ or $(-\infty, \infty)$, where $a < b$.

> **Definitions**: Given any interval I, a function f is...
>
> **increasing** on I if, for all x_1 and x_2 in I, $x_1 < x_2 \Rightarrow f(x_1) < f(x_2)$
> **decreasing** on I if, for all x_1 and x_2 in I, $x_1 < x_2 \Rightarrow f(x_1) > f(x_2)$
> **monotonic** on I if f is increasing or decreasing on I

Graphically, f is increasing (decreasing) if, as we move from left to right along the graph of f, the height of the graph increases (decreases).

These same ideas make sense if we consider $h(t)$ to be the height (in feet) of a rocket at time t seconds. We naturally say that the rocket is rising or that its height is increasing if the height $h(t)$ increases over a period of time, as t increases.

Example 1. List the intervals on which the function graphed below is increasing or decreasing.

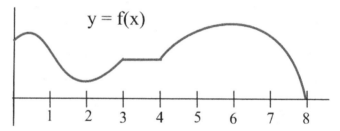

Solution. f is increasing on the intervals $[0, 0.3]$ (approximately), $[2, 3]$ and $[4, 6]$. f is decreasing on (approximately) $[0.3, 2]$ and $[6, 8]$. On the interval $[3, 4]$ the function is not increasing or decreasing—it is **constant**. It is also valid to say that f is increasing on the intervals $[0.5, 0.8]$ and $(0.5, 0.8)$ as well as many others, but we usually talk about the longest intervals on which f is monotonic. ◄

Practice 1. List the intervals on which the function graphed below is increasing or decreasing.

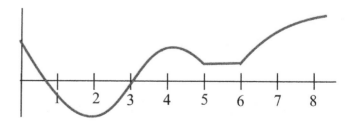

If we have an accurate graph of a function, then it is relatively easy to determine where f is monotonic, but if the function is defined by a formula, then a little more work is required. The next two theorems relate the values of the derivative of f to the monotonicity of f. The first theorem says that if we know where f is monotonic, then we also know something about the values of f'. The second theorem says that if we know about the values of f' then we can draw conclusions about where f is monotonic.

First Shape Theorem:

For a function f that is differentiable on an interval (a, b):

- if f is increasing on (a, b) then $f'(x) \geq 0$ for all x in (a, b)

- if f is decreasing on (a, b) then $f'(x) \leq 0$ for all x in (a, b)

- if f is constant on (a, b), then $f'(x) = 0$ for all x in (a, b)

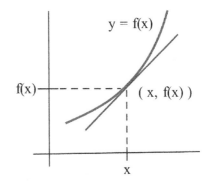

Proof. Most people find a picture such as the one in the margin to be a convincing justification of this theorem: if the graph of f increases near a point $(x, f(x))$, then the tangent line is also increasing, and the slope of the tangent line is positive (or perhaps zero at a few places). A more precise proof, however, requires that we use the definitions of the derivative of f and of "increasing" (given above).

Case I: Assume that f is increasing on (a, b). We know that f is differentiable, so if x is any number in the interval (a, b) then

$$f'(x) = \lim_{h \to 0} \frac{f(x + h) - f(x)}{h}$$

and this limit exists and is a finite value. If h is any small enough **positive** number so that $x + h$ is also in the interval (a, b), then $x < x + h \Rightarrow f(x) < f(x + h)$ (by the definition of "increasing"). We know that the numerator, $f(x + h) - f(x)$, and the denominator, h, are both positive, so the limiting value, $f'(x)$, must be positive or zero: $f'(x) \geq 0$.

Case II: Assume that f is decreasing on (a, b). If $x < x + h$, then $f(x) > f(x + h)$ (by the definition of "decreasing"). So the numerator of the limit, $f(x + h) - f(x)$, will be negative but the denominator, h, will still be positive, so the limiting value, $f'(x)$, must be negative or zero: $f'(x) \leq 0$.

The proof of this part is very similar to the "increasing" proof.

Case III: The derivative of a constant is 0, so if f is constant on (a, b) then $f'(x) = 0$ for all x in (a, b).   □

The previous theorem is easy to understand, but you need to pay attention to exactly what it says and what it does **not** say. It is possible for a differentiable function that is increasing on an interval to have horizontal tangent lines at some places in the interval (see margin). It is

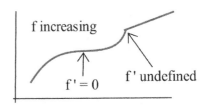

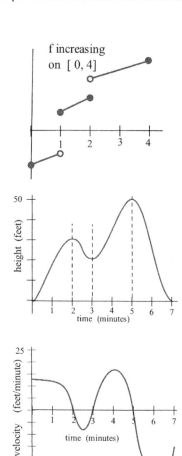

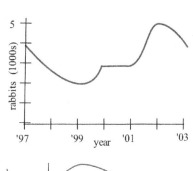

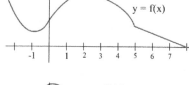

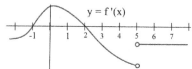

also possible for a continuous function that is increasing on an interval to have an undefined derivative at some places in the interval. Finally, it is possible for a function that is increasing on an interval to fail to be continuous at some places in the interval (see margin).

The First Shape Theorem has a natural interpretation in terms of the height $h(t)$ and upward velocity $h'(t)$ of a helicopter at time t. If the height of the helicopter is increasing ($h(t)$ is an increasing function), then the helicopter has a positive or zero upward velocity: $h'(t) \geq 0$. If the height of the helicopter is not changing, then its upward velocity is 0: $h'(t) = 0$.

Example 2. A figure in the margin shows the height of a helicopter during a period of time. Sketch the graph of the upward velocity of the helicopter, $\dfrac{dh}{dt}$.

Solution. The graph of $v(t) = \dfrac{dh}{dt}$ appears in the margin. Notice that $h(t)$ has a local maximum when $t = 2$ and $t = 5$, and that $v(2) = 0$ and $v(5) = 0$. Similarly, $h(t)$ has a local minimum when $t = 3$, and $v(3) = 0$. When h is increasing, v is positive. When h is decreasing, v is negative. ◀

Practice 2. A figure in the margin shows the population of rabbits on an island during a 6-year period. Sketch the graph of the rate of population change, $\dfrac{dR}{dt}$, during those years.

Example 3. A graph of f appears in the margin; sketch a graph of f'.

Solution. It is a good idea to look first for the points where $f'(x) = 0$ or where f is not differentiable (the critical points of f). These locations are usually easy to spot, and they naturally break the problem into several smaller pieces. The only numbers at which $f'(x) = 0$ are $x = -1$ and $x = 2$, so the only places the graph of $f'(x)$ will cross the x-axis are at $x = -1$ and $x = 2$: we can therefore plot the points $(-1, 0)$ and $(2, 0)$ on the graph of f'. The only place where f is not differentiable is at the "corner" above $x = 5$, so the graph of f' will not be defined for $x = 5$. The rest of the graph of f is relatively easy to sketch:

- if $x < -1$ then $f(x)$ is decreasing so $f'(x)$ is negative

- if $-1 < x < 2$ then $f(x)$ is increasing so $f'(x)$ is positive

- if $2 < x < 5$ then $f(x)$ is decreasing so $f'(x)$ is negative

- if $5 < x$ then $f(x)$ is decreasing so $f'(x)$ is negative

A graph of f' appears on the previous page: $f(x)$ is continuous at $x = 5$, but not differentiable at $x = 5$ (indicated by the "hole"). ◀

Practice 3. A graph of f appears in the margin. Sketch a graph of f'. (The graph of f has a "corner" at $x = 5$.)

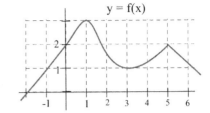

The next theorem is almost the converse of the First Shape Theorem and explains the relationship between the values of the derivative and the graph of a function from a different perspective. It says that if we know something about the values of f', then we can draw some conclusions about the shape of the graph of f.

Second Shape Theorem:

For a function f that is differentiable on an interval I:

- if $f'(x) > 0$ for all x in the interval I, then f is increasing on I

- if $f'(x) < 0$ for all x in the interval I, then f is decreasing on I

- if $f'(x) = 0$ for all x in the interval I, then f is constant on I

Proof. This theorem follows directly from the Mean Value Theorem, and the last part is just a restatement of the First Corollary of the Mean Value Theorem.

Case I: Assume that $f'(x) > 0$ for all x in I and pick any points a and b in I with $a < b$. Then, by the Mean Value Theorem, there is a point c between a and b so that $\dfrac{f(b) - f(a)}{b - a} = f'(c) > 0$ and we can conclude that $f(b) - f(a) > 0$, which means that $f(b) > f(a)$. Because $a < b \Rightarrow f(a) < f(b)$, we know that f is increasing on I.

Case II: Assume that $f'(x) < 0$ for all x in I and pick any points a and b in I with $a < b$. Then there is a point c between a and b so that $\dfrac{f(b) - f(a)}{b - a} = f'(c) < 0$, and we can conclude that $f(b) - f(a) = (b - a)f'(c) < 0$ so $f(b) < f(a)$. Because $a < b \Rightarrow f(a) > f(b)$, we know f is decreasing on I. □

Practice 4. Rewrite the Second Shape Theorem as a statement about the height $h(t)$ and upward velocity $h'(t)$ of a helicopter at time t seconds.

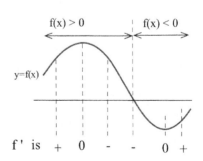

The value of a function f at a number x tells us the height of the graph of f above or below the point $(x, 0)$ on the x-axis. The value of f' at a number x tells us whether the graph of f is increasing or decreasing (or neither) as the graph passes through the point $(x, f(x))$ on the graph of f. If $f(x)$ is positive, it is possible for $f'(x)$ to be positive, negative, zero or undefined: the value of $f(x)$ has absolutely nothing to do with the value of f'. The margin figure illustrates some of the possible combinations of values for f and f'.

Practice 5. Graph a continuous function that satisfies the conditions on f and f' given below:

x	-2	-1	0	1	2	3
$f(x)$	1	-1	-2	-1	0	2
$f'(x)$	-1	0	1	2	-1	1

The Second Shape Theorem can be particularly useful if we need to graph a function f defined by a formula. Between any two consecutive critical numbers of f, the graph of f is monotonic (why?). If we can find all of the critical numbers of f, then the domain of f will be broken naturally into a number of pieces on which f will be monotonic.

Example 4. Use information about the values of f' to help graph $f(x) = x^3 - 6x^2 + 9x + 1$.

Solution. $f'(x) = 3x^2 - 12x + 9 = 3(x-1)(x-3)$ so $f'(x) = 0$ only when $x = 1$ or $x = 3$; f' is a polynomial, so it is always defined. The only critical numbers, $x = 1$ and $x = 3$, break the real number line into three pieces on which f is monotonic: $(-\infty, 1)$, $(1, 3)$ and $(3, \infty)$.

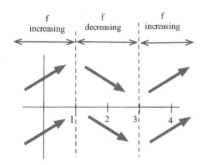

- $x < 1 \Rightarrow f'(x) = 3(\text{negative})(\text{negative}) > 0 \Rightarrow f$ increasing

- $1 < x < 3 \Rightarrow f'(x) = 3(\text{positive})(\text{negative}) < 0 \Rightarrow f$ is decreasing

- $3 < x \Rightarrow f'(x) = 3(\text{positive})(\text{positive}) > 0 \Rightarrow f$ is increasing

Although we don't yet know the value of f anywhere, we do know a lot about the shape of the graph of f: as we move from left to right along the x-axis, the graph of f increases until $x = 1$, then decreases until $x = 3$, after which the graph increases again (see margin). The graph of f "turns" when $x = 1$ and $x = 3$. To plot the graph of f, we still need to evaluate f at a few values of x, but only at a **very** few values: $f(1) = 5$, and $(1, 5)$ is a local maximum of f; $f(3) = 1$, and $(3, 1)$ is a local minimum of f. A graph of f appears in the margin. ◀

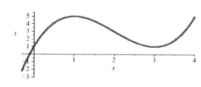

Practice 6. Use information about the values of f' to help graph the function $f(x) = x^3 - 3x^2 - 24x + 5$.

Example 5. Use the graph of f' in the margin to sketch the shape of the graph of f. Why isn't the graph of f' enough to completely determine the graph of f?

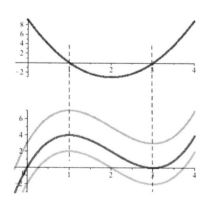

Solution. Several functions that have the derivative we want appear in the margin, and each provides a correct answer. By the Second Corollary to the Mean Value Theorem, we know there is a whole family of "parallel" functions that share the derivative we want, and each of these functions provides a correct answer. If we had additional information about the function — such as a point it passes through — then only one member of the family would satisfy the extra condition and there would be only one correct answer.   ◀

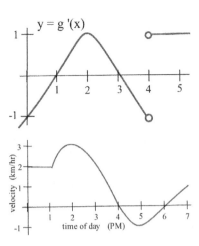

Practice 7. Use the graph of g' provided in the margin to sketch the shape of a graph of g.

Practice 8. A weather balloon is released from the ground and sends back its upward velocity measurements (see margin). Sketch a graph of the height of the balloon. When was the balloon highest?

Using the Derivative to Test for Extremes

The first derivative of a function tells about the general shape of the function, and we can use that shape information to determine whether an extreme point is a (local) maximum or minimum or neither.

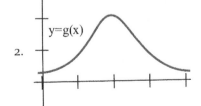

> **First Derivative Test for Local Extremes:**
> Let f be a continuous function with $f'(c) = 0$ or $f'(c)$ undefined.
>
> - If f'(left of c) > 0 and f'(right of c) < 0 then $(c, f(c))$ is a local maximum.
>
> - If f'(left of c) < 0 and f'(right of c) > 0 then $(c, f(c))$ is a local minimum.
>
> - If f'(left of c) > 0 and f'(right of c) > 0 then $(c, f(c))$ is **not** a local extreme.
>
> - If f'(left of c) < 0 and f'(right of c) < 0 then $(c, f(c))$ is **not** a local extreme.

Practice 9. Find all extremes of $f(x) = 3x^2 - 12x + 7$ and use the First Derivative Test to classify them as maximums, minimums or neither.

3.3 Problems

In Problems 1–3, sketch the graph of the **derivative** of each function.

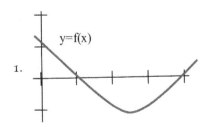

1.

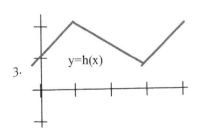

2.

3.

Problems 4–6 show the graph of the height of a helicopter; sketch a graph of its upward velocity.

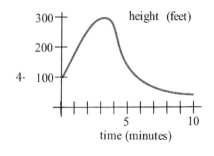

4.

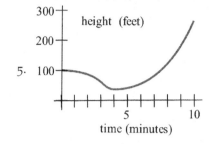

5.

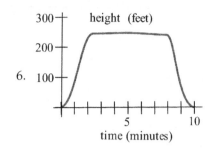

6.

7. In the figure below, match the graphs of the functions with those of their derivatives.

8. Match the graphs showing the heights of rockets with those showing their velocities.

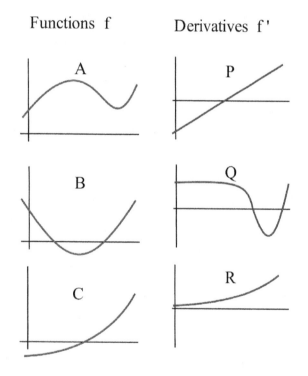

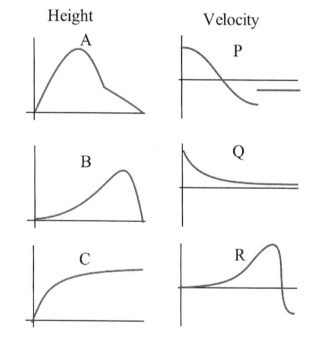

9. Use the Second Shape Theorem to show that $f(x) = \ln(x)$ is monotonic increasing on the interval $(0, \infty)$.

10. Use the Second Shape Theorem to show that $g(x) = e^x$ is monotonic increasing on the entire real number line.

11. A student is working with a complicated function f and has shown that the derivative of f is always positive. A minute later the student also claims that $f(x) = 2$ when $x = 1$ and when $x = \pi$. Without checking the student's work, how can you be certain that it contains an error?

12. The figure below shows the graph of the **derivative** of a continuous function f.

   (a) List the critical numbers of f.

   (b) What values of x result in a local maximum?

   (c) What values of x result in a local minimum?

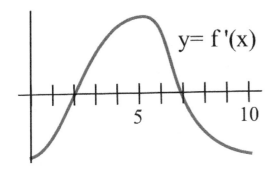

13. The figure below shows the graph of the **derivative** of a continuous function g.

   (a) List the critical numbers of g.

   (b) What values of x result in a local maximum?

   (c) What values of x result in a local minimum?

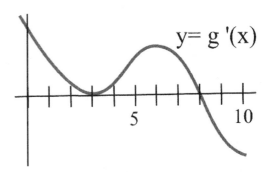

Problems 14–16 show the graphs of the upward velocities of three helicopters. Use the graphs to determine when each helicopter was at a (relative) maximum or minimum height.

14.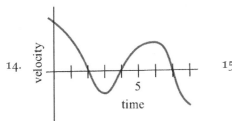

15.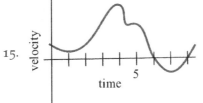

16.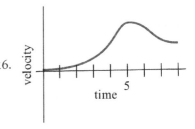

In 17–22, use information from the derivative of each function to help you graph the function. Find all local maximums and minimums of each function.

17. $f(x) = x^3 - 3x^2 - 9x - 5$

18. $g(x) = 2x^3 - 15x^2 + 6$

19. $h(x) = x^4 - 8x^2 + 3$

20. $s(t) = t + \sin(t)$

21. $r(t) = \dfrac{2}{t^2 + 1}$

22. $f(x) = \dfrac{x^2 + 3}{x}$

23. $f(x) = 2x + \cos(x)$ so $f(0) = 1$. Without graphing the function, you can be certain that f has how many **positive** roots?

24. $g(x) = 2x - \cos(x)$ so $g(0) = -1$. Without graphing the function, you can be certain that g has how many **positive** roots?

25. $h(x) = x^3 + 9x - 10$ has a root at $x = 1$. Without graphing h, show that h has no other roots.

26. Sketch the graphs of monotonic decreasing functions that have exactly (a) no roots (b) one root and (c) two roots.

27. Each of the following statements is false. Give (or sketch) a counterexample for each statement.

   (a) If f is increasing on an interval I, then $f'(x) > 0$ for all x in I.

   (b) If f is increasing and differentiable on I, then $f'(x) > 0$ for all x in I.

   (c) If cars A and B always have the same speed, then they will always be the same distance apart.

28. (a) Find several different functions f that all have the same derivative $f'(x) = 2$.

 (b) Determine a function f with derivative $f'(x) = 2$ that also satisfies $f(1) = 5$.

 (c) Determine a function g with $g'(x) = 2$ for which the graph of g goes through $(2, 1)$.

29. (a) Find several different functions h that all have the same derivative $h'(x) = 2x$.

 (b) Determine a function f with derivative $f'(x) = 2x$ that also satisfies $f(3) = 20$.

 (c) Determine a function g with $g'(x) = 2x$ for which the graph of g goes through $(2, 7)$.

30. Sketch functions with the given properties to help determine whether each statement is true or false.

 (a) If $f'(7) > 0$ and $f'(x) > 0$ for all x near 7, then $f(7)$ is a local maximum of f on $[1, 7]$.

 (b) If $g'(7) < 0$ and $g'(x) < 0$ for all x near 7, then $g(7)$ is a local minimum of g on $[1, 7]$.

 (c) If $h'(1) > 0$ and $h'(x) > 0$ for all x near 1, then $h(1)$ is a local minimum of h on $[1, 7]$.

 (d) If $r'(1) < 0$ and $r'(x) < 0$ for all x near 1, then $r(1)$ is a local maximum of r on $[1, 7]$.

 (e) If $s'(7) = 0$, then $s(7)$ is a local maximum of s on $[1, 7]$.

3.3 Practice Answers

1. g is increasing on $[2, 4]$ and $[6, 8]$; g is decreasing on $[0, 2]$ and $[4, 5]$; g is constant on $[5, 6]$.

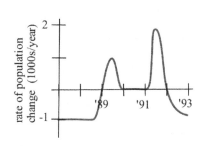

2. The graph in the margin shows the rate of population change, $\dfrac{dR}{dt}$.

3. A graph of f' appears below. Notice how the graph of f' is 0 where f has a maximum or minimum.

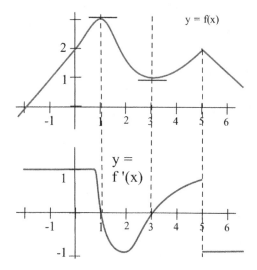

4. The Second Shape Theorem for helicopters:

 • If the upward velocity h' is positive during time interval I then the height h is increasing during time interval I.

 • If the upward velocity h' is negative during time interval I then the height h is decreasing during time interval I.

 • If the upward velocity h' is zero during time interval I then the height h is constant during time interval I.

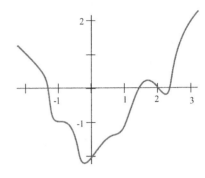

5. A graph satisfying the conditions in the table appears in the margin.

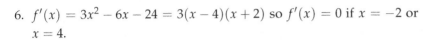

x	-2	-1	0	1	2	3
$f(x)$	1	-1	-2	-1	0	2
$f'(x)$	-1	0	1	2	-1	1

6. $f'(x) = 3x^2 - 6x - 24 = 3(x-4)(x+2)$ so $f'(x) = 0$ if $x = -2$ or $x = 4$.

   - $x < -2 \Rightarrow f'(x) = 3(\text{negative})(\text{negative}) > 0 \Rightarrow f$ increasing
   - $-2 < x < 4 \Rightarrow f'(x) = 3(\text{negative})(\text{positive}) < 0 \Rightarrow f$ decreasing
   - $x > 4 \Rightarrow f'(x) = 3(\text{positive})(\text{positive}) > 0 \Rightarrow f$ increasing

   Thus f has a relative maximum at $x = -2$ and a relative minimum at $x = 4$. A graph of f appears in the margin.

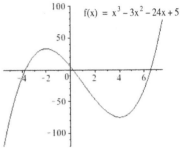

$f(x) = x^3 - 3x^2 - 24x + 5$

7. The figure below left shows several possible graphs for g. Each has the correct shape to give the graph of g'. Notice that the graphs of g are "parallel" (differ by a constant).

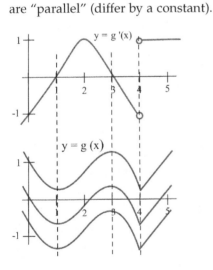

$y = g'(x)$

$y = g(x)$

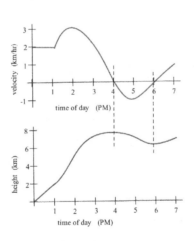

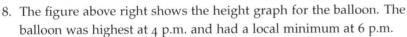

8. The figure above right shows the height graph for the balloon. The balloon was highest at 4 p.m. and had a local minimum at 6 p.m.

9. $f'(x) = 6x - 12$ so $f'(x) = 0$ only if $x = 2$.

   - $x < 2 \Rightarrow f'(x) < 0 \Rightarrow f$ decreasing
   - $x > 2 \Rightarrow f'(x) > 0 \Rightarrow f$ increasing

   From this we can conclude that f has a minimum when $x = 2$ and has a shape similar to graph provided in the margin.

   We could also have noticed that the graph of the quadratic function $f(x) = 3x^2 - 12x + 7$ must be an upward-opening parabola.

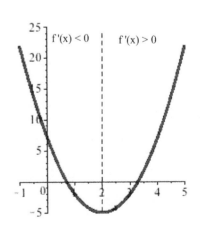

$f'(x) < 0$ $f'(x) > 0$

3.4   *The Second Derivative and the Shape of f*

The first derivative of a function provides information about the shape of the function, so the second derivative of a function provides information about the shape of the first derivative, which in turn will provide additional information about the shape of the original function f.

In this section we investigate how to use the second derivative (and the shape of the first derivative) to reach conclusions about the shape of the original function. The first derivative tells us whether the graph of f is increasing or decreasing. The second derivative will tell us about the "concavity" of f: whether f is curving upward or downward.

Concavity

Graphically, a function is **concave up** if its graph is curved with the opening upward (see margin); similarly, a function is **concave down** if its graph opens downward. The concavity of a function can be important in applied problems and can even affect billion-dollar decisions.

An Epidemic: Suppose you, as an official at the CDC, must decide whether current methods are effectively fighting the spread of a disease — or whether more drastic measures are required. In the margin figure, $f(x)$ represents the number of people infected with the disease at time x in two different situations. In both cases the number of people with the disease, $f(\text{now})$, and the rate at which new people are getting sick, $f'(\text{now})$, are the same. The difference is the concavity of f, and that difference might have a big effect on your decision. In (a), f is concave down at "now," and it appears that the current methods are starting to bring the epidemic under control; in (b), f is concave up, and it appears that the epidemic is growing out of control.

Usually it is easy to determine the concavity of a function by examining its graph, but we also need a definition that does not require a graph of the function, a definition we can apply to a function described by a formula alone.

Definition: Let f be a differentiable function.

- f is **concave up** at a if the graph of f is above the tangent line L to f for all x close to (but not equal to) a:

$$f(x) > L(x) = f(a) + f'(a)(x - a)$$

- f is **concave down** at a if the graph of f is below the tangent line L to f for all x close to (but not equal to) a:

$$f(x) < L(x) = f(a) + f'(a)(x - a)$$

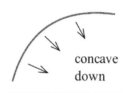

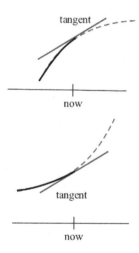

The margin figure shows the concavity of a function at several points. The next theorem provides an easily applied test for the concavity of a function given by a formula.

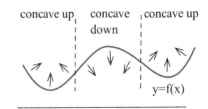

concave up | concave down | concave up

$y=f(x)$

The Second Derivative Condition for Concavity:
Let f be a twice differentiable function on an interval I.

(a) $f''(x) > 0$ on I $\Rightarrow$ $f'(x)$ increasing on I $\Rightarrow$ f concave up on I

(b) $f''(x) < 0$ on I $\Rightarrow$ $f'(x)$ decreasing on I $\Rightarrow$ f concave down on I

(c) $f''(a) = 0$ $\Rightarrow$ no information
  ($f(x)$ may be concave up or concave down or neither at a)

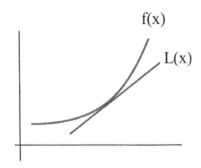

$f(x)$

$L(x)$

Proof. (a) Assume that $f''(x) > 0$ for all x in I, and let a be any point in I. We want to show that f is concave up at a, so we need to prove that the graph of f (see margin) is above the tangent line to f at a: $f(x) > L(x) = f(a) + f'(a)(x-a)$ for x close to a. Assume that x is in I and apply the Mean Value Theorem to f on the interval with endpoints a and x: there is a number c between a and x so that

$$f'(c) = \frac{f(x) - f(a)}{x - a} \Rightarrow f(x) = f(a) + f'(c)(x-a)$$

Because $f'' > 0$ on I, we know that $f'' > 0$ between a and x, so the Second Shape Theorem tells us that f' is increasing between a and x. We will consider two cases: $x > a$ and $x < a$.

- If $x > a$ then $x - a > 0$ and c is in the interval $[a, x]$ so $a < c$. Because f' is increasing, $a < c \Rightarrow f'(a) < f'(c)$. Multiplying each side of this last inequality by the positive quantity $x - a$ yields $f'(a)(x-a) < f'(c)(x-a)$. Adding $f(a)$ to each side of this last inequality, we have:

$$L(x) = f(a) + f'(a)(x-a) < f(a) + f'(c)(x-a) = f(x)$$

- If $x < a$ then $x - a < 0$ and c is in the interval $[x, a]$ so $c < a$. Because f' is increasing, $c < a \Rightarrow f'(c) < f'(a)$. Multiplying each side of this last inequality by the negative quantity $x - a$ yields $f'(c)(x-a) > f'(a)(x-a)$ so:

$$f(x) = f(a) + f'(c)(x-a) > f(a) + f'(a)(x-a) = L(x)$$

In each case we see that $f(x)$ is above the tangent line $L(x)$.

(b) The proof of this part is similar.

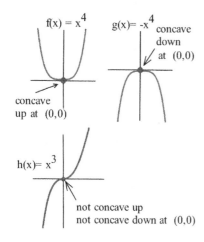

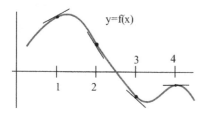

y=f(x)

(c) Let $f(x) = x^4$, $g(x) = -x^4$ and $h(x) = x^3$ (see margin). The second derivative of each of these functions is zero at $a = 0$, and at $(0,0)$ they all have the same tangent line: $L(x) = 0$ (the x-axis). However, at $(0,0)$ they all have different concavity: f is concave up, while g is concave down and h is neither concave up nor concave down.   □

Practice 1. Use the graph of f in the lower margin figure to finish filling in the table with "+" for positive, "−" for negative or "0."

x	$f(x)$	$f'(x)$	$f''(x)$	concavity
1	+	+	−	down
2	+			
3	−			
4				

"Feeling" the Second Derivative

Earlier we saw that if a function $f(t)$ represents the position of a car at time t, then $f'(t)$ gives the velocity and $f''(t)$ the acceleration of the car at the instant t.

If we are driving along a straight, smooth road, then what we *feel* is the acceleration of the car:

- a large positive acceleration feels like a "push" toward the back of the car

- a large negative acceleration (a deceleration) feels like a "push" toward the front of the car

- an acceleration of 0 for a period of time means the velocity is constant and we do not feel pushed in either direction

In a moving vehicle it is possible to measure these "pushes," the acceleration, and from that information to determine the velocity of the vehicle, and from the velocity information to determine the position. Inertial guidance systems in airplanes use this tactic: they measure front–back, left–right and up–down acceleration several times a second and then calculate the position of the plane. They also use computers to keep track of time and the rotation of the earth under the plane. After all, in six hours the Earth has made a quarter of a revolution, and Dallas has rotated more than 5,000 miles!

Example 1. The upward acceleration of a rocket was $a(t) = 30$ m/sec^2 during the first six seconds of flight, $0 \le t \le 6$. The velocity of the rocket at $t = 0$ was 0 m/sec and the height of the rocket above the ground at $t = 0$ was 25 m. Find a formula for the height of the rocket at time t and determine the height at $t = 6$ seconds.

Solution. $v'(t) = a(t) = 30 \Rightarrow v(t) = 30t + K$ for some constant K. We also know $v(0) = 0$ so $30(0) + K = 0 \Rightarrow K = 0$ and this $v(t) = 30t$.

Similarly, $h'(t) = v(t) = 30t \Rightarrow h(t) = 15t^2 + C$ for some constant C. We know that $h(0) = 25$ so $15(0)^2 + C = 25 \Rightarrow C = 25$. Thus $h(t) = 15t^2 + 25$ so $h(6) = 15(6)^2 + 25 = 565$ m.   ◀

f'' and Extreme Values of f

The concavity of a function can also help us determine whether a critical point is a maximum or minimum or neither. For example, if a point is at the bottom of a concave-up function then that point is a minimum.

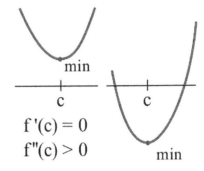

The Second Derivative Test for Extremes:

Let f be a twice differentiable function.

(a) If $f'(c) = 0$ and $f''(c) < 0$
   then f is concave down and has a local maximum at $x = c$.

(b) If $f'(c) = 0$ and $f''(c) > 0$
   then f is concave up and has a local minimum at $x = c$.

(c) If $f'(c) = 0$ and $f''(c) = 0$ then f may have a local maximum,
   a local minimum or neither at $x = c$.

Proof. (a) Assume that $f'(c) = 0$. If $f''(c) < 0$ then f is concave down at $x = c$ so the graph of f will be below the tangent line $L(x)$ for values of x near c. The tangent line, however, is given by $L(x) = f(c) + f'(c)(x - c) = f(c) + 0(x - c) = f(c)$, so if x is close to c then $f(x) < L(x) = f(c)$ and f has a local maximum at $x = c$.

(b) The proof for a local minimum of f is similar.

(c) If $f'(c) = 0$ and $f''(c) = 0$, then we cannot immediately conclude anything about local maximums or minimums of f at $x = c$. The functions $f(x) = x^4$, $g(x) = -x^4$ and $h(x) = x^3$ all have their first and second derivatives equal to zero at $x = 0$, but f has a local minimum at 0, g has a local maximum at 0, and h has neither a local maximum nor a local minimum at $x = 0$.   □

The Second Derivative Test for Extremes is very useful when f'' is easy to calculate and evaluate. Sometimes, however, the First Derivative Test—or simply a graph of the function—provides an easier way to determine if the function has a local maximum or a local minimum: it depends on the function and on which tools you have available.

Practice 2. $f(x) = 2x^3 - 15x^2 + 24x - 7$ has critical numbers $x = 1$ and $x = 4$. Use the Second Derivative Test for Extremes to determine whether $f(1)$ and $f(4)$ are maximums or minimums or neither.

Inflection Points

Maximums and minimums typically occur at places where the second derivative of a function is positive or negative, but the places where the second derivative is 0 are also of interest.

> **Definition:**
>
> An **inflection point** is a point on the graph of a function where the concavity of the function changes, from concave up to concave down or from concave down to concave up.

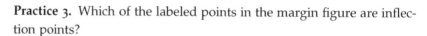

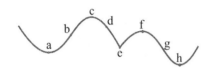

Practice 3. Which of the labeled points in the margin figure are inflection points?

To find the inflection points of a function we can use the second derivative of the function. If $f''(x) > 0$, then the graph of f is concave up at the point $(x, f(x))$ so $(x, f(x))$ is not an inflection point. Similarly, if $f''(x) < 0$ then the graph of f is concave down at the point $(x, f(x))$ and the point is not an inflection point. The only points left that can possibly be inflection points are the places where $f''(x) = 0$ or where $f''(x)$ does not exist (in other words, where f' is not differentiable). To find the inflection points of a function we need only check the points where $f''(x)$ is 0 or undefined. If $f''(c) = 0$ or is undefined, then the point $(c, f(c))$ **may or may not** be an inflection point—we need to check the concavity of f on each side of $x = c$. The functions in the next example illustrate what can happen at such a point.

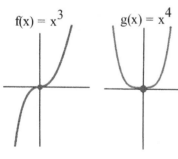

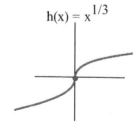

Example 2. Let $f(x) = x^3$, $g(x) = x^4$ and $h(x) = \sqrt[3]{x}$ (see margin). For which of these functions is the point $(0, 0)$ an inflection point?

Solution. Graphically, it is clear that the concavity of $f(x) = x^3$ and $h(x) = \sqrt[3]{x}$ changes at $(0, 0)$, so $(0, 0)$ is an inflection point for f and h. The function $g(x) = x^4$ is concave up everywhere, so $(0, 0)$ is not an inflection point of g.

$f(x) = x^3 \Rightarrow f'(x) = 3x^2 \Rightarrow f''(x) = 6x$ so the only point at which $f''(x) = 0$ or is undefined (f' is not differentiable) is at $x = 0$. If $x < 0$ then $f''(x) < 0$ so f is concave down; if $x > 0$ then $f''(x) > 0$ so f is concave up. Thus at $x = 0$ the concavity of f changes so the point $(0, f(0)) = (0, 0)$ is an inflection point of $f(x) = x^3$.

$g(x) = x^4 \Rightarrow g'(x) = 4x^3 \Rightarrow g''(x) = 12x^2$ so the only point at which $g''(x) = 0$ or is undefined is at $x = 0$. But $g''(x) > 0$ (so g is concave up) for any $x \neq 0$. Thus the concavity of g never changes, so the point $(0, g(0)) = (0, 0)$ is not an inflection point of $g(x) = x^4$.

$h(x) = \sqrt[3]{x} = x^{\frac{1}{3}} \Rightarrow h'(x) = \frac{1}{3}x^{-\frac{2}{3}} \Rightarrow h''(x) = -\frac{2}{9}x^{-\frac{5}{3}}$ so h'' is not defined if $x = 0$ (and $h''(x) \neq 0$ elsewhere); $h''(\text{negative number}) > 0$

and h''(positive number) < 0, so h changes concavity at $(0,0)$ and $(0,0)$ is an inflection point of $h(x) = \sqrt[3]{x}$.   ◀

Practice 4. Find all inflection points of $f(x) = x^4 - 12x^3 + 30x^2 + 5x - 7$.

Example 3. Sketch a graph of a function with $f(2) = 3$, $f'(2) = 1$ and an inflection point at $(2,3)$.

Solution. Two solutions appear in the margin.   ◀

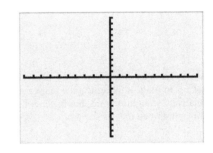

Using f' and f'' to Graph f

Today you can easily graph most functions of interest using a graphing calculator — and create even nicer graphs using an app on your phone or a Web-based graphing utility. Earlier generations of calculus students did not have these tools, so they relied on calculus to help them draw graphs of unfamiliar functions by hand. While you can create a graph in seconds that your predecessors may have labored over for half an hour or longer, you can still use calculus to help you select an appropriate graphing "window," and to be confident that your window has not missed any points of interest on the graph of a function.

Example 4. Create a graph of $f(x) = xe^{-9x^2}$ that shows all local and global extrema and all inflection points.

Solution. If you graph $f(x)$ on a calculator using the standard window ($-10 \le x \le 10$ and $-10 \le y \le 10$) you will likely see nothing other than the coordinate axes (see margin). You might consult a table of values for the function to help adjust the window, but this trial-and-error technique will still not guarantee that you have displayed all points of interest. Computing the first derivative of f, we get:

$$f'(x) = x\left[-18xe^{-9x^2}\right] + e^{-9x^2} \cdot 1 = \left[1 - 18x^2\right]e^{-9x^2}$$

which is defined for all values of x; $f'(x) = 0 \Rightarrow 1 - 18x^2 = 0 \Rightarrow x^2 = \frac{1}{18} \Rightarrow x = \pm\frac{1}{3\sqrt{2}}$, so the only critical numbers are $x = -\frac{1}{3\sqrt{2}}$ and $x = \frac{1}{3\sqrt{2}}$. Computing the second derivative of f, we get:

$$f''(x) = \left[1 - 18x^2\right] \cdot \left[-18xe^{-9x^2}\right] + e^{-9x^2} \cdot [-36x]$$
$$= \left[324x^3 - 54x\right]e^{-9x^2} = 54x\left[6x^2 - 1\right]e^{-9x^2}$$

We can check that $f''\left(-\frac{1}{3\sqrt{2}}\right) = 12\sqrt{\frac{2}{e}} > 0$, so f must have a local minimum at $x = -\frac{1}{3\sqrt{2}}$; similarly, $f''\left(\frac{1}{3\sqrt{2}}\right) = -12\sqrt{\frac{2}{e}} < 0$, so f must have a local maximum at $x = \frac{1}{3\sqrt{2}}$.

Furthermore, $f''(x) = 0$ only when $x = 0$ or when $6x^2 - 1 = 0 \Rightarrow$ $x = \pm\frac{1}{\sqrt{6}}$, so these three values are candidates for locations of inflection points of f. Noting that:

$$-1 < -\frac{1}{\sqrt{6}} < -\frac{1}{3\sqrt{2}} < 0 < \frac{1}{3\sqrt{2}} < \frac{1}{\sqrt{6}} < 1$$

and that $f''(-1) = -270e^{-9} < 0$ and $f''\left(-\frac{1}{3\sqrt{2}}\right) = 12\sqrt{\frac{2}{e}} > 0$, we observe that f is concave down to the left of $x = -\frac{1}{\sqrt{6}}$ and concave up to the right of $x = -\frac{1}{\sqrt{6}}$, so f does in fact have an inflection point at $x = -\frac{1}{\sqrt{6}}$. Likewise, $f''\left(\frac{1}{3\sqrt{2}}\right) = -12\sqrt{\frac{2}{e}} < 0$ and $f''(1) = 270e^{-9} > 0$, so $f''(x)$ switches sign at $x = 0$ and at $x = \frac{1}{\sqrt{6}}$, and therefore $f(x)$ changes concavity at those points as well.

We have now identified two local extrema of f and three inflection points of f. Equally important, we have used calculus to show that these five points of interest are the *only* places where extrema or inflection points can occur. If we create a graph of f that includes these five points, our graph is guaranteed to include all "interesting" features of the graph of f. A window with $-1 \le x \le 1$ and $-0.2 < y < 0.2$ (because the local extreme values are $f\left(\pm\frac{1}{2\sqrt{3}}\right) \approx \pm 0.14$) should provide a graph (see margin) that includes all five points of interest. ◀

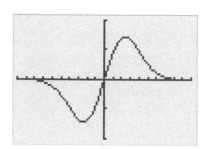

Practice 5. Compute the first and second derivatives of the function $g(x) = x^4 + 4x^3 - 90x^2 + 13$, locate all extrema and inflection points of $g(x)$, and create a graph of $g(x)$ that shows these points of interest.

Most problems in calculus textbooks are set up to make solving these equations relatively straightforward, but in general this will not be the case.

Even with calculus, we will typically need calculators or computers to help solve the equations $f'(x) = 0$ and $f''(x) = 0$ that we use to find critical numbers and candidates for inflection points.

3.4 Problems

In Problems 1–2, each statement describes a quantity $f(t)$ changing over time. For each statement, tell what f represents and whether the first and second derivatives of f are positive or negative.

1. (a) "Unemployment rose again, but the rate of increase is smaller than last month."

   (b) "Our profits declined again, but at a slower rate than last month."

   (c) "The population is still rising and at a faster rate than last year."

2. (a) "The child's temperature is still rising, but more slowly than it was a few hours ago."

   (b) "The number of whales is decreasing, but at a slower rate than last year."

   (c) "The number of people with the flu is rising and at a faster rate than last month."

3. Sketch the graphs of functions that are defined and concave up everywhere and have exactly:
   (a) no roots. (b) 1 root. (c) 2 roots. (d) 3 roots.

4. On which intervals is the function graphed below:

   (a) concave up?   (b) concave down?

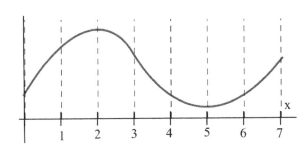

5. On which intervals is the function graphed below:

   (a) concave up?   (b) concave down?

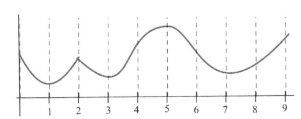

Problems 6–10 give a function and values of x so that $f'(x) = 0$. Use the Second Derivative Test to determine whether each point $(x, f(x))$ is a local maximum, a local minimum or neither.

6. $f(x) = 2x^3 - 15x^2 + 6$; $x = 0, 5$

7. $g(x) = x^3 - 3x^2 - 9x + 7$; $x = -1, 3$

8. $h(x) = x^4 - 8x^2 - 2$; $x = -2, 0, 2$

9. $f(x) = \sin^5(x)$; $x = \frac{\pi}{2}, \pi, \frac{3\pi}{2}$

10. $f(x) = x \cdot \ln(x)$; $x = \frac{1}{e}$

11. At which values of x labeled in the figure below is the point $(x, f(x))$ an inflection point?

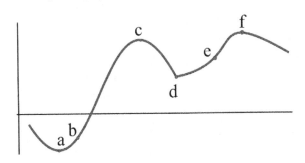

12. At which values of x labeled in the figure below is the point $(x, g(x))$ an inflection point?

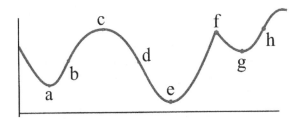

13. How many inflection points can a:

   (a) quadratic polynomial have?

   (b) cubic polynomial have?

   (c) polynomial of degree n have?

14. Fill in the table with "+," "−,"or "0" for the function graphed below.

x	$f(x)$	$f'(x)$	$f''(x)$
0			
1			
2			
3			

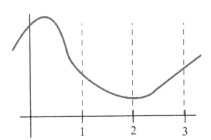

15. Fill in the table with "+," "−,"or "0" for the function graphed below.

x	$g(x)$	$g'(x)$	$g''(x)$
0			
1			
2			
3			

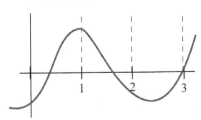

16. Sketch functions f for x-values near 1 so that $f(1) = 2$ and:

    (a) $f'(1) > 0$, $f''(1) > 0$

    (b) $f'(1) > 0$, $f''(1) < 0$

    (c) $f'(1) < 0$, $f''(1) > 0$

    (d) $f'(1) > 0$, $f''(1) = 0$, $f''(1^-) < 0$, $f''(1^+) > 0$

    (e) $f'(1) > 0$, $f''(1) = 0$, $f''(1^-) > 0$, $f''(1^+) < 0$

17. Some people like to think of a concave-up graph as one that will "hold water" and of a concave-down graph as one which will "spill water." That description is accurate for a concave-down graph, but it can fail for a concave-up graph. Sketch the graph of a function that is concave up on an interval but will not "hold water."

18. The function $f(x) = \dfrac{1}{2\pi} e^{-\frac{(x-c)^2}{2b^2}}$ defines the **Gaussian distribution** used extensively in statistics and probability; its graph (see below) is a "bell-shaped" curve.

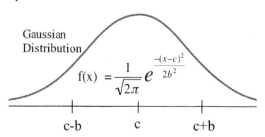

Gaussian Distribution

$$f(x) = \frac{1}{\sqrt{2\pi}} e^{\frac{-(x-c)^2}{2b^2}}$$

c-b   c   c+b

    (a) Show that f has a maximum at $x = c$. (The value c is called the **mean** of this distribution.)

    (b) Show that f has inflection points where $x = c + b$ and $x = c - b$. (The value b is called the **standard deviation** of this distribution.)

In Problems 19–36, locate all critical numbers, local extrema and inflection points of the given function, and use these results to sketch a graph of the function showing all points of interest.

19. $f(x) = x^3 - 21x^2 + 144x - 350$

20. $g(x) = \dfrac{1}{6}x^3 + x^2 - \dfrac{45}{2}x + 100$

21. $f(x) = e^{7x} - 5x$

22. $g(x) = e^{7x} - 5x$

23. $f(x) = e^{-3x} + x$

24. $g(x) = e^{-3x} - x$

25. $f(x) = xe^{-3x}$

26. $g(x) = xe^{5x}$

27. $f(x) = x^{\frac{4}{3}} - x^{\frac{1}{3}}$

28. $g(x) = 6x^{\frac{4}{3}} + 3x^{\frac{1}{3}}$

29. $f(x) = \ln\left(1 + x^2\right)$

30. $g(x) = \ln\left(x^2 - 6x + 10\right)$

31. $f(x) = \sqrt[3]{x^2 + 2x + 2}$

32. $g(x) = \sqrt{x^2 + 2x + 2}$

33. $f(x) = x^{\frac{2}{3}}\left(1 - x\right)^{\frac{1}{3}}$

34. $g(x) = x^{\frac{1}{3}}\left(1 - x\right)^{\frac{2}{3}}$

35. $f(\theta) = \sin(\theta) + \sin^2(\theta)$

36. $g(\theta) = \cos(\theta) - \sin^2(\theta)$

3.4 Practice Answers

1. See the margin figure for reference.

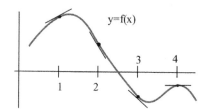

y=f(x)

x	$f(x)$	$f'(x)$	$f''(x)$	concavity
1	+	+	−	down
2	+	−	−	down
3	−	−	+	up
4	−	0	−	down

2. $f'(x) = 6x^2 - 30x + 24$, which is defined for all x. $f'(x) = 0$ if $x = 1$ or $x = 4$ (critical values). $f''(x) = 12x - 30$ so $f''(1) = -18 < 0$ tells us that f is concave down at the critical value $x = 1$, so $(1, f(1)) = (1, 4)$ is a relative maximum; and $f''(4) = 18 > 0$ tells us that f is concave up at the critical value $x = 4$, so $(4, f(4)) = (4, -23)$ is a relative minimum. A graph of f appears in the margin.

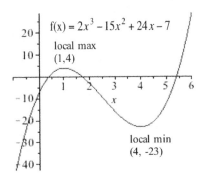

3. The points labeled b and g are inflection points.

4. $f'(x) = 4x^3 - 36x^2 + 60x + 5 \Rightarrow f''(x) = 12x^2 - 72x + 60 = 12(x^2 - 6x + 5) = 12(x - 1)(x - 5)$ so the only candidates to be inflection points are $x = 1$ and $x = 5$.

   - If $x < 1$ then $f''(x) = 12(\text{neg})(\text{neg}) > 0$
   - If $1 < x < 5$ then $f''(x) = 12(\text{pos})(\text{neg}) < 0$
   - If $5 < x$ then $f''(x) = 12(\text{pos})(\text{pos}) > 0$

   f changes concavity at $x = 1$ and $x = 5$, so $x = 1$ and $x = 5$ are both inflection points. A graph of f appears in the margin.

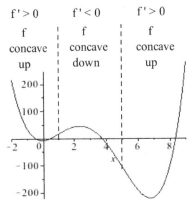

$f(x) = x^4 - 12x^3 + 30x^2 + 5x - 7$

5. $g(x) = x^4 + 4x^3 - 90x^2 + 13 \Rightarrow g'(x) = 4x^3 + 12x^2 - 180x \Rightarrow g''(x) = 12x^2 + 24x - 180$; because $g'(x)$ and $g''(x)$ are polynomials, they exist everywhere. The critical numbers for $g(x)$ occur where $g'(x) = 0 \Rightarrow 4x^3 + 12x^2 - 180x = 4x(x^2 + 3x - 45) = 4x(x + 9)(x - 5) = 0 \Rightarrow x = -9$, $x = 0$ or $x = 5$. Using the Second Derivative Test: $g''(-9) = 576 > 0$, so $g(x)$ has a local minimum at $x = -9$; $g''(0) = -180 < 0$, so $g(x)$ has a local maximum at $x = 0$; and $g''(5) = 240 > 0$, so $g(x)$ has a local minimum at $x = 5$.

Candidates for inflection points occur where $g''(x) = 0$:

$$12x^2 + 24x - 180 = 12(x^2 + 2x - 15) = 12(x - 3)(x + 5) = 0$$
$$\Rightarrow x = -5 \text{ or } x = 3$$

Observing that $g''(x) > 0$ for $x < -5$, $g''(x) < 0$ for $-5 < x < 3$ and $g''(x) > 0$ for $x > 3$ confirms that both candidates are in fact inflection points. A graphing window with $-12 \le x \le 8$ (this is only one reasonable possibility) should include all points of interest. Checking that $g(-9) = -3632$, $g(0) = 13$ and $g(5) = -1112$ suggests that a graphing window with $-4000 \le y \le 1000$ should work (see margin).

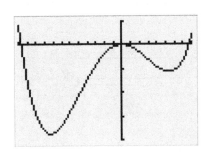

3.5 Applied Maximum and Minimum Problems

We have used derivatives to find maximums and minimums of functions given by formulas, but it is very unlikely that someone will simply hand you a function and ask you to find its extreme value(s). Typically, someone will describe a problem and ask your help to maximize or minimize a quantity: "What is the largest volume of a package that the post office will accept?"; "What is the quickest way to get from here to there?"; or "What is the least expensive way to accomplish some task?" These problems often involve restrictions — or **constraints** — and sometimes neither the problem nor the constraints are clearly stated.

Before we can use calculus or other mathematical techniques to solve these "**max/min**" problems, we need to understand the situation at hand and translate the problem into mathematical form. After solving the problem using calculus (or other mathematical techniques) we need to check that our mathematical solution really solves the original problem. Often, the most challenging part of this procedure is understanding the problem and translating it into mathematical form.

In this section we examine some problems that require understanding, translation, solution and checking. Most will not be as complicated as those a working scientist, engineer or economist needs to solve, but they represent a step toward developing the required skills.

Example 1. The company you own has a large supply of 8-inch by 15-inch rectangular pieces of tin, and you decide to use them to make boxes by cutting a square from each corner and folding up the sides (see margin). For example, if you cut a 1-inch square from each corner, the resulting 6-inch by 13-inch by 1-inch box has a volume of 78 cubic inches. The amount of money you can charge for a box depends on how much the box holds, so you want to make boxes with the largest possible volume. What size square should you cut from each corner?

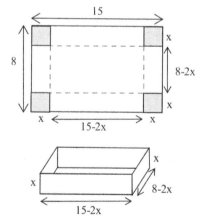

Solution. To help understand the problem, first drawing a diagram can be very helpful. Then we need to translate it into a mathematical problem:

- identify the variables

- label the variable and constant parts of the diagram

- write the quantity to be maximized as a function of the variables

If we label the side of the square to be removed as x inches, then the box is x inches high, $8 - 2x$ inches wide and $15 - 2x$ inches long, so the volume is:

$$(\text{length})(\text{width})(\text{height}) = (15 - 2x)(8 - 2x) \cdot x$$
$$= 4x^3 - 46x^2 + 120x \text{ cubic inches}$$

Now we have a mathematical problem, to maximize the function $V(x) = 4x^3 - 46x^2 + 120x$, so we use existing calculus techniques, computing $V'(x) = 12x^2 - 92x + 120$ to find the critical points.

- Set $V'(x) = 0$ and solve by factoring or using the quadratic formula:

$$V'(x) = 12x^2 - 92x + 120 = 4(3x - 5)(x - 6) = 0 \Rightarrow x = \frac{5}{3} \text{ or } x = 6$$

so $x = \frac{5}{3}$ and $x = 6$ are critical points of V.

- $V'(x)$ is a polynomial so it is defined everywhere and there are no critical points resulting from an undefined derivative.

- What are the endpoints for x in this problem? A square cannot have a negative length, so $x \geq 0$. We cannot remove more than half of the width, so $8 - 2x \geq 0 \Rightarrow x \leq 4$. Together, these two inequalities say that $0 \leq x \leq 4$, so the endpoints are $x = 0$ and $x = 4$. (Note that the value $x = 6$ is not in this interval, so $x = 6$ cannot maximize the volume and we do not consider it further.)

The maximum volume must occur at the critical point $x = \frac{5}{3}$ or at one of the endpoints ($x = 0$ and $x = 4$): $V(0) = 0$, $V(\frac{5}{3}) = \frac{2450}{27} \approx 90.74$ cubic inches, and $V(4) = 0$, so the maximum volume of the box occurs when we remove a $\frac{5}{3}$-inch by $\frac{5}{3}$-inch square from each corner, resulting in a box $\frac{5}{3}$ inches high, $8 - 2(\frac{5}{3}) = \frac{14}{3}$ inches wide and $15 - 2(\frac{5}{3}) = \frac{35}{3}$ inches long. ◀

Practice 1. If you start with 7-inch by 15-inch pieces of tin, what size square should you remove from each corner so the box will have as large a volume as possible? [Hint: $12x^2 - 88x + 105 = (2x - 3)(6x - 35)$]

We were fortunate in the previous Example and Practice problem because the functions we created to describe the volume were functions of only one variable. In other situations, the function we get will have more than one variable, and we will need to use additional information to rewrite our function as a function of a single variable. Typically, the constraints will contain the additional information we need.

Example 2. We want to fence a rectangular area in our backyard for a garden. One side of the garden is along the edge of the yard, which is already fenced, so we only need to build a new fence along the other three sides of the rectangle (see margin). If a neighbor gives us 80 feet of fencing left over from a home-improvement project, what dimensions should the garden have in order to enclose the largest possible area using all of the available material?

Solution. As a first step toward understanding the problem, we draw a diagram or picture of the situation. Next, we identify the variables:

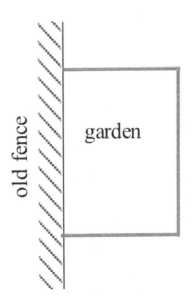

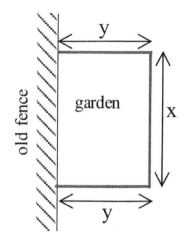

in this case, the length (call it x) and width (call it y) of the garden. The margin figure shows a labeled diagram, which we can use to write a formula for the function that we want to maximize:

$$A = \text{area of the rectangle} = (\text{length})(\text{width}) = x \cdot y$$

Unfortunately, our function A involves two variables, x and y, so we need to find a relationship between them (an equation containing both x and y) that we can solve for wither x or y. The constraint says that we have 80 feet of fencing available, so $x + 2y = 80 \Rightarrow y = 40 - \frac{x}{2}$. Then:

$$A = x \cdot y = x \left(40 - \frac{x}{2}\right) = 40x - \frac{x^2}{2}$$

which is a function of a single variable (x). We want to maximize A.

$A'(x) = 40 - x$ so the only way $A'(x) = 0$ is to have $x = 40$, and $A(x)$ is differentiable for all x so the only critical number (other than the endpoints) is $x = 40$. Finally, $0 \leq x \leq 80$ (why?) so we also need to check $x = 0$ and $x = 80$: the maximum area must occur at $x = 0$, $x = 40$ or $x = 80$.

$$A(0) = 40(0) - \frac{0^2}{2} = 0 \text{ square feet}$$

$$A(40) = 40(40) - \frac{40^2}{2} = 800 \text{ square feet}$$

$$A(80) = 40(80) - \frac{80^2}{2} = 0 \text{ square feet}$$

so the largest rectangular garden has an area of 800 square feet, with dimensions $x = 40$ feet by $y = 40 - \frac{40}{2} = 20$ feet. ◀

Practice 2. Suppose you decide to create the rectangular garden in a **corner** of your yard. Then two sides of the garden are bounded by the existing fence, so you only need to use the available 80 feet of fencing to enclose the other two sides. What are the dimensions of the new garden of largest area? What are the dimensions if you have F feet of new fencing available?

Example 3. You need to reach home as quickly as possible, but you are in a rowboat on a lake 4 miles from shore and your home is 2 miles up the shore (see margin). If you can row at 3 miles per hour and walk at 5 miles per hour, toward which point on the shore should you row? What if your home is 7 miles up the coast?

Solution. The margin figure shows a labeled diagram with the variable x representing the distance along the shore from point A, the nearest point on the shore to your boat, to point P, the point you row toward.

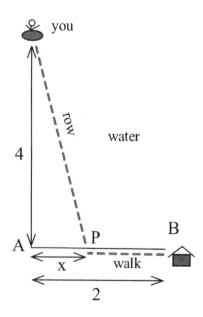

The total time — rowing and walking — is:

T = total time

= (rowing time from boat to P) + (walking time from P to B)

$$= \frac{\text{distance from boat to } P}{\text{rate rowing boat}} + \frac{\text{distance from } P \text{ to } B}{\text{rate walking along shore}}$$

$$= \frac{\sqrt{x^2 + 4^2}}{3} + \frac{2 - x}{5} = \frac{\sqrt{x^2 + 16}}{3} + \frac{2 - x}{5}$$

It is not reasonable to row to a point below A and then walk home, so $x \geq 0$. Similarly, we can conclude that $x \leq 2$, so our interval is $0 \leq x \leq 2$ and the endpoints are $x = 0$ and $x = 2$.

To find the other critical numbers of T between $x = 0$ and $x = 2$, we need the derivative of T:

$$T'(x) = \frac{1}{3} \cdot \frac{1}{2} \left(x^2 + 16 \right)^{-\frac{1}{2}} (2x) - \frac{1}{5} = \frac{x}{3\sqrt{x^2 + 16}} - \frac{1}{5}$$

This derivative is defined for all values of x (and in particular for all values in the interval $0 \leq x \leq 2$). To find where $T'(x) = 0$ we solve:

$$\frac{x}{3\sqrt{x^2 + 16}} - \frac{1}{5} = 0 \;\Rightarrow\; 5x = 3\sqrt{x^2 + 16}$$

$$\Rightarrow\; 25x^2 = 9x^2 + 144$$

$$\Rightarrow\; 16x^2 = 144 \;\Rightarrow\; x^2 = 9 \;\Rightarrow\; x = \pm 3$$

Neither of these numbers, however, is in our interval $0 \leq x \leq 2$, so neither of them gives a minimum time. The only critical numbers for T on this interval are the endpoints, $x = 0$ and $x = 2$:

$$T(0) = \frac{\sqrt{0 + 16}}{3} + \frac{2 - 0}{5} = \frac{4}{3} + \frac{2}{5} \approx 1.73 \text{ hours}$$

$$T(2) = \frac{\sqrt{2^2 + 16}}{3} + \frac{2 - 2}{5} = \frac{\sqrt{20}}{3} \approx 1.49 \text{ hours}$$

The quickest route has P 2 miles down the coast: you should row directly toward home.

If your home is 7 miles down the coast, then the interval for x is $0 \leq x \leq 7$, which has endpoints $x = 0$ and $x = 7$. Our function for the travel time is now:

$$T(x) = \frac{\sqrt{x^2 + 16}}{3} + \frac{7 - x}{5} \;\Rightarrow\; T'(x) = \frac{x}{3\sqrt{x^2 + 16}} - \frac{1}{5}$$

so the only point in our interval where $T'(x) = 0$ is at $x = 3$ and the derivative is defined for all values in this interval. So the only critical numbers for T are $x = 0$, $x = 3$ and $x = 7$:

$$T(0) = \frac{\sqrt{0 + 16}}{3} + \frac{7 - 0}{5} = \frac{4}{3} + \frac{7}{5} \qquad \approx 2.73 \text{ hours}$$

$$T(3) = \frac{\sqrt{3^2 + 16}}{3} + \frac{7 - 3}{5} = \frac{\sqrt{65}}{3} + \frac{4}{5} \qquad \approx 2.47 \text{ hours}$$

$$T(7) = \frac{\sqrt{7^2 + 16}}{3} + \frac{7 - 7}{5} = \frac{5}{3} \qquad \approx 2.68 \text{ hours}$$

The quickest way home is to aim for a point P that is 3 miles down the shore, row directly to P, and then walk along the shore to home. ◀

One challenge of max/min problems is that they may require geometry, trigonometry or other mathematical facts and relationships.

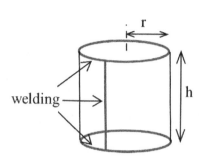

Example 4. Find the height and radius of the least expensive closed cylinder that has a volume of 1,000 cubic inches. Assume that the materials needed to construct the cylinder are free, but that it costs 80¢ per inch to weld the top and bottom onto the cylinder and to weld the seam up the side of the cylinder (see margin).

Solution. If we let r be the radius of the cylinder and h be its height, then the volume is $V = \pi r^2 h = 1000$. The quantity we want to minimize is cost, and

$$C = (\text{top seam cost}) + (\text{bottom seam cost}) + (\text{side seam cost})$$
$$= (\text{total seam length}) \left(80 \, \frac{¢}{\text{inch}} \right)$$
$$= (2\pi r + 2\pi r + h) \, (80) = 320\pi r + 80h$$

Unfortunately, C is a function of two variables, r and h, but we can use the information in the constraint ($V = \pi r^2 h = 1000$) to solve for h and then substitute this expression for h into the formula for C:

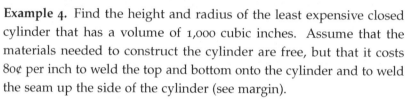

$$1000 = \pi r^2 h \Rightarrow h = \frac{1000}{\pi r^2} \Rightarrow C = 320\pi r + 80h = 320\pi r + 80 \left(\frac{1000}{\pi r^2} \right)$$

which is a function of a single variable. Differentiating:

$$C'(r) = 320\pi - \frac{160000}{\pi r^3}$$

which is defined except when $r = 0$ (a value that does not make sense in the original problem) and there are no restrictions on r (other than $r > 0$) so there are no endpoints to check. Thus C will be at a minimum when $C'(r) = 0$:

$$320\pi - \frac{160000}{\pi r^3} = 0 \Rightarrow r^3 = \frac{500}{\pi^2} \Rightarrow r = \sqrt[3]{\frac{500}{\pi^2}}$$

so $r \approx 3.7$ inches and $h = \frac{1000}{\pi r^2} = \frac{1000}{\pi \left(\sqrt[3]{\frac{500}{\pi^2}} \right)^2} \approx 23.3$ inches. ◀

Practice 3. Find the height and radius of the least expensive closed cylinder that has a volume of 1,000 cubic inches, assuming that the only cost for this cylinder is the price of the materials: the material for the top and bottom costs 5¢ per square inch, while the material for the sides costs 3¢ per square inch (see margin).

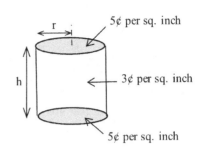

5¢ per sq. inch

3¢ per sq. inch

5¢ per sq. inch

Example 5. Find the dimensions of the least expensive rectangular box that is three times as long as it is wide and which holds 100 cubic centimeters of water. The material for the bottom costs 7¢ per cm², the sides cost 5¢ per cm² and the top costs 2¢ per cm².

Solution. Label the box so that w = width, l = length and h = height. Then our cost function C is:

C = (bottom cost) + (cost of front and back) + (cost of ends) + (top cost)

$\quad$ = (bottom area)(7) + (front and back area)(5) + (ends area)(5) + (top area)(2)

$\quad$ = $(wl)(7) + (2lh)(5) + (2wh)(5) + (wl)(2)$

$\quad$ = $7wl + 10lh + 10wh + 2wl$

$\quad$ = $9wl + 10lh + 10wh$

Unfortunately, C is a function of three variables (w, l and h) but we can use the information from the constraints to eliminate some of the variables: the box is "three times as long as it is wide" so $l = 3w$ and

$$C = 9wl + 10lh + 10wh = 9w(3w) + 10(3w)h + 10wh = 27w^2 + 40wh$$

We also know the volume V is 100 in³ and $V = lwh = 3w^2h$ (because $l = 3w$), so $h = \dfrac{100}{3w^2}$. Then:

$$C = 27w^2 + 40wh = 27w^2 + 40w\left(\frac{100}{3w^2}\right) = 27w^2 + \frac{4000}{3w}$$

which is a function of a single variable. Differentiating:

$$C'(w) = 54w - \frac{4000}{3w^2}$$

which is defined everywhere except $w = 0$ (yielding a box of volume 0) and there is no constraint interval, so C is minimized when $C'(w) = 0 \Rightarrow w = \sqrt[3]{\dfrac{4000}{162}} \approx 2.91$ inches $\Rightarrow l = 3w \approx 8.73$ inches $\Rightarrow h = \frac{100}{3w^2} \approx 3.94$ inches. The minimum cost is approximately \$6.87. ◀

Problems described in words are usually more difficult to solve because we first need to understand and "translate" a real-life problem into a mathematical problem. Unfortunately, those skills only seem to come with practice. With practice, however, you will start to recognize patterns for understanding, translating and solving these problems, and you will develop the skills you need. So read carefully, draw pictures, think hard — and do the best you can.

3.5 Problems

1. (a) You have 200 feet of fencing to enclose a rectangular vegetable garden. What should the dimensions of your garden be in order to enclose the largest area?

   (b) Show that if you have P feet of fencing available, the garden of greatest area is a square.

   (c) What are the dimensions of the largest rectangular garden you can enclose with P feet of fencing if one edge of the garden borders a straight river and does not need to be fenced?

   (d) Just thinking — calculus will not help: What do you think is the shape of the largest garden that can be enclosed with P feet of fencing if we do not require the garden to be rectangular? What if one edge of the garden borders a (straight) river?

2. (a) You have 200 feet of fencing available to construct a rectangular pen with a fence divider down the middle (see below). What dimensions of the pen enclose the largest total area?

   (b) If you need two dividers, what dimensions of the pen enclose the largest area?

   (c) What are the dimensions in parts (a) and (b) if one edge of the pen borders on a river and does not require any fencing?

3. You have 120 feet of fencing to construct a pen with four equal-sized stalls.

   (a) If the pen is rectangular and shaped like the one shown below, what are the dimensions of the pen of largest area and what is that area?

   (b) The square pen below uses 120 feet of fencing but encloses a larger area (400 ft²) than the best

design in part (a). Design a pen that uses only 120 feet of fencing and has four equal-sized stalls but encloses more than 400 ft². (Hint: Don't use rectangles and squares.)

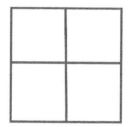

4. (a) You need to form a 10-inch by 15-inch piece of tin into a box (with no top) by cutting a square from each corner and folding up the sides. How much should you cut so the resulting box has the greatest volume?

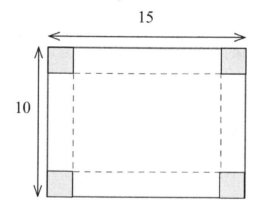

   (b) If the piece of tin is A inches by B inches, how much should you cut from each corner so the resulting box has the greatest volume?

5. Find the dimensions of a box with largest volume formed from a 10-inch by 10-inch piece of cardboard cut and folded as shown below.

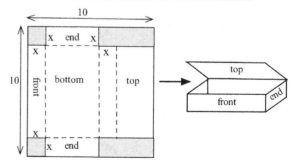

6. (a) You must construct a square-bottomed box with no top that will hold 100 cubic inches of water. If the bottom and sides are made from the same material, what are the dimensions of the box that uses the least material? (Assume that no material is wasted.)

   (b) Suppose the box in part (a) uses different materials for the bottom and the sides. If the bottom material costs 5¢ per square inch and the side material costs 3¢ per square inch, what are the dimensions of the least expensive box that will hold 100 cubic inches of water?

   (This is a "classic" problem with many variations. We could require that the box be twice as long as it is wide, or that the box have a top, or that the ends cost a different amount than the front and back, or even that it costs a certain amount to weld each edge. You should be able to set up the cost equations for these variations.)

7. (a) Determine the dimensions of the least expensive cylindrical can that will hold 100 cubic inches if the materials cost 2¢, 5¢ and 3¢ per square inch, respectively, for the top, bottom and sides.

   (b) How do the dimensions of the least expensive can change if the bottom material costs more than 5¢ per square inch?

8. You have 100 feet of fencing to build a pen in the shape of a circular sector, the "pie slice" shown below. The area of such a sector is $\frac{rs}{2}$.

   (a) What value of r maximizes the enclosed area?

   (b) What central angle maximizes the area?

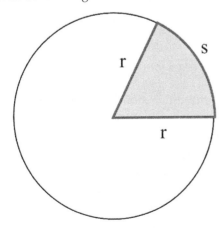

9. You are a lifeguard standing at the edge of the water when you notice a swimmer in trouble (see figure below) 40 m out in the water from a point 60 m down the beach. Assuming you can run at a speed of 8 meters per second and swim at a rate of 2 meters per second, how far along the shore should you run before diving into the water in order to reach the swimmer as quickly as possible?

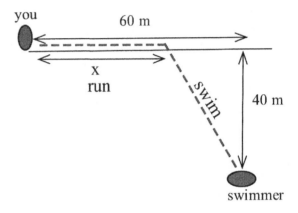

10. You have been asked to determine the least expensive route for a telephone cable that connects Andersonville with Beantown (see figure below).

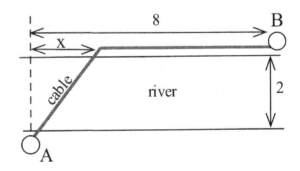

   (a) If it costs $5000 per mile to lay the cable on land and $8000 per mile to lay the cable across the river (with the cost of the cable included), find the least expensive route.

   (b) What is the least expensive route if the cable costs $7000 per mile in addition to the cost to lay it?

11. You have been asked to determine where a water works should be built along a river between Chesterville and Denton (see below) to minimize the total cost of the pipe to the towns.

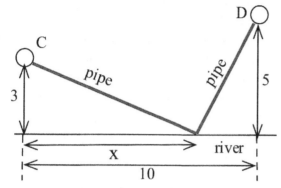

(a) Assume that the same size (and cost) pipe is used to each town. (This part can be done quickly without using calculus.)

(b) Assume instead that the pipe to Chesterville costs \$3000 per mile and to Denton it costs \$7000 per mile.

12. Light from a bulb at A is reflected off a flat mirror to your eye at point B (see below). If the time (and length of the path) from A to the mirror and then to your eye is a minimum, show that the angle of incidence equals the angle of reflection. (Hint: This is similar to the previous problem.)

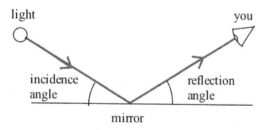

13. U.S. postal regulations state that the sum of the length and girth (distance around) of a package must be no more than 108 inches (see below).

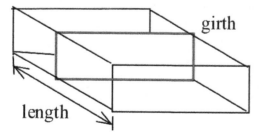

(a) Find the dimensions of the acceptable box with a square end that has the largest volume.

(b) Find the dimensions of the acceptable box that has the largest volume if its end is a rectangle twice as long as it is wide.

(c) Find the dimensions of the acceptable box with a circular end that has the largest volume.

14. Just thinking — you don't need calculus for this problem: A spider and a fly are located on opposite corners of a cube (see below). What is the shortest path along the surface of the cube from the spider to the fly?

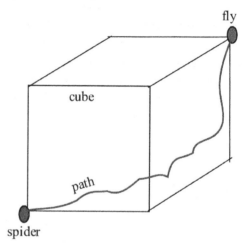

15. Two sides of a triangle are 7 and 10 inches long. What is the length of the third side so the area of the triangle will be greatest? (This problem can be done without using calculus. How? If you do use calculus, consider the angle θ between the two sides.)

16. Find the shortest distance from the point $(2,0)$ to the curve:

(a) $y = 3x - 1$ (b) $y = x^2$

(c) $x^2 + y^2 = 1$ (d) $y = \sin(x)$

17. Find the dimensions of the rectangle with the largest area if the base must be on the x-axis and its other two corners are on the graph of:

(a) $y = 16 - x^2$, $-4 \le x \le 4$

(b) $x^2 + y^2 = 1$

(c) $|x| + |y| = 1$

(d) $y = \cos(x)$, $-\frac{\pi}{2} \le x \le \frac{\pi}{2}$

18. The strength of a wooden beam is proportional to the product of its width and the square of its height (see figure below). What are the dimensions of the strongest beam that can be cut from a log with diameter:

   (a) 12 inches?

   (b) d inches?

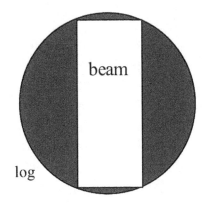

19. You have a long piece of 12-inch-wide metal that you plan to fold along the center line to form a V-shaped gutter (see below). What angle θ will yield a gutter that holds the most water (that is, has the largest cross-sectional area)?

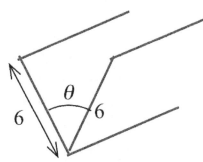

20. You have a long piece of 8-inch-wide metal that you plan to make into a gutter by bending up 3 inches on each side (see below). What angle θ will yield a gutter that holds the most water?

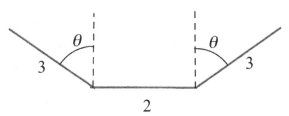

21. You have a 6-inch-diameter paper disk that you want to form into a drinking cup by removing a pie-shaped wedge (sector) and then forming the remaining paper into a cone (see below). Find the height and top radius of the cone so the that the volume of the cup is as large as possible.

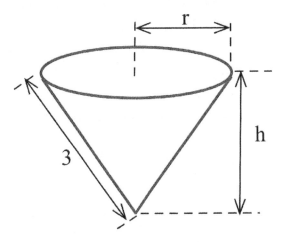

22. (a) What value of b minimizes the sum of the squares of the vertical distances from $y = 2x + b$ to the points $(1,1)$, $(1,2)$ and $(2,2)$?

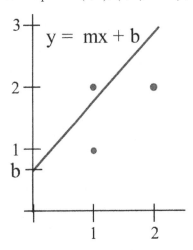

   (b) What slope m minimizes the sum of the squares of the vertical distances from the line $y = mx$ to the points $(1,1)$, $(1,2)$ and $(2,2)$?

   (c) What slope m minimizes the sum of the squares of the vertical distances from the line $y = mx$ to the points $(2,1)$, $(4,3)$, $(-2,-2)$ and $(-4,-2)$?

23. You own a small airplane that holds a maximum of 20 passengers. It costs you $100 per flight from St. Thomas to St. Croix for gas and wages plus an additional $6 per passenger for the extra gas required by the extra weight. The charge per passenger is $30 each if 10 people charter your plane (10 is the minimum number you will fly), and this charge is reduced by $1 per passenger for each passenger over 10 who travels (that is, if 11 fly they each pay $29, if 12 fly they each pay $28, etc.). What number of passengers on a flight will maximize your profit?

24. Prove: If f and g are differentiable functions and if the vertical distance between f and g is greatest at $x = c$, then $f'(c) = g'(c)$ and the tangent lines to f and g are parallel when $x = c$.

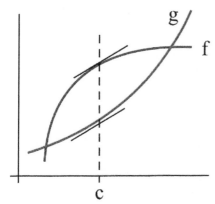

25. Profit = revenue − expenses. Assuming revenue and expenses are differentiable functions, show that when profit is maximized, then marginal revenue $\left(\dfrac{dR}{dx}\right)$ equals marginal expense $\left(\dfrac{dE}{dx}\right)$.

26. Dean Simonton claims the "productivity levels" of people in various fields can be described as a function of their "career age" t by $p(t) = e^{-at} - e^{-bt}$ where a and b are constants depending on the field, and career age is approximately 20 less than the actual age of the individual.

    (a) Based on this model, at what ages do mathematicians ($a = 0.03$, $b = 0.05$), geologists ($a = 0.02$, $b = 0.04$) and historians ($a = 0.02$, $b = 0.03$) reach their maximum productivity?

    (b) Simonton says, "With a little calculus we can show that the curve ($p(t)$) maximizes at $t =$

$\dfrac{1}{b-a} \ln\left(\dfrac{b}{a}\right)$." Use calculus to show that Simonton is correct.

Note: Models of this type have uses for describing the behavior of groups, but it is dangerous — and usually invalid — to apply group descriptions or comparisons to individuals in a group. (*Scientific Genius* by Dean Simonton, Cambridge University Press, 1988, pp. 69–73)

27. After the table was wiped and the potato chips dried off, the question remained: "Just how far could a can of cola be tipped before it fell over?"

    (a) For a full can or an empty can the answer was easy: the center of gravity (CG) of the can is at the middle of the can, half as high as the height of the can, and we can tilt the can until the CG is directly above the bottom rim (see below left). Find θ if the height of the can is 12 cm and the diameter is 5 cm.

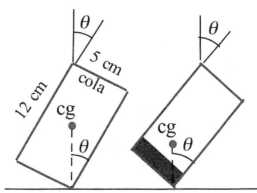

    (b) For a partly filled can, more thinking was needed. Some ideas you will see in Chapter 5 tell us that the CG of a can holding x cm of cola is $C(x) = \dfrac{360 + 9.6x^2}{60 + 19.2x}$ cm above the bottom of the can. Find the height x of cola that will make the CG as low as possible.

    (c) Assuming that the cola is frozen solid (so the top of the cola stays parallel to the bottom of the can), how far can we tilt a can containing x cm of cola? (See above right.)

    (d) If the can contained x cm of liquid cola, could we tilt it farther or less far than the frozen cola before it would fall over?

28. Just thinking—calculus will not help with this one.

(a) Four towns are located at the corners of a square. What is the shortest length of road we can construct so that it is possible to travel along the road from any town to any other town?

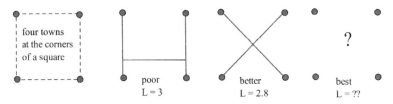

The problem of finding the shortest path connecting several points in the plane is called the "Steiner problem." It is important for designing computer chips and telephone networks to be as efficient as possible.

(b) What is the shortest connecting path for five towns located on the corners of a pentagon?

Generalized Max/Min Problems

The previous max/min problems were mostly numerical problems: the amount of fencing in Problem 2 was 200 feet, the lengths of the piece of tin in Problem 4 were 10 and 15, and the parabola in Problem 17(a) was $y = 16 - x^2$. In working those problems, you might have noticed some patterns among the numbers in the problem and the numbers in your answers, and you might have wondered if the pattern was a coincidence or if there really was a general pattern at work. Rather than trying several numerical examples to see if the "pattern" holds, mathematicians, engineers, scientists and others sometimes resort to generalizing the problem. We free the problem from the particular numbers by replacing the numbers with letters, and then we solve the generalized problem. In this way, relationships between the values in the problem and those in the solution can become more obvious. Solutions to these generalized problems are also useful if you want to program a computer to quickly provide numerical answers.

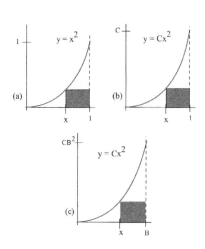

29. (a) Find the dimensions of the rectangle with the greatest area that can be built so the base of the rectangle is on the x-axis between 0 and 1 ($0 \leq x \leq 1$) and one corner of the rectangle is on the curve $y = x^2$ (see above right). What is the area of this rectangle?

(b) Generalize the problem in part (a) for the parabola $y = Cx^2$ with $C > 0$ and $0 \leq x \leq 1$.

(c) Generalize for the parabola $y = Cx^2$ with $C > 0$ and $0 \leq x \leq B$.

30. (a) Find the dimensions of the rectangle with the greatest area that can be built so the base of the rectangle is on the x-axis between 0 and 1 and one corner of the rectangle is on the curve $y = x^3$. What is the area of this rectangle?

(b) Generalize the problem in part (a) for the curve $y = Cx^3$ with $C > 0$ and $0 \leq x \leq 1$.

(c) Generalize for the curve $y = Cx^3$ with $C > 0$ and $0 \leq x \leq B$.

(d) Generalize for the curve $y = Cx^n$ with $C > 0$, n a positive integer, and $0 \leq x \leq B$.

31. (a) The base of a right triangle is 50 and the height is 20. Find the dimensions and area of the rectangle with the greatest area that can be enclosed in the triangle if the base of the rectangle must lie on the base of the triangle.

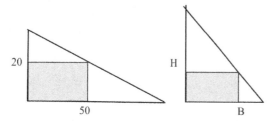

(b) The base of a right triangle is B and the height is H. Find the dimensions and area of the rectangle with the greatest area that can be enclosed in the triangle if the base of the rectangle must lie on the base of the triangle.

(c) State your general conclusion from part (b) in words.

32. (a) You have T dollars to buy fencing material to enclose a rectangular plot of land. The fence for the top and bottom costs $5 per foot and for the sides it costs $3 per foot. Find the dimensions of the plot with the largest area. For this largest plot, how much money was used for the top and bottom, and for the sides?

(b) You have T dollars to buy fencing material to enclose a rectangular plot of land. The fence for the top and bottom costs $A per foot and for the sides it costs $B per foot. Find the dimensions of the plot with the largest area. For this largest plot, how much money was used for the top and bottom (together), and for the sides (together)?

(c) You have T dollars to buy fencing material to enclose a rectangular plot of land. The fence costs $A per foot for the top, $B/foot for the bottom, $C/ft for the left side and $D/ft for the right side. Find the dimensions of the plot with the largest area. For this largest plot, how much money was used for the top and bottom (together), and for the sides (together)?

33. Determine the dimensions of the least expensive cylindrical can that will hold V cubic inches if the top material costs $A per square inch, the bottom material costs $B per square inch, and the side material costs $C per square inch.

34. Find the location of C in the figure below so that the sum of the distances from A to C and from C to B is a minimum.

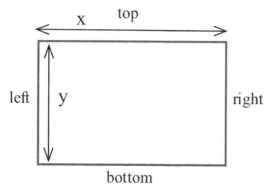

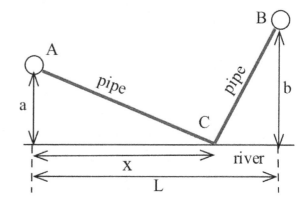

3.5 Practice Answers

1. $V(x) = x(15 - 2x)(7 - 2x) = 4x^3 - 44x^2 + 105x$ so:

$$V'(x) = 12x^2 - 88x + 105 = (2x - 3)(6x - 35)$$

which is defined for all x: the only critical numbers are the endpoints $x = 0$ and $x = \frac{7}{2}$ and where $V'(x) = 0$: $x = \frac{3}{2}$ and $x = \frac{35}{6}$ (but $\frac{35}{6}$ is not in the interval $[0, \frac{7}{2}]$ so it is not practical). The maximum volume must occur when $x = 0$, $x = \frac{3}{2}$ or $x = \frac{7}{2}$:

$$V(0) = 0 \cdot (15 - 2 \cdot 0) \cdot (7 - 2 \cdot 0) = 0$$

$$V\left(\frac{3}{2}\right) = \frac{3}{2} \cdot \left(15 - 2 \cdot \frac{3}{2}\right) \cdot \left(7 - 2 \cdot \frac{3}{2}\right) = \frac{3}{2}(12)(4) = 72$$

$$V\left(\frac{7}{2}\right) = \frac{7}{2} \cdot \left(15 - 2 \cdot \frac{7}{2}\right) \cdot \left(7 - 2 \cdot \frac{7}{2}\right) = \frac{7}{2}(8)(0) = 0$$

The maximum-volume box will result from cutting a 1.5-by-1.5 inch square from each corner. A graph of $V(x)$ appears in the margin.

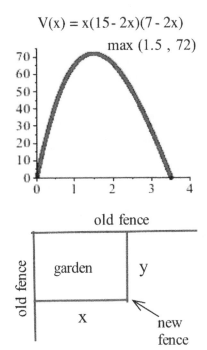

2. (a) We have 80 feet of fencing (see margin). Our assignment is to maximize the area of the garden: $A = x \cdot y$ (two variables). Fortunately, we have the constraint that $x + y = 80$, so $y = 80 - x$ and our assignment reduces to maximizing a function of one variable:

$$A = x \cdot y = x \cdot (80 - x) = 80x - x^2 \Rightarrow A'(x) = 80 - 2x$$

so $A'(x) = 0 \Rightarrow x = 40$. Because $A''(x) = -2 < 0$, the graph of A is concave down, hence A has a maximum at $x = 40$. The maximum area is $A(40) = 40 \cdot 40 = 1600$ ft^2 when $x = 40$ feet and $y = 40$ feet. The maximum-area garden is a square.

(b) This is similar to part (a) except we have F feet of fencing instead of 80 feet: $x + y = F \Rightarrow y = F - x$ and we want to maximize $A = xy = x(F - x) = Fx - x^2$. Differentiating, $A'(x) = F - 2x$ so $A'(x) = 0 \Rightarrow x = \frac{F}{2} \Rightarrow y = \frac{F}{2}$. The maximum area is $A\left(\frac{F}{2}\right) = \frac{F^2}{4}$ square feet when the garden is a square with half of the new fence used on each of the two new sides.

3. The cost C is given by:

$$C = 5(\text{area of top}) + 3(\text{area of sides}) + 5(\text{area of bottom})$$
$$= 5(\pi r^2) + 3(2\pi rh) + 5(\pi r^2)$$

so our assignment is to minimize $C = 10\pi r^2 + 6\pi rh$, a function of two variables (r and h). Fortunately, we also have the constraint that volume $= 1000$ in$^3 = \pi r^2 h \Rightarrow h = \frac{1000}{\pi r^2}$. So:

$$C = 10\pi r^2 + 6\pi r \left(\frac{1000}{\pi r^2}\right) = 10\pi r^2 + \frac{6000}{r} \Rightarrow C'(r) = 20\pi r - \frac{6000}{r^2}$$

which exists for $r \neq 0$ ($r = 0$ is not in the domain of $C(r)$).

$$C'(r) = 0 \Rightarrow 20\pi r - \frac{6000}{r^2} = 0 \Rightarrow 20\pi r^3 = 6000 \Rightarrow r = \sqrt[3]{\frac{6000}{20\pi}} \approx 4.57 \text{ inches}$$

When $r = 4.57$, $h = \frac{1000}{\pi(4.57)^2} \approx 15.24$ inches. Examining the second derivative, $C''(r) = 20\pi + \frac{12000}{r^3} > 0$ for all $r > 0$ so C is concave up and we have found the minimum cost.

3.6   Asymptotic Behavior of Functions

When you turn on an automobile or a light bulb or a computer, many things happen. Some of them are uniquely part of the start-up process of the system. These "transient" things occur only during start up, and then the system settles down to its steady-state operation. This start-up behavior can be very important, but sometimes we want to investigate the steady-state — or long-term — behavior: how does the system behave "after a long time?" In this section we investigate and describe the long-term behavior of functions and the systems they model: how does a function behave "when x (or $-x$) is arbitrarily large?"

Limits as x Becomes Arbitrarily Large ("Approaches Infinity")

The same type of questions we considered about a function f as x approached a finite number can also be asked about f as x "becomes arbitrarily large" (or "increases without bound") — that is, eventually becomes larger than any fixed number.

Example 1. What happens to the values of $f(x) = \dfrac{5x}{2x+3}$ and $g(x) = \dfrac{\sin(7x+1)}{3x}$ as x becomes arbitrarily large (increases without bound)?

x	$\frac{5x}{2x+3}$	$\frac{\sin(7x+1)}{3x}$
10	2.17	0.031702
100	2.463	−0.001374
1000	2.4962	0.000333
10,000	2.4996	0.000001

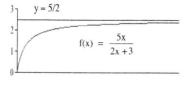

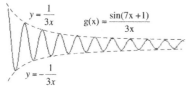

Solution. One approach is numerical: evaluate $f(x)$ and $g(x)$ for some "large" values of x and see if there is a pattern to the values of $f(x)$ and $g(x)$. The margin table shows the values of $f(x)$ and $g(x)$ for several large values of x. When x is very large, it appears that the values of $f(x)$ are close to 2.5 $= \frac{5}{2}$ and the values of $g(x)$ are close to 0. In fact, we can guarantee that the values of $f(x)$ are as close to $\frac{5}{2}$ as someone wants by taking x to be "big enough." The values of $f(x) = \dfrac{5x}{2x+3}$ may or may not ever equal $\frac{5}{2}$ (they never do), but if x is "large," then $f(x)$ is "very close to" $\frac{5}{2}$. Similarly, we can guarantee that the values of $g(x)$ are as close to 0 as someone wants by taking x to be "big enough." The graphs of f and g for "large" values of x appear in the margin.   ◀

Practice 1. What happens to the values of $f(x) = \dfrac{3x+4}{x-2}$ and $g(x) = \dfrac{\cos(5x)}{2x+7}$ as x becomes arbitrarily large?

We can express the answers to Example 1 using limits. "As x becomes arbitrarily large, the values of $\dfrac{5x}{2x+3}$ approach $\frac{5}{2}$" can be written:

$$\lim_{x\to\infty} \frac{5x}{2x+3} = \frac{5}{2}$$

and "the values of $\dfrac{\sin(7x+1)}{3x}$ approach 0" can be written:

$$\lim_{x\to\infty} \frac{\sin(7x+1)}{3x} = 0$$

We read $\lim\limits_{x\to\infty}$ as "the limit as x approaches infinity," meaning "the limit as x becomes arbitrarily large" or "as x increases without bound."

The notation "$x \to -\infty$," read as "x approaches negative infinity," means that the values of $-x$ become arbitrarily large.

Practice 2. Rewrite your answers to Practice 1 using limit notation.

The expression $\lim\limits_{x\to\infty} f(x)$ asks about the behavior of $f(x)$ as the values of x get larger and larger without any bound. One way to determine this behavior is to look at the values of $f(x)$ for some values of x that are very "large." If the values of the function get arbitrarily close to a single number as x gets larger and larger, then we will say that number is the limit of the function as x approaches infinity.

Practice 3. Fill in the table for $f(x) = \dfrac{6x+7}{3-2x}$ and $g(x) = \dfrac{\sin(3x)}{x}$ and use those values to estimate $\lim\limits_{x\to\infty} f(x)$ and $\lim\limits_{x\to\infty} g(x)$.

x	$\frac{6x+7}{3-2x}$	$\frac{\sin(3x)}{x}$
10		
200		
500		
20,000		

Example 2. How large must x be to guarantee that $f(x) = \dfrac{1}{x} < 0.1$? That $f(x) < 0.001$? That $f(x) < E$ (with $E > 0$)?

Solution. If $x > 10$, then $\dfrac{1}{x} < \dfrac{1}{10} = 0.1$. If $x > 1000$, then $\dfrac{1}{x} < \dfrac{1}{1000} = 0.001$. In general, if E is any positive number, then we can guarantee that $|f(x)| < E$ by picking only values of $x > \dfrac{1}{E} > 0$: if $x > \dfrac{1}{E}$, then $\dfrac{1}{x} < E$. From this we can conclude that $\lim\limits_{x\to\infty} \dfrac{1}{x} = 0$. ◄

Practice 4. How large must x be to guarantee that $f(x) = \dfrac{1}{x^2} < 0.1$? That $f(x) < 0.001$? That $f(x) < E$ (with $E > 0$)? Evaluate $\lim\limits_{x\to\infty} \dfrac{1}{x^2}$.

The Main Limit Theorem (Section 1.2) about limits of combinations of functions still holds true if the limits as "$x \to a$" are replaced with limits as "$x \to \infty$" but we will not prove those results.

Polynomials arise regularly in applications, and we often need the limit, as "$x \to \infty$," of ratios of polynomials or functions containing powers of x. In these situations the following technique is often helpful:

During this discussion—and throughout this book—we do not treat "infinity" or "∞" as a number, but only as a useful notation. "Infinity" is not part of the real number system, and we use the common notation "$x \to \infty$" and the phrase "x approaches infinity" only to mean that "x becomes arbitrarily large."

A more formal definition of the limit as "$x \to \infty$" appears at the end of this section.

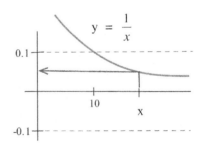

• factor the highest power of x in the denominator from both the numerator and the denominator

• cancel the common factor from the numerator and denominator

The limit of the new denominator is a constant, so the limit of the resulting ratio is easier to determine.

Example 3. Determine $\lim\limits_{x \to \infty} \dfrac{7x^2 + 3x - 4}{3x^2 - 5}$ and $\lim\limits_{x \to \infty} \dfrac{9x + 2}{3x^2 - 5x + 1}$.

Solution. Factoring x^2 out of the numerator and the denominator of the first rational function results in:

$$\lim_{x \to \infty} \frac{7x^2 + 3x - 4}{3x^2 - 5} = \lim_{x \to \infty} \frac{x^2\left(7 + \frac{3}{x} - \frac{4}{x^2}\right)}{x^2\left(3 - \frac{5}{x^2}\right)} = \lim_{x \to \infty} \frac{7 + \frac{3}{x} - \frac{4}{x^2}}{3 - \frac{5}{x^2}} = \frac{7}{3}$$

where we used the facts that $\frac{3}{x} \to 0$, $\frac{4}{x^2} \to 0$ and $\frac{5}{x^2} \to 0$ as $x \to \infty$. Similarly:

$$\lim_{x \to \infty} \frac{9x + 2}{3x^2 - 5x + 1} = \lim_{x \to \infty} \frac{x^2\left(\frac{9}{x} + \frac{2}{x^2}\right)}{x^2\left(3 - \frac{5}{x} + \frac{1}{x^2}\right)} = \lim_{x \to \infty} \frac{\frac{9}{x} + \frac{2}{x^2}}{3 - \frac{5}{x} + \frac{1}{x^2}} = \frac{0}{3} = 0$$

because $\frac{k}{x} \to 0$ and $\frac{c}{x^2} \to 0$ as $x \to \infty$ for any constants k and c.   ◀

If we need to evaluate a more difficult limit as $x \to \infty$, it is often useful to algebraically manipulate the function into the form of a ratio and then use the previous technique.

If the values of the function oscillate and do not approach a single number as x becomes arbitrarily large, then the function does not have a limit as x approaches ∞: the limit **does not exist**.

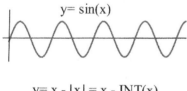

y= sin(x)

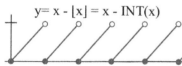

y= x - ⌊x⌋ = x - INT(x)

Example 4. Evaluate $\lim\limits_{x \to \infty} \sin(x)$ and $\lim\limits_{x \to \infty} x - \lfloor x \rfloor$

Solution. As $x \to \infty$, $f(x) = \sin(x)$ and $g(x) = x - \lfloor x \rfloor$ do not have limits. As x grows without bound, the values of $f(x) = \sin(x)$ oscillate between -1 and $+1$ (see margin), and these values do not approach a single number. Similarly, $g(x) = x - \lfloor x \rfloor$ continues to take on all values between 0 and 1, and these values never approach a single number.   ◀

Using Calculators to Help Find Limits as "$x \to \infty$" or "$x \to -\infty$"

Calculators only store a limited number of digits for each quantity. This becomes a severe limitation when we deal with extremely large quantities.

Example 5. The value of $f(x) = (x + 1) - x$ is clearly equal to 1 for all values of x, and your calculator will give the right answer if you use it to evaluate $f(4)$ or $f(5)$. Now use it to evaluate f for a big value of x, say $x = 10^{40}$: $f(10^{40}) = (10^{40} + 1) - 10^{40} = 1$, but most calculators do not store 40 digits of a number, and they will respond that $f(10^{40}) = 0$, which is **wrong**. In this example the calculator's error is obvious, but similar errors can occur in less obvious ways when using calculators for computations involving very large numbers.

> **You should be careful with — and somewhat suspicious of — the answers your calculator gives you.**

Calculators can still be helpful for examining limits as $x \to \infty$ and $x \to -\infty$ as long as we don't place too much faith in their responses.

Even if you have forgotten some of the properties of the natural logarithm function $\ln(x)$ and the cube root function $\sqrt[3]{x}$, a little experimentation on your calculator can help convince you that $\displaystyle\lim_{x \to \infty} \frac{\ln(x)}{\sqrt[3]{x}} = 0$.

The Limit Is Infinite

The function $f(x) = \dfrac{1}{x^2}$ is undefined at $x = 0$, but we can still ask about the behavior of $f(x)$ for values of x "close to" 0. The margin figure indicates that if x is very small (close to 0) then $f(x)$ is very large. As the values of x get closer to 0, the values of $f(x)$ grow larger and can be made as large as we want by picking x to be close enough to 0. Even though the values of f are not approaching any one number, we use the "infinity" notation to indicate that the values of f are growing without bound, and write: $\displaystyle\lim_{x \to 0} \frac{1}{x^2} = \infty$.

The values of $\dfrac{1}{x^2}$ do not *equal* "infinity": the notation $\displaystyle\lim_{x \to 0} \frac{1}{x^2} = \infty$ means that the values of $\dfrac{1}{x^2}$ can be made arbitrarily large by picking values of x very close to 0.

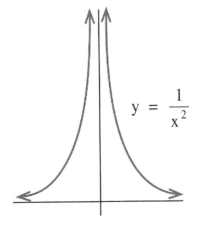

$$y = \frac{1}{x^2}$$

The limit, as $x \to 0$, of $\dfrac{1}{x}$ is slightly more complicated. If x is close to 0, then the value of $f(x) = \dfrac{1}{x}$ can be a large positive number or a large negative number, depending on the sign of x. The function $f(x) = \dfrac{1}{x}$ does not have a (two-sided) limit as x approaches 0, but we can still investigate one-sided limits:

$$\lim_{x \to 0^+} \frac{1}{x} = \infty \quad \text{and} \quad \lim_{x \to 0^-} \frac{1}{x} = -\infty$$

Example 6. Determine $\displaystyle\lim_{x \to 3^+} \frac{x - 5}{x - 3}$ and $\displaystyle\lim_{x \to 3^-} \frac{x - 5}{x - 3}$.

Solution. As $x \to 3^+$, $x - 5 \to -2$ and $x - 3 \to 0$. Because the denominator is approaching 0, we cannot use the Main Limit Theorem, and we need to examine the function more carefully. When $x \to 3^+$, we know that $x > 3$ so $x - 3 > 0$. So if x is close to 3 and slightly larger than 3, then the ratio of $x - 5$ to $x - 3$ is:

$$\frac{\text{a number close to } -2}{\text{small positive number}} = \text{large negative number}$$

As $x > 3$ gets closer to 3:

$$\frac{x-5}{x-3} = \frac{\text{a number closer to } -2}{\text{positive and closer to } 0} = \text{larger negative number}$$

By taking $x > 3$ even closer to 3, the denominator gets closer to 0 but remains positive, so the ratio gets arbitrarily large and negative:
$$\lim_{x \to 3^+} \frac{x-5}{x-3} = -\infty.$$

As $x \to 3^-$, $x - 5 \to -2$ and $x - 3 \to 0$ as before, but now we know that $x < 3$ so $x - 3 < 0$. So if x is close to 3 and slightly smaller than 3, then the ratio of $x - 5$ to $x - 3$ is:

$$\frac{\text{a number close to } -2}{\text{small negative number}} = \text{large positive number}$$

so $\lim\limits_{x \to 3^-} \dfrac{x-5}{x-3} = \infty.$ ◀

Practice 5. Find: (a) $\lim\limits_{x \to 2^+} \dfrac{7}{2-x}$ (b) $\lim\limits_{x \to 2^+} \dfrac{3x}{2x-4}$ (c) $\lim\limits_{x \to 2^+} \dfrac{3x^2 - 6x}{x - 2}$.

Horizontal Asymptotes

The limits of f, as "$x \to \infty$" and "$x \to -\infty$," provide information about horizontal asymptotes of f.

> **Definition:** The line $y = K$ is a **horizontal asymptote** of f if:
>
> $$\lim_{x \to \infty} f(x) = K \quad \text{or} \quad \lim_{x \to -\infty} f(x) = K$$

Example 7. Find any horizontal asymptotes of $f(x) = \dfrac{2x + \sin(x)}{x}$.

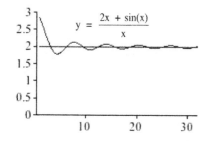

Solution. Computing the limit as $x \to \infty$:

$$\lim_{x \to \infty} \frac{2x + \sin(x)}{x} = \lim_{x \to \infty} \left[\frac{2x}{x} + \frac{\sin(x)}{x} \right] = \lim_{x \to \infty} \left[2 + \frac{\sin(x)}{x} \right]$$
$$= 2 + \lim_{x \to \infty} \frac{\sin(x)}{x} = 2 + 0 = 2$$

so the line $y = 2$ is a horizontal asymptote of f. The limit, as "$x \to -\infty$," is also 2 so $y = 2$ is the *only* horizontal asymptote of f. The graphs of f and $y = 2$ appear in the margin. A function may or may not cross its asymptote. ◀

You likely explored horizontal asymptotes in a previous course using terms like "end behavior" and investigating only rational functions. The tools of calculus allow us to make the the notion of "end behavior" more precise and investigate a wider variety of functions.

Vertical Asymptotes

As with horizontal asymptotes, you have likely studied vertical asymptotes before (at least for rational functions). We can now define vertical asymptotes using infinite limits.

> **Definition**: The vertical line $x = a$ is a **vertical asymptote** of the graph of f if either or both of the one-sided limits of f, as $x \to a^-$ or $x \to a^+$, is infinite.

If our function f is the ratio of a polynomial $P(x)$ and a polynomial $Q(x)$, $f(x) = \dfrac{P(x)}{Q(x)}$, then the only **candidates** for vertical asymptotes are the values of x where $Q(x) = 0$. However, the fact that $Q(a) = 0$ is **not** enough to guarantee that the line $x = a$ is a vertical asymptote of f; we also need to evaluate $P(a)$.

If $Q(a) = 0$ and $P(a) \neq 0$, then the line $x = a$ must be a vertical asymptote of f. If $Q(a) = 0$ and $P(a) = 0$, then the line $x = a$ may or may not be a vertical asymptote.

Example 8. Find the vertical asymptotes of $f(x) = \dfrac{x^2 - x - 6}{x^2 - x}$ and $g(x) = \dfrac{x^2 - 3x}{x^2 - x}$.

Solution. Factoring the numerator and denominator of $f(x)$ yields $f(x) = \dfrac{(x - 3)(x + 2)}{x(x - 1)}$ so the only values of x that make the denominator 0 are $x = 0$ and $x = 1$, and these are the only candidates to be vertical asymptotes. Because $\lim\limits_{x \to 0^+} f(x) = +\infty$ and $\lim\limits_{x \to 1^+} f(x) = -\infty$, both $x = 0$ and $x = 1$ are vertical asymptotes of f.

Factoring the numerator and denominator of $g(x)$ yields $\dfrac{x(x - 3)}{x(x - 1)}$ so the only candidate to be vertical asymptotes are $x = 0$ and $x = 1$. Because $\lim\limits_{x \to 1^+} g(x) = \lim\limits_{x \to 1^+} \dfrac{x(x - 3)}{x(x - 1)} = \lim\limits_{x \to 1^+} \dfrac{x - 3}{x - 1} = -\infty$ the line $x = 1$ must be a vertical asymptote of g. But $\lim\limits_{x \to 0} g(x) = \lim\limits_{x \to 0} \dfrac{x - 3}{x - 1} = 3 \neq \pm\infty$ so $x = 0$ is **not** a vertical asymptote of g.   ◀

Practice 6. Find the vertical asymptotes of $f(x) = \dfrac{x^2 + x}{x^2 + x - 2}$ and $g(x) = \dfrac{x^2 - 1}{x - 1}$.

Other Asymptotes as "$x \to \infty$" and "$x \to -\infty$"

If the limit of $f(x)$, as $x \to \infty$ or $x \to -\infty$, is a constant K, then the graph of f gets arbitrarily close to the horizontal line $y = K$, in which case we call $y = K$ a horizontal asymptote of f. Some functions, however, approach lines that are not horizontal.

Example 9. Find all asymptotes of $f(x) = \dfrac{x^2 + 2x + 1}{x} = x + 2 + \dfrac{1}{x}$.

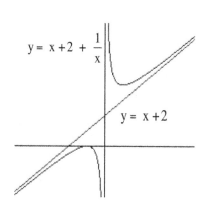

$y = x + 2 + \dfrac{1}{x}$

$y = x + 2$

Solution. If x is a large positive (or negative) number, then $\dfrac{1}{x}$ is very close to 0, and the graph of $f(x)$ is very close to the line $y = x + 2$ (see margin). The line $y = x + 2$ is an asymptote of the graph of f.

If x is a large positive number, then $\dfrac{1}{x}$ is positive, and the graph of f is slightly above the graph of $y = x + 2$. If x is a large negative number, then $\dfrac{1}{x}$ is negative, and the graph of f will be slightly below the graph of $y = x + 2$. The $\dfrac{1}{x}$ piece of f never equals 0, so the graph of f never crosses or touches the graph of the asymptote $y = x + 2$.

The graph of f also has a vertical asymptote at $x = 0$ because $\displaystyle\lim_{x \to 0^+} f(x) = \infty$ and $\displaystyle\lim_{x \to 0^-} f(x) = -\infty$. ◀

Practice 7. Find all asymptotes of $g(x) = \dfrac{2x^2 - x - 1}{x + 1} = 2x - 3 + \dfrac{2}{x + 1}$.

Some functions even have **nonlinear asymptotes**: asymptotes that are not straight lines. The graphs of these functions approach some nonlinear function when the values of x become arbitrarily large.

Example 10. Find all asymptotes of $f(x) = \dfrac{x^4 + 3x^3 + x^2 + 4x + 5}{x^2 + 1} = x^2 + 3x + \dfrac{x + 5}{x^2 + 1}$.

Solution. When x is very large, positive or negative, then $\dfrac{x + 5}{x^2 + 1}$ is very close to 0 and the graph of f is very close to the graph of $g(x) = x^2 + 3x$. The function $g(x) = x^2 + 3x$ is a nonlinear asymptote of f. The denominator of f is never 0 and f has no vertical asymptotes. ◀

Practice 8. Find all asymptotes of $f(x) = \dfrac{x^3 + 2\sin(x)}{x} = x^2 + \dfrac{2\sin(x)}{x}$.

If we can write $f(x)$ as a sum of two functions, $f(x) = g(x) + r(x)$, with $\displaystyle\lim_{x \to \pm\infty} r(x) = 0$, then the graph of f is asymptotic to the graph of g, and g is an asymptote of f. In this situation:

• if $g(x) = K$, then f has a horizontal asymptote $y = K$

• if $g(x) = ax + b$, then f has a linear asymptote $y = ax + b$

• otherwise f has a nonlinear asymptote $y = g(x)$

Formal Definition of $\lim_{x \to \infty} f(x) = K$

The following definition states precisely what we mean by the phrase "we can guarantee that the values of $f(x)$ are arbitrarily close to K by restricting the values of x to be sufficiently large."

Definition: $\lim_{x \to \infty} f(x) = K$ means that, for every given $\epsilon > 0$, there is a number N so that:

    if x is larger than N

    then $f(x)$ is within ϵ units of K.

Equivalently: $|f(x) - K| < \epsilon$ whenever $x > N$.

Example 11. Show that $\lim_{x \to \infty} \dfrac{x}{2x+1} = \dfrac{1}{2}$.

Solution. Typically, we need to do two things. First we need to find a value of N, often depending on ϵ. Then we need to show that the value of N we found satisfies the conditions of the definition.

Assume that $|f(x) - K|$ is less than ϵ and solve for $x > 0$:

$$\epsilon > \left| \frac{x}{2x+1} \right| = \left| \frac{2x - (2x+1)}{2(2x+1)} \right| = \left| \frac{-1}{4x+2} \right| = \frac{1}{4x+2}$$

$$\Rightarrow 4x + 2 > \frac{1}{\epsilon} \Rightarrow x > \frac{1}{4}\left(\frac{1}{\epsilon} - 2 \right)$$

So, given any $\epsilon > 0$, take $N = \dfrac{1}{4}\left(\dfrac{1}{\epsilon} - 2 \right)$.

Now we can just reverse the order of the steps above to show that this N satisfies the limit definition. If $x > 0$ and $x > \dfrac{1}{4}\left(\dfrac{1}{\epsilon} - 2 \right)$ then:

$$4x + 2 > \frac{1}{\epsilon} \Rightarrow \epsilon > \frac{1}{4x+2} = \left| \frac{x}{2x+1} - \frac{1}{2} \right| = |f(x) - K|$$

We have shown that "for every given ϵ, there is an N" that satisfies the definition. ◀

3.6 Problems

1. The margin figure shows $f(x)$ and $g(x)$ for $0 \le x \le 5$. Define a new function $h(x) = \dfrac{f(x)}{g(x)}$.

   (a) At what value of x does $h(x)$ have a root?

   (b) Determine the limits of $h(x)$ as $x \to 1^+$, $x \to 1^-$, $x \to 3^+$ and $x \to 3^-$.

   (c) Where does $h(x)$ have a vertical asymptote?

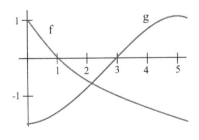

2. The figure below shows $f(x)$ and $g(x)$ on the interval $0 \le x \le 5$. Let $h(x) = \dfrac{f(x)}{g(x)}$.

   (a) At what value(s) of x does $h(x)$ have a root?

   (b) Where does $h(x)$ have vertical asymptotes?

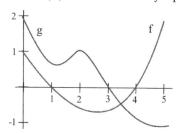

3. The figure below shows $f(x)$ and $g(x)$ for $0 \le x \le 5$. Let $h(x) = \dfrac{f(x)}{g(x)}$. Determine the limits of $h(x)$ as $x \to 2^+$, $x \to 2^-$, $x \to 4^+$ and $x \to 4^-$.

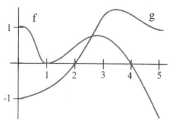

For Problems 4–24, calculate the limit of each expression as "$x \to \infty$."

4. $\dfrac{6}{x+2}$

5. $\dfrac{28}{3x-5}$

6. $\dfrac{7x+12}{3x-2}$

7. $\dfrac{4-3x}{x+8}$

8. $\dfrac{5\sin(2x)}{2x}$

9. $\dfrac{\cos(3x)}{5x-1}$

10. $\dfrac{2x-3\sin(x)}{5x-1}$

11. $\dfrac{4+x\cdot\sin(x)}{2x-4}$

12. $\dfrac{x^2-5x+2}{x^2+8x-4}$

13. $\dfrac{2x^2-9}{3x^2+10x}$

14. $\dfrac{\sqrt{x+5}}{\sqrt{4x-2}}$

15. $\dfrac{5x^2-7x+2}{2x^3+4x}$

16. $\dfrac{x+\sin(x)}{x-\sin(x)}$

17. $\dfrac{7x^2+x\cdot\sin(x)}{3-x^2+\sin(7x^2)}$

18. $\dfrac{7x^{143}+734x-2}{x^{150}-99x^{83}+25}$

19. $\dfrac{\sqrt{9x^2+16}}{2+\sqrt{x^2+1}}$

20. $\sin\left(\dfrac{3x+5}{2x-1}\right)$

21. $\cos\left(\dfrac{7x+4}{x^2+x+1}\right)$

22. $\ln\left(\dfrac{3x^2+5x}{x^2-4}\right)$

23. $\ln(x+8) - \ln(x-5)$

24. $\ln(3x+8) - \ln(2x+5)$

25. Salt water with a concentration of 0.2 pounds of salt per gallon flows into a large tank that initially contains 50 gallons of pure water.

   (a) If the flow rate of salt water into the tank is 4 gallons per minute, what is the volume $V(t)$ of water and the amount $A(t)$ of salt in the tank t minutes after the flow begins?

   (b) Show that the salt concentration $C(t)$ at time t is $C(t) = \dfrac{0.8t}{4t+50}$.

   (c) What happens to the concentration $C(t)$ after a "long" time?

   (d) Redo parts (a)–(c) for a large tank that initially contains 200 gallons of pure water.

26. Under certain laboratory conditions, an agar plate contains $B(t) = 100\left(2 - e^{-t}\right)$ bacteria t hours after the start of the experiment.

   (a) How many bacteria are on the plate at the start of the experiment ($t = 0$)?

   (b) Show that the population is always increasing. (Show $B'(t) > 0$ for all $t > 0$.)

   (c) What happens to the population $B(t)$ after a "long" time?

   (d) Redo parts (a)–(c) for $B(t) = A\left(2 - e^{-t}\right)$.

For Problems 27–41, calculate the limits.

27. $\displaystyle\lim_{x\to0} \dfrac{x+5}{x^2}$

28. $\displaystyle\lim_{x\to3} \dfrac{x-1}{(x-3)^2}$

29. $\displaystyle\lim_{x\to5} \dfrac{x-7}{(x-5)^2}$

30. $\displaystyle\lim_{x\to2^+} \dfrac{x-1}{x-2}$

31. $\displaystyle\lim_{x\to2^-} \dfrac{x-1}{x-2}$

32. $\displaystyle\lim_{x\to3^+} \dfrac{x-1}{x-2}$

33. $\displaystyle\lim_{x\to4^+} \dfrac{x+3}{4-x}$

34. $\displaystyle\lim_{x\to1^-} \dfrac{x^2+5}{1-x}$

35. $\displaystyle\lim_{x\to3^+} \dfrac{x^2-4}{x^2-2x-3}$

36. $\displaystyle\lim_{x\to2} \dfrac{x^2-x-2}{x^2-4}$

37. $\displaystyle\lim_{x\to 0}\frac{x-2}{1-\cos(x)}$

38. $\displaystyle\lim_{x\to\infty}\frac{x^3+7x-4}{x^2+11x}$

39. $\displaystyle\lim_{x\to 5}\frac{\sin(x-5)}{(x-5)}$

40. $\displaystyle\lim_{x\to 0}\frac{x+1}{\sin^2(x)}$

41. $\displaystyle\lim_{x\to 0^+}\frac{1+\cos(x)}{1-e^x}$

In Problems 42–59, write an **equation** of each asymptote for each function and state whether it is a vertical, horizontal or slant asymptote.

42. $f(x)=\dfrac{x+2}{x-1}$

43. $f(x)=\dfrac{x-3}{x^2}$

44. $f(x)=\dfrac{x-1}{x^2-x}$

45. $f(x)=\dfrac{x+5}{x^2-4x+3}$

46. $f(x)=\dfrac{x+\sin(x)}{3x-3}$

47. $f(x)=\dfrac{x^2-4}{x+1}$

48. $f(x)=\dfrac{\cos(x)}{x^2}$

49. $f(x)=2+\dfrac{3-x}{x-1}$

50. $f(x)=\dfrac{x\cdot\sin(x)}{x^2-x}$

51. $f(x)=\dfrac{2x^2+x+5}{x}$

52. $f(x)=\dfrac{x^2+x}{x+1}$

53. $f(x)=\dfrac{1}{x-2}+\sin(x)$

54. $f(x)=x+\dfrac{x}{x^2+1}$

55. $f(x)=x^2+\dfrac{x}{x^2+1}$

56. $f(x)=x^2+\dfrac{x}{x+1}$

57. $f(x)=\dfrac{x\cdot\cos(x)}{x-3}$

58. $f(x)=\dfrac{x^3-x^2+2x-1}{x-1}$

59. $f(x)=\sqrt{\dfrac{x^2+3x+2}{x+3}}$

3.6 Practice Answers

1. As x becomes arbitrarily large, the values of $f(x)$ approach 3 and the values of $g(x)$ approach 0.

2. $\displaystyle\lim_{x\to\infty}\frac{3x+4}{x-2}=3$ and $\displaystyle\lim_{x\to\infty}\frac{\cos(5x)}{2x+7}=0$

3. The completed table appears in the margin.

4. If $x>\sqrt{10}\approx 3.162$ then $f(x)=\dfrac{1}{x^2}<0.1$.

   If $x>\sqrt{1000}\approx 31.62$ then $f(x)=\dfrac{1}{x^2}<0.001$.

   If $x>\sqrt{\dfrac{1}{E}}=\dfrac{1}{\sqrt{E}}$ then $f(x)=\dfrac{1}{x^2}<E$.

x	$\frac{6x+7}{3-2x}$	$\frac{\sin(3x)}{x}$
10	−3.94117647	−0.09880311
200	−3.04030227	0.00220912
500	−3.00160048	0.00017869
20,000	−3.00040003	0.00004787
↓	↓	↓
	−3	0

5. (a) As $x\to 2^+$, $2-x\to 0$, and $x>2$ so $2-x<0$: $2-x$ takes on small negative values.

$$\frac{7}{2-x}=\frac{7}{\text{small negative number}}=\text{large negative number}$$

so we represent the limit as: $\displaystyle\lim_{x\to 2^+}\frac{7}{2-x}=-\infty$.

(b) As $x\to 2^+$, $2x-4\to 0$, and $x>2$ so $2x-4>0$: $2x-4$ takes on small positive values. And as $x\to 2^+$, $3x\to 6$ so:

$$\frac{3x}{2x-4}=\frac{\text{number near 6}}{\text{small positive number}}=\text{large positive number}$$

so we represent the limit as: $\displaystyle\lim_{x\to 2^+}\frac{3x}{2x-4}=+\infty$.

(c) As $x \to 2^+$, $3x^2 - 6x \to 0$ and $x - 2 \to 0$ so we need to do more work. Factoring the numerator as $3x^2 - 6x = 3x(x - 2)$:

$$\lim_{x \to 2^+} \frac{3x^2 - 6x}{x - 2} = \lim_{x \to 2^+} \frac{3x(x - 2)}{x - 2} = \lim_{x \to 2^+} 3x = 6$$

where we were able to cancel the $x - 2$ factor because the limit involves values of x close to (but not equal to) 2.

6. (a) $f(x) = \dfrac{x^2 + x}{x^2 + x - 2} = \dfrac{x(x + 1)}{(x - 1)(x + 2)}$ so f has vertical asymptotes at $x = 1$ and $x = -2$.

(b) $g(x) = \dfrac{x^2 - 1}{x - 1} = \dfrac{(x + 1)(x - 1)}{x - 1}$ so the value $x = 1$ is not in the domain of g. If $x \neq 1$, then $g(x) = x + 1$: g has a "hole" when $x = 1$ and no vertical asymptotes.

7. $g(x) = 2x - 3 + \dfrac{2}{x + 1}$ has a vertical asymptote at $x = -1$ and no horizontal asymptotes, but $\lim\limits_{x \to \infty} \dfrac{2}{x + 1} = 0$ so g has the linear asymptote $y = 2x - 3$.

8. $f(x) = x^2 + \dfrac{2\sin(x)}{x}$ is not defined at $x = 0$, so f has a vertical asymptote or a "hole" there; $\lim\limits_{x \to 0} x^2 + \dfrac{2\sin(x)}{x} = 0 + 2 = 2$ so f has a "hole" when $x = 0$. Because $\lim\limits_{x \to \infty} \dfrac{2\sin(x)}{x} = 0$, f has the nonlinear asymptote $y = x^2$ (but no horizontal asymptotes).

3.7 L'Hôpital's Rule

When taking limits of slopes of secant lines, $m_{\text{sec}} = \dfrac{f(x+h) - f(x)}{h}$ as $h \to 0$, we frequently encountered one difficulty: both the numerator and the denominator approached 0. And because the denominator approached 0, we could not apply the Main Limit Theorem. In many situations, however, we managed to get past this "$\frac{0}{0}$" difficulty by using algebra or geometry or trigonometry to rewrite the expression and then take the limit. But there was no common approach or pattern. The algebraic steps we used to evaluate $\lim\limits_{h \to 0} \dfrac{(2+h)^2 - 4}{h}$ seem quite different from the trigonometric steps needed for $\lim\limits_{h \to 0} \dfrac{\sin(2+h) - \sin(2)}{h}$.

In this section we consider a single technique, called l'Hôpital's Rule, that enables us to quickly and easily evaluate many limits of the form "$\frac{0}{0}$" as well as several other challenging indeterminate forms.

Although discovered by Johann Bernoulli, this rule was named for the Marquis de l'Hôpital (pronounced low-pee-TALL), who published it in his 1696 calculus textbook, *Analysis of the Infinitely Small for the Understanding of Curved Lines.*

A Linear Example

The graphs of two linear functions appear in the margin and we want to find $\lim\limits_{x \to 5} \dfrac{f(x)}{g(x)}$. Unfortunately, $\lim\limits_{x \to 5} f(x) = 0$ and $\lim\limits_{x \to 5} g(x) = 0$ so we cannot apply the Main Limit Theorem. We do know, however, that f and g are linear, so we can calculate their slopes, and we know that they both lines go through the point $(5,0)$ so we can find their equations: $f(x) = -2(x-5)$ and $g(x) = 3(x-5)$.

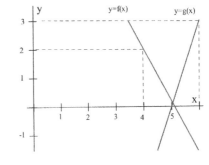

Now the limit is easier to compute:

$$\lim_{x \to 5} \frac{f(x)}{g(x)} = \lim_{x \to 5} \frac{-2(x-5)}{3(x-5)} = \lim_{x \to 5} \frac{-2}{3} = -\frac{2}{3} = \frac{\text{slope of } f}{\text{slope of } g}$$

In fact, this pattern works for any two linear functions: If f and g are linear functions with slopes $m \neq 0$ and $n \neq 0$ and a common root at $x = a$, then $f(x) - f(a) = m(x-a)$ and $g(x) - g(a) = n(x-a)$ so $f(x) = m(x-a)$ and $g(x) = n(x-a)$. Then:

$$\lim_{x \to a} \frac{f(x)}{g(x)} = \lim_{x \to a} \frac{m(x-a)}{n(x-a)} = \lim_{x \to a} \frac{m}{n} = \frac{m}{n} = \frac{\text{slope of } f}{\text{slope of } g}$$

A more powerful result—that the same pattern holds true for differentiable functions even if they are not linear—is called l'Hôpital's Rule.

L'Hôpital's Rule ("$\frac{0}{0}$" Form)

If $\quad$ f and g are differentiable at $x = a$,
$\qquad$ $f(a) = 0$, $g(a) = 0$ and $g'(a) \neq 0$

then $\quad \lim\limits_{x \to a} \dfrac{f(x)}{g(x)} = \dfrac{f'(a)}{g'(a)} = \dfrac{\text{slope of } f \text{ at } a}{\text{slope of } g \text{ at } a}$

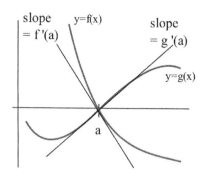

slope = f'(a)   y=f(x)

slope = g'(a)

y=g(x)

a

Unfortunately, we have ignored some subtle difficulties, such as $g(x)$ or $g'(x)$ possibly being 0 when x is close to, but not equal to, a. Because of these issues, a full-fledged proof of l'Hôpital's Rule is omitted.

Idea for a proof: Even though f and g may not be linear functions, they *are* differentiable. So at the point $x = a$ they are "almost linear" in the sense that we can approximate them quite well using their tangent lines at that point (see margin).

Because $f(a) = g(a) = 0$, $f(x) \approx f(a) + f'(a)(x - a) = f'(a)(x - a)$ and $g(x) \approx g(a) + g'(a)(x - a) = g'(a)(x - a)$. So:

$$\lim_{x \to a} \frac{f(x)}{g(x)} \approx \lim_{x \to a} \frac{f'(a)(x - a)}{g'(a)(x - a)} = \lim_{x \to a} \frac{f'(a)}{g'(a)} = \frac{f'(a)}{g'(a)}$$

Example 1. Determine $\lim\limits_{x \to 0} \dfrac{x^2 + \sin(5x)}{3x}$ and $\lim\limits_{x \to 1} \dfrac{\ln(x)}{e^x - e}$.

Solution. We could evaluate the first limit without l'Hôpital's Rule, but let's use it anyway. We can match the pattern of l'Hôpital's Rule by letting $a = 0$, $f(x) = x^2 + \sin(5x)$ and $g(x) = 3x$. Then $f(0) = 0$, $g(0) = 0$, and f and g are differentiable with $f'(x) = 2x + 5\cos(5x)$ and $g'(x) = 3$, so:

$$\lim_{x \to 0} \frac{x^2 + \sin(5x)}{3x} = \frac{f'(0)}{g'(0)} = \frac{2 \cdot 0 + 5\cos(5 \cdot 0)}{3} = \frac{5}{3}$$

For the second limit, let $a = 1$, $f(x) = \ln(x)$ and $g(x) = e^x - e$. Then $f(1) = 0$, $g(1) = 0$, f and g are differentiable for x near 1 (when $x > 0$), and $f'(x) = \dfrac{1}{x}$ and $g'(x) = e^x$. Then:

$$\lim_{x \to 1} \frac{\ln(x)}{e^x - e} = \frac{f'(1)}{g'(1)} = \frac{\frac{1}{1}}{e^1} = \frac{1}{e}$$

Here no simplification was possible, so we needed l'Hôpital's Rule. ◄

Practice 1. Evaluate $\lim\limits_{x \to 0} \dfrac{1 - \cos(5x)}{3x}$ and $\lim\limits_{x \to 2} \dfrac{x^2 + x - 6}{x^2 + 2x - 8}$.

Strong Version of l'Hôpital's Rule

We can strengthen L'Hôpital's Rule to include cases when $g'(a) = 0$, and the indeterminate form "$\frac{\infty}{\infty}$" when f and g increase without bound.

L'Hôpital's Rule (Strong "$\frac{0}{0}$" and "$\frac{\infty}{\infty}$" Forms)

If f and g are differentiable on an open interval I containing a, $g'(x) \neq 0$ on I except possibly at a, and
$$\lim_{x \to a} \frac{f(x)}{g(x)} = \text{"}\frac{0}{0}\text{" or "}\frac{\infty}{\infty}\text{"}$$
then $\lim\limits_{x \to a} \dfrac{f(x)}{g(x)} = \lim\limits_{x \to a} \dfrac{f'(x)}{g'(x)}$ if the limit on the right exists.

(Here "a" can represent a finite number or "∞.")

Example 2. Evaluate $\lim\limits_{x \to \infty} \dfrac{e^{7x}}{5x}$.

Solution. As "$x \to \infty$," both e^{7x} and $5x$ increase without bound, so we have an "$\frac{\infty}{\infty}$" indeterminate form and can use the Strong Version of l'Hôpital's Rule: $\lim\limits_{x \to \infty} \dfrac{e^{7x}}{5x} = \lim\limits_{x \to \infty} \dfrac{7e^{7x}}{5} = \infty.$ ◀

The limit of $\dfrac{f'}{g'}$ may also be an indeterminate form, in which case we can apply l'Hôpital's Rule again to the ratio $\dfrac{f'}{g'}$. We can continue using l'Hôpital's Rule at each stage as long as we have an indeterminate quotient.

Example 3. Compute $\lim\limits_{x \to 0} \dfrac{x^3}{x - \sin(x)}$.

Solution. As $x \to 0$, $f(x) = x^3 \to 0$ and $g(x) = x - \sin(x) \to 0$ so:

$$\lim_{x \to 0} \frac{x^3}{x - \sin(x)} = \lim_{x \to 0} \frac{3x^2}{1 - \cos(x)} = \lim_{x \to 0} \frac{6x}{\sin(x)} = \lim_{x \to 0} \frac{6}{\cos(x)} = 6$$

where we have used l'Hôpital's Rule three times in succession. (At each stage, you should verify the conditions for l'Hôpital's Rule hold.)   ◀

Practice 2. Use l'Hôpital's Rule to find $\lim\limits_{x \to \infty} \dfrac{x^2 + e^x}{x^3 + 8x}$.

Which Function Grows Faster?

Sometimes we want to compare the asymptotic behavior of two systems or functions for large values of x. L'Hôpital's Rule can be useful in such situations. For example, if we have two algorithms for sorting names, and each algorithm takes longer and longer to sort larger collections of names, we may want to know which algorithm will accomplish the task more efficiently for really large collections of names.

Example 4. Algorithm A requires $n \cdot \ln(n)$ steps to sort n names and algorithm B requires $n^{1.5}$ steps. Which algorithm will be better for sorting very large collections of names?

Solution. We can compare the ratio of the number of steps each algorithm requires, $\dfrac{n \cdot \ln(n)}{n^{1.5}}$, and then take the limit of this ratio as n grows arbitrarily large: $\lim\limits_{n \to \infty} \dfrac{n \cdot \ln(n)}{n^{1.5}}$.

If this limit is infinite, we say that $n \cdot \ln(n)$ "grows faster" than $n^{1.5}$. If the limit is 0, we say that $n^{1.5}$ grows faster than $n \cdot \ln(n)$.

Because $n \cdot \ln(n)$ and $n^{1.5}$ both grow arbitrarily large when n becomes large, we can simplify the ratio to $\dfrac{\ln(n)}{n^{0.5}}$ and then use l'Hôpital's Rule:

$$\lim_{n \to \infty} \frac{\ln(n)}{n^{0.5}} = \lim_{n \to \infty} \frac{\frac{1}{n}}{0.5n^{-0.5}} = \lim_{n \to \infty} \frac{2}{\sqrt{n}} = 0$$

We conclude that $n^{1.5}$ grows faster than $n \cdot \ln(n)$ so algorithm A requires fewer steps for really large sorts.   ◄

Practice 3. Algorithm A requires e^n operations to find the shortest path connecting n towns, while algorithm B requires $100 \cdot \ln(n)$ operations for the same task and algorithm C requires n^5 operations. Which algorithm is best for finding the shortest path connecting a very large number of towns? The worst?

Other Indeterminate Forms

We call "$\frac{0}{0}$" an **indeterminate form** because knowing that f approaches 0 and g approaches 0 is not enough to determine the limit of $\frac{f}{g}$, even if that limit exists. The ratio of a "small" number divided by a "small" number can be almost anything as three simple "$\frac{0}{0}$" examples show:

$$\lim_{x \to 0} \frac{3x}{x} = 3 \quad \text{while} \quad \lim_{x \to 0} \frac{x^2}{x} = 0 \quad \text{and} \quad \lim_{x \to 0} \frac{5x}{x^3} = \infty$$

Similarly, "$\frac{\infty}{\infty}$" is an indeterminate form because knowing that f and g both grow arbitrarily large is not enough to determine the value of the limit of $\frac{f}{g}$ or even if the limit exists:

$$\lim_{x \to \infty} \frac{3x}{x} = 3 \quad \text{while} \quad \lim_{x \to \infty} \frac{x^2}{x} = \infty \quad \text{and} \quad \lim_{x \to \infty} \frac{5x}{x^3} = 0$$

In addition to the indeterminate quotient forms "$\frac{0}{0}$" and "$\frac{\infty}{\infty}$" there are several other "indeterminate forms." In each case, the resulting limit depends not only on each function's limit but also on how quickly each function approaches its limit.

- **Product**: If f approaches 0 and g grows arbitrarily large, the product $f \cdot g$ has the indeterminate form "$0 \cdot \infty$."

- **Exponent**: If f and g both approach 0, the function f^g has the indeterminate form "0^0."

- **Exponent**: If f approaches 1 and g grows arbitrarily large, the function f^g has the indeterminate form "1^∞."

- **Exponent**: If f grows arbitrarily large and g approaches 0, the function f^g has the indeterminate form "∞^0."

- **Difference**: If f and g both grow arbitrarily large, the function $f - g$ has the indeterminate form "$\infty - \infty$."

Unfortunately, l'Hôpital's Rule can only be used directly with an indeterminate quotient ($\frac{0}{0}$ or "$\frac{\infty}{\infty}$"), but we can algebraically manipulate these other forms into quotients and *then* apply l'Hôpital's Rule.

Example 5. Evaluate $\lim\limits_{x \to 0^+} x \cdot \ln(x)$.

Solution. This limit involves an indeterminate product (of the form "$0 \cdot -\infty$") but we need a quotient in order to apply l'Hôpital's Rule. If we rewrite the product $x \cdot \ln(x)$ as a quotient:

$$\lim_{x \to 0^+} x \cdot \ln(x) = \lim_{x \to 0^+} \frac{\ln(x)}{\frac{1}{x}} = \lim_{x \to 0^+} \frac{\frac{1}{x}}{\frac{-1}{x^2}} = \lim_{x \to 0^+} -x = 0$$

results from applying the "$\frac{\infty}{\infty}$" version of l'Hôpital's Rule.   ◄

> To use l'Hôpital's Rule on a product $f \cdot g$ with indeterminate form "$0 \cdot \infty$," first rewrite $f \cdot g$ as a quotient: $\dfrac{f}{\frac{1}{g}}$ or $\dfrac{g}{\frac{1}{f}}$. Then apply l'Hôpital's Rule.

Example 6. Evaluate $\lim\limits_{x \to 0^+} x^x$.

Solution. This limit involves the indeterminate form 0^0. We can convert it to a product by recalling a property of exponential and logarithmic functions: for any positive number a, $a = e^{\ln(a)}$ so:

$$f^g = e^{\ln(f^g)} = e^{g \cdot \ln(f)}$$

Applying this to x^x:

$$\lim_{x \to 0^+} x^x = \lim_{x \to 0^+} e^{\ln(x^x)} = \lim_{x \to 0^+} e^{x \cdot \ln(x)}$$

This last limit involves the indeterminate product $x \cdot \ln(x)$. From the previous example we know that $\lim\limits_{x \to 0^+} x \cdot \ln(x) = 0$ so we can conclude that:

$$\lim_{x \to 0^+} x^x = \lim_{x \to 0^+} e^{x \cdot \ln(x)} = e^{\lim_{x \to 0^+} x \cdot \ln(x)} = e^0 = 1$$

because the function $f(u) = e^u$ is continuous everywhere.   ◄

> To use l'Hôpital's Rule on an expression involving exponents, f^g with the indeterminate form "0^0," "1^∞" or "∞^0," first convert it to an expression involving an indeterminate product by recognizing that $f^g = e^{g \cdot \ln(f)}$ and then determining the limit of $g \cdot \ln(f)$. The final result is $e^{\text{limit of } g \cdot \ln(f)}$.

Example 7. Evaluate $\lim\limits_{x\to\infty}\left(1+\dfrac{a}{x}\right)^x$.

Solution. This expression has the form 1^∞ so we first use logarithms to convert the problem into a limit involving a product:

$$\lim_{x\to\infty}\left(1+\frac{a}{x}\right)^x = \lim_{x\to\infty} e^{x\cdot\ln\left(1+\frac{a}{x}\right)}$$

so now we need to compute $\lim\limits_{x\to\infty} x\cdot\ln\left(1+\dfrac{a}{x}\right)$. This limit has the form "$\infty\cdot 0$" so we now convert the product to a quotient:

$$\lim_{x\to\infty} x\cdot\ln\left(1+\frac{a}{x}\right) = \lim_{x\to\infty}\frac{\ln\left(1+\frac{a}{x}\right)}{\frac{1}{x}}$$

This last limit has the form "$\frac{0}{0}$" so we can finally apply l'Hôpital's Rule:

$$\lim_{x\to\infty}\frac{\ln\left(1+\frac{a}{x}\right)}{\frac{1}{x}} = \lim_{x\to\infty}\frac{\frac{-a}{x^2}}{\frac{-1}{x^2}} = \lim_{x\to\infty}\frac{a}{1+\frac{a}{x}} = \frac{a}{1} = a$$

and conclude that:

$$\lim_{x\to\infty}\left(1+\frac{a}{x}\right)^x = \lim_{x\to\infty} e^{x\cdot\ln\left(1+\frac{a}{x}\right)} = e^{\lim_{x\to\infty} x\cdot\ln\left(1+\frac{a}{x}\right)} = e^a$$

where we have again used the continuity of the function $f(u)=e^u$. ◀

3.7 Problems

In Problems 1–15, evaluate each limit. Be sure to justify any use of l'Hôpital's Rule.

1. $\lim\limits_{x\to 1}\dfrac{x^3-1}{x^2-1}$

2. $\lim\limits_{x\to 2}\dfrac{x^4-16}{x^5-32}$

3. $\lim\limits_{x\to 0}\dfrac{\ln(1+3x)}{5x}$

4. $\lim\limits_{x\to\infty}\dfrac{e^x}{x^3}$

5. $\lim\limits_{x\to 0}\dfrac{x\cdot e^x}{1-e^x}$

6. $\lim\limits_{x\to 0}\dfrac{2^x-1}{x}$

7. $\lim\limits_{x\to\infty}\dfrac{\ln(x)}{x}$

8. $\lim\limits_{x\to\infty}\dfrac{\ln(x)}{\sqrt{x}}$

9. $\lim\limits_{x\to\infty}\dfrac{\ln(x)}{x^p}$ $(p>0)$

10. $\lim\limits_{x\to 0}\dfrac{e^{3x}-e^{2x}}{4x}$

11. $\lim\limits_{x\to 0}\dfrac{1-\cos(3x)}{x^2}$

12. $\lim\limits_{x\to 0}\dfrac{1-\cos(2x)}{x}$

13. $\lim\limits_{x\to a}\dfrac{x^m-a^m}{x^n-a^n}$

14. $\lim\limits_{x\to 0}\dfrac{\cos(a+x)-\cos(a)}{x}$

15. $\lim\limits_{x\to 0}\dfrac{1-\cos(x)}{x\cdot\cos(x)}$

16. Find a value for p so that $\lim\limits_{x\to\infty}\dfrac{3x}{px+7}=2$.

17. Find a value for p so that $\lim\limits_{x\to 0}\dfrac{e^{px}-1}{3x}=5$.

18. The limit $\lim\limits_{x\to\infty}\dfrac{\sqrt{3x+5}}{\sqrt{2x-1}}$ has the indeterminate form "$\frac{\infty}{\infty}$." Why doesn't l'Hôpital's Rule work with this limit? (Hint: Apply l'Hôpital's Rule twice and see what happens.) Evaluate the limit without using l'Hôpital's Rule.

19. (a) Evaluate $\lim\limits_{x\to\infty}\dfrac{e^x}{x}$, $\lim\limits_{x\to\infty}\dfrac{e^x}{x^2}$ and $\lim\limits_{x\to\infty}\dfrac{e^x}{x^5}$.

    (b) An algorithm is "exponential" if it requires $a\cdot e^{bn}$ steps (a, and b are positive constants). An algorithm is "polynomial" if it requires $c\cdot n^d$ steps. Show that polynomial algorithms require fewer steps than exponential ones for large values of n.

20. The problem $\lim\limits_{x \to 0} \dfrac{x^2}{3x^2 + x}$ appeared on a test. One student determined the limit was an indeterminate "$\frac{0}{0}$" form and applied l'Hôpital's Rule to get:

$$\lim_{x \to 0} \frac{x^2}{3x^2 + x} = \lim_{x \to 0} \frac{2x}{6x + 1} = \lim_{x \to 0} \frac{2}{6} = \frac{1}{3}$$

Another student also determined the limit was an indeterminate "$\frac{0}{0}$" form and wrote:

$$\lim_{x \to 0} \frac{x^2}{3x^2 + x} = \lim_{x \to 0} \frac{2x}{6x + 1} = \frac{0}{0 + 1} = 0$$

Which student is correct? Why?

In Problems 21–30, evaluate each limit. Be sure to justify any use of l'Hôpital's Rule.

21. $\lim\limits_{x \to 0^+} \sin(x) \cdot \ln(x)$

22. $\lim\limits_{x \to \infty} x^3 e^{-x}$

23. $\lim\limits_{x \to 0^+} \sqrt{x} \cdot \ln(x)$

24. $\lim\limits_{x \to 0^+} x^{\sin(x)}$

25. $\lim\limits_{x \to \infty} \left(1 - \dfrac{3}{x^2}\right)^x$

26. $\lim\limits_{x \to 0} \left(1 - \cos(3x)\right)^x$

27. $\lim\limits_{x \to 0} \left(\dfrac{1}{x} - \dfrac{1}{\sin(x)}\right)$

28. $\lim\limits_{x \to \infty} \left[x - \ln(x)\right]$

29. $\lim\limits_{x \to \infty} \left(\dfrac{x + 5}{x}\right)^{\frac{1}{x}}$

30. $\lim\limits_{x \to \infty} \left(1 + \dfrac{3}{x}\right)^{\frac{2}{x}}$

3.7 Practice Answers

1. Both numerator and denominator in the first limit are differentiable and both equal 0 when $x = 0$, so we apply l'Hôpital's Rule:

$$\lim_{x \to 0} \frac{1 - \cos(5x)}{3x} = \lim_{x \to 0} \frac{5 \sin(5x)}{3} = \frac{0}{3} = 0$$

Both numerator and denominator in the second limit are differentiable and both equal 0 when $x = 0$, so we apply l'Hôpital's Rule:

$$\lim_{x \to 2} \frac{x^2 + x - 6}{x^2 + 2x - 8} = \lim_{x \to 2} \frac{2x + 1}{2x + 2} = \frac{5}{6}$$

2. Both numerator and denominator are differentiable and both become arbitrarily large as $x \to \infty$, so we apply l'Hôpital's Rule:

$$\lim_{x \to \infty} \frac{x^2 + e^x}{x^3 + 8x} = \lim_{x \to \infty} \frac{2x + e^x}{3x^2 + 8} = \lim_{x \to \infty} \frac{2 + e^x}{6x} = \lim_{x \to \infty} \frac{e^x}{6} = \infty$$

Note that we needed to apply l'Hôpital's Rule three times and that each stage involved an "$\frac{\infty}{\infty}$" indeterminate form.

3. Comparing A with e^n operations to B with $100 \cdot \ln(n)$ operations we can apply l'Hôpital's Rule:

$$\lim_{n \to \infty} \frac{e^n}{100 \ln(n)} = \lim_{n \to \infty} \frac{e^n}{\frac{1}{n}} = \lim_{n \to \infty} \frac{n \cdot e^n}{100} = \infty$$

to show that B requires fewer operations than A.

Comparing B with $100 \ln(n)$ operations to C with n^5 operations, we again apply l'Hôpital's Rule:

$$\lim_{n \to \infty} \frac{100 \ln(n)}{n^5} = \lim_{n \to \infty} \frac{\frac{100}{n}}{5n^4} = \lim_{n \to \infty} \frac{20}{n^5} = 0$$

to show that B requires fewer operations than C. So B requires the fewest operations of the three algorithms.

Comparing A to C we must apply l'Hôpital's Rule repeatedly:

$$\lim_{n\to\infty}\frac{e^n}{n^5} = \lim_{n\to\infty}\frac{e^n}{5n^4} = \lim_{n\to\infty}\frac{e^n}{20n^3} = \lim_{n\to\infty}\frac{e^n}{60n^2}$$

$$= \lim_{n\to\infty}\frac{e^n}{120n} = \lim_{n\to\infty}\frac{e^n}{120} = \infty$$

So A requires more operations than C and thus A requires the most operations of the three algorithms.

A

Answers

Important Note about Precision of Answers: In many of the problems in this book you are required to read information from a graph and to calculate with that information. You should take reasonable care to read the graphs as accurately as you can (a small straightedge is helpful), but even skilled and careful people make slightly different readings of the same graph. That is simply one of the drawbacks of graphical information. When answers are given to graphical problems, the answers should be viewed as the best approximations we could make, and they usually include the word "approximately" or the symbol "$\approx$" meaning "approximately equal to." Your answers should be close to the given answers, but you should not be concerned if they differ a little. (Yes those are vague terms, but it is all we can say when dealing with graphical information.)

Section 0.1

1. approx. 1, 0, −1

3. (a) $\approx \dfrac{70° - 150°}{10 \text{ min} - 0 \text{ min}} = -8 \dfrac{°}{\text{min}}$

   (b) $\approx 6 \dfrac{°}{\text{min}}$ cooling; $5 \dfrac{°}{\text{min}}$ cooling

   (c) $\approx 5.5 \dfrac{°}{\text{min}}$ cooling; $10 \dfrac{°}{\text{min}}$ cooling

   (d) When $t = 6$ min.

5. We estimate that the area is approximately (very approximately) 9 cm^2.

7. **Method 1**: Measure the diameter of the can, fill it half full of water, measure the height of the water and calculate the volume. Submerge the bulb, measure the height of the water again, and calculate the new volume. The volume of the bulb is the difference of the two calculated volumes.

   Method 2: Fill the can with water and weigh it. Submerge the bulb (displacing a volume of water equal to the volume of the bulb), remove the bulb, and weigh the can again. By subtracting, find the weight of the displaced water and use the fact that the density of water is 1 gram per 1 cubic centimeter to determine the volume of the bulb.

Section 0.2

1. (a) $-\frac{3}{4}$ (b) $\frac{1}{2}$ (c) 0 (d) 2 (e) undefined

3. (a) $\frac{4}{3}$ (b) $-\frac{9}{5}$ (c) $x + 2$ (if $x \neq 2$) (d) $4 + h$ (if $h \neq 0$)
   (e) $a + x$ (if $a \neq x$)

5. (a) $t = 5$: $\frac{5000}{1500} = \frac{10}{3}$; $t = 10$: $\frac{5000}{3000} = \frac{5}{3}$; $t = 20$:
   $\frac{5000}{6000} = \frac{5}{6}$ (b) any $t > 0$: $\frac{5000}{300t} = \frac{50}{3t}$ (c) Decreasing, since the numerator remains constant at 5000 while the denominator increases.

7. The restaurant is 4 blocks south and 2 blocks east. The distance is $\sqrt{4^2 + 2^2} = \sqrt{20} \approx 4.47$ blocks.

9. (a) $\sqrt{20^2 - 4^2} = \sqrt{384} \approx 19.6$ feet (b) $\frac{\sqrt{384}}{4} \approx 4.9$
   (c) $\tan(\theta) = \frac{\sqrt{384}}{4} \approx 4.9$ so $\theta \approx 1.37$ ($\approx 78.5°$)

11. (a) The equation of the line through $P = (2, 3)$ and $Q = (8, 11)$ is $y - 3 = \frac{8}{6}(x - 2) \Rightarrow 6y - 8x = 2$. Substituting $x = 2a + 8(1 - a) = 8 - 6a$ and $y = 3a + 11(1 - a) = 11 - 8a$ into the equation for the line, we get:

   $6(11 - 8a) - 8(8 - 6a) = 66 - 48a - 64 + 48a = 2$

   for all a, so the point with $x = 2a + 8(1 - a)$ and $y = 3a + 11(1 - a)$ is on the line through P and

Q for any a. Furthermore, $2 \le 8 - 6a \le 8$ for $0 \le a \le 1$, so the point in question must be on the line **segment** PQ.

(b) $\text{dist}(P, Q) = \sqrt{6^2 + 8^2} = 10$, while:

$$\begin{aligned}
\text{dist}(P, R) &= \sqrt{(8 - 6a - 2)^2 + (11 - 8a - 3)^2} \\
&= \sqrt{(6 - 6a)^2 + (8 - 8a)^2} \\
&= \sqrt{6^2(1 - a)^2 + 8^2(1 - a)^2} \\
&= \sqrt{100(1 - a)^2} = 10 \cdot |1 - a| \\
&= |1 - a| \cdot \text{dist}(P, Q)
\end{aligned}$$

13. (a) $m_1 \cdot m_2 = (1)(-1) = -1$

(b)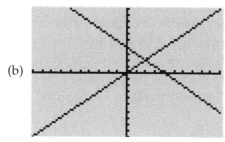

(c) Because 20 units of x-values are physically wider on the screen than 20 units of y-values.

(d) Set the window so that:

$$(\text{xmax} - \text{xmin}) \approx 1.7(\text{ymax} - \text{ymin})$$

15. (a) $y - 5 = 3(x - 2)$ or $y = 3x - 1$
    (b) $y - 2 = -2(x - 3)$ or $y = 8 - 2x$
    (c) $y - 4 = -\frac{1}{2}(x - 1)$ or $y = -\frac{1}{2}x + \frac{9}{2}$

17. (a) $y - 5 = \frac{3}{2}(x - 2)$ or $y = \frac{3}{2}x + 2$
    (b) $y - 2 = \frac{3}{2}(x + 1)$ or $y = \frac{3}{2}x + \frac{7}{2}$
    (c) $x = 3$

19. Distance between the centers $= \sqrt{6^2 + 8^2} = 10$.
    (a) $10 - 2 - 4 = 4$ (b) $10 - 2 - 7 = 1$ (c) 0 (they intersect) (d) $15 - 10 - 3 = 2$ (e) $12 - 10 - 1 = 1$

21. Find $\text{dist}(P, C) = \sqrt{(x - h)^2 + (y - k)^2}$ and compare the value to r:

P is $\begin{cases} \text{inside the circle} & \text{if} \quad \text{dist}(P, C) < r \\ \text{on the circle} & \text{if} \quad \text{dist}(P, C) = r \\ \text{outside the circle} & \text{if} \quad \text{dist}(P, C) > r \end{cases}$

23. A point $P = (x, y)$ lies on the circle if and only if its distance from $C = (h, k)$ is r: $\text{dist}(P, C) = r$. So P is on the circle if and only if:

$$\sqrt{(x - h)^2 + (y - k)^2} = r$$

if and only if:

$$(x - h)^2 + (y - k)^2 = r^2$$

25. (a) $-\frac{5}{12}$ (b) undefined (vertical line) (c) $\frac{12}{5}$ (d) 0 (horizontal line)

27. (a) ≈ 2.22 (b) ≈ 2.24 (c) (by inspection) 3 units, which occurs at the point $(5, 3)$

29. (a) If $B \ne 0$, we can solve for y: $y = -\frac{A}{B}x + \frac{C}{B}$, so the slope is $m = -\frac{A}{B}$.

    (b) The required slope is $\frac{B}{A}$ (the negative reciprocal of $-\frac{A}{B}$) and the y-intercept is 0, so the equation is $y = \frac{B}{A}x$ or $Bx - Ay = 0$.

    (c) Solve the equations $Ax + By = C$ and $Bx - Ay = 0$ simultaneously to get:

    $$x = \frac{AC}{A^2 + B^2} \quad \text{and} \quad y = \frac{BC}{A^2 + B^2}$$

    (d) The distance from this point to the origin is:

    $$\begin{aligned}
    &\sqrt{\left(\frac{AC}{A^2 + B^2}\right)^2 + \left(\frac{BC}{A^2 + B^2}\right)^2} \\
    &= \sqrt{\frac{A^2C^2}{(A^2 + B^2)^2} + \frac{B^2C^2}{(A^2 + B^2)^2}} \\
    &= \sqrt{\frac{(A^2 + B^2)C^2}{(A^2 + B^2)^2}} \\
    &= \sqrt{\frac{C^2}{A^2 + B^2}} = \frac{|C|}{\sqrt{A^2 + B^2}}
    \end{aligned}$$

Section 0.3

1. A: a, B: c, C: d, D:b

3. A: b, B: c, C: d, D: a

5. (a) C (b) A (c) B

7. (a) $f(1) = 4$, $g(1)$ is undefined, $H(1) = -1$

(b)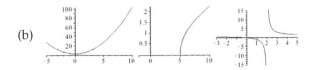

(c) $f(3x) = (3x)^2 + 3 = 9x^2 + 3$, $g(3x) = \sqrt{3x - 5}$ (for $x \geq \frac{5}{3}$), $H(3x) = \frac{3x}{3x-2}$

(d) $f(x + h) = (x + h)^2 + 3 = x^2 + 2xh + h^2 + 3$, $g(x + h) = \sqrt{x + h - 5}$, $H(x + h) = \frac{x+h}{x+h-2}$

9. (a) $m = 2$ (b) $m = 2x + 3 + h$ (c) If $x = 1.3$, then $m = 5.6 + h$; if $x = 1.1$, then $m = 5.2 + h$; if $x = 1.002$, then $m = 5.004 + h$.

11. $\frac{f(a+h)-f(a)}{h} = 2a + h - 2$ (if $h \neq 0$). If $a = 1$: h. If $a = 2$: $2 + h$. If $a = 3$: $4 + h$. If $a = x$: $2x + h - 2$. $\frac{g(a+h)-g(a)}{h} = \frac{\sqrt{a+h}-\sqrt{a}}{h}$. If $a = 1$: $\frac{\sqrt{1+h}-1}{h}$. If $a = 2$: $\frac{\sqrt{2+h}-\sqrt{2}}{h}$. If $a = 3$: $\frac{\sqrt{3+h}-\sqrt{3}}{h}$. If $a = x$: $\frac{\sqrt{x+h}-\sqrt{x}}{h}$.

13. (a) Approx. 250 miles, 375 miles. (b) Approx. 200 miles/hour. (c) By flying along a circular arc about 375 miles from the airport (or by landing at another airport).

15. (a)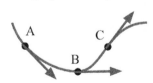

(b) Largest: $x = 2$; smallest: $x = 4$.

(c) Largest: at $x = 5$; smallest at $x = 3$.

17. The path of the slide is a straight line tangent to the graph of the path at the point of fall:

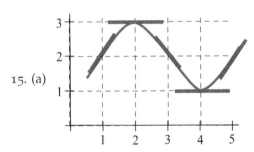

19. (a) $s(1) = 2$, $s(3) = \frac{4}{3}$, $s(4) = \frac{5}{4}$ (b) $s(x) = \frac{x+1}{x}$

21. Approximate values:

x	$f(x)$	$g(x)$
0	1.0	1.0
1	2.0	1.0
2	2.0	-1.0
3	1.0	0.0
4	1.5	0.5

23. On your own.

Section 0.4

1. (a) ≈ -18, ≈ -2.2 (b) If $T = 11°C$, $WCI_{11} =$

$$\begin{cases} 11 & \text{if } 0 \leq v \leq 6.5 \\ 22.55 - 5.29\sqrt{v} + 0.279v & \text{if } 6.5 < v \leq 72 \\ -2.2 & \text{if } v > 72 \end{cases}$$

3. $g(0) = 3$, $g(1) = 1$, $g(2) = 2$, $g(3) = 3$, $g(4) = 1$, $g(5) = 1$.

$$g(x) = \begin{cases} 3 - x & \text{if } x < 1 \\ x & \text{if } 1 \leq x \leq 3 \\ 1 & \text{if } x > 3 \end{cases}$$

5. (a) $f(f(1)) = 1$, $f(g(2)) = 2$, $f(g(0)) = 2$, $f(g(1)) = 3$

(b) $g(f(2)) = 0$, $g(f(3)) = 1$, $g(g(0)) = 0$, $g(f(0)) = 0$

(c) $f(h(3)) = 3$, $f(h(4)) = 2$, $h(g(0)) = 0$, $h(g(1)) = -1$

7. (a)

x	-1	0	1	2	3	4
$f(x)$	3	3	-1	0	1	1
$g(x)$	-2	0	1	2	3	4
$h(x)$	-3	-2	-1	0	1	2

(b) $f(g(1)) = -1$, $f(h(1)) = 3$, $h(f(1)) = -3$, $f(f(2)) = 3$, $g(g(3.5)) = 3$

(c)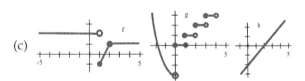

9. If $L(d)$ represents location on day d:

$$L(d) = \begin{cases} \text{England} & \text{if } d = \text{Monday or Tuesday} \\ \text{France} & \text{if } d = \text{Wednesday} \\ \text{Germany} & \text{if } d = \text{Thursday or Friday} \\ \text{Italy} & \text{if } d = \text{Saturday} \end{cases}$$

11. Assuming the left portion is part of a parabola:

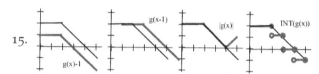

$$f(x) = \begin{cases} x^2 & \text{if } x < 2 \\ x - 1 & \text{if } x > 2 \end{cases}$$

13. (a) $B(1) = 1 \cdot f(1) = 1 \cdot \frac{1}{1} = 1$, $B(2) = 2 \cdot \frac{1}{2} = 1$, $B(3) = 3 \cdot \frac{1}{3} = 1$. (b) $B(x) = x \cdot f(x) = x \cdot \frac{1}{x} = 1$ (if $x > 0$)

15.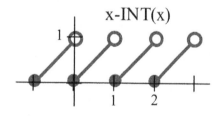

17. (a) $f(g(x)) = 6x + 2 + 3A$, $g(f(x)) = g(3x + 2) = 6x + 4 + A$. If $f(g(x)) = g(f(x))$, then $A = 1$.

(b) $f(g(x)) = 3Bx - 1$, $g(f(x)) = 3Bx + 2B - 1$. If $f(g(x)) = g(f(x))$, then $B = 0$.

19. Graph of $f(x) = x - \lfloor x \rfloor$:

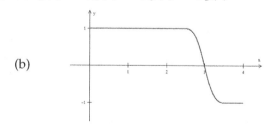

x-INT(x)

21. $f(x) = \lfloor 1.3 + 0.5 \sin(x) \rfloor$ works. The value of A in $f(x) = \lfloor A + 0.5 \sin(x) \rfloor$ determines the relative lengths of the long and short parts of the pattern.

23. (a) $g(1) = 1$, $g(2) = 1$, $g(3) = 0$, $g(4) = -1$.

(b)

25. ≈ 0.739 starting with $x = 1, 2, 10$ or any value

27. $f(1) = \frac{2}{2} = 1$. $f(0.5) = 1.25$, $f(1.25) = 1.025$, $f(1.025) \approx 1.0003049$, $f(1.0003049) \approx 1.000000046, \ldots f(4) = 2.125$, $f(2.125) \approx 1.297794$, $f(1.297794) \approx 1.034166$, $f(1.034166) \approx 1.000564, \ldots$

29. (a) $f(2) = \frac{14}{3} \approx 4.7$, $f(\frac{14}{3}) = \frac{50}{9} \approx 5.6$, $f(\frac{50}{9}) = \frac{158}{27} \approx 5.85$, $f(\frac{158}{27}) = \frac{482}{81} \approx 5.95$ $\ldots f(4) = \frac{16}{3} \approx 5.3$, $f(\frac{16}{3}) = \frac{52}{9} \approx 5.8$, $f(\frac{52}{9}) = \frac{160}{27} \approx 5.93$, $f(\frac{160}{27}) = \frac{484}{81} \approx 5.975$ $\ldots f(6) = 6$. (b) $c = 6$ (c) Solve $c = g(c) = \frac{c}{3} + A$ to get $3c = c + 3A \Rightarrow 2c = 3A \Rightarrow c = \frac{3A}{2}$ is a fixed point of g.

31. On your own.

Section 0.5

1. (a) $x = 2, 4$ (b) $x = -2, -1, 0, 1, 2, 3, 4, 5$ (c) $x = -2, -1, 1, 3$

3. (a) all x (all real numbers) (b) $x > \sqrt[3]{-2}$ (c) all x

5. (a) $x = -2, -3, 3$ (b) no values of x (c) $x \geq 0$

7. (a) If $x \neq 2$ and $x \neq -3$, then $x^2 + x - 6 \neq 0$. True. (b) If an object does not have 3 sides, then it is not a triangle. True.

9. (a) If your car does not get at least 24 miles per gallon, then it is not tuned properly. (b) If you cannot have dessert, then you did not eat your vegetables.

11. (a) If you will not vote for me, then you do not love your country. (b) If not only outlaws have guns, then guns are not outlawed. (poor English) If someone legally has a gun, then guns are not illegal.

13. (a) Both $f(x)$ and $g(x)$ are not positive. (b) x is not positive. ($x \leq 0$) (c) 8 is not a prime number.

15. (a) For some numbers a and b, $|a + b| \neq |a| + |b|$. (b) Some snake is not poisonous. (c) Some dog can climb trees.

17. If x is an integer, then $2x$ is an even integer. True. Converse: If $2x$ is an even integer, then x is an integer. True. (It is not likely that these were the statements you thought of; there are lots of other examples.)

19. (a) False. If $a = 3$, $b = 4$, then $(a + b)^2 = 7^2 = 49$, but $a^2 + b^2 = 3^2 + 4^2 = 9 + 16 = 25$. (b) False. If $a = -2$, $b = -3$, then $a > b$, but $a^2 = 4 < 9 = b^2$. (c) True.

21. (a) True. (b) False. If $f(x) = x + 1$ and $g(x) = x + 2$, Then $f(x) \cdot g(x) = x^2 + 3x + 2$ is not a linear function. (c) True.

23. (a) If a and b are prime numbers, then $a + b$ is prime. False: take $a = 3$ and $b = 5$.

    (b) If a and b are prime numbers, then $a + b$ is not prime. False: take $a = 2$ and $b = 3$.

    (c) If x is a prime number, then x is odd. False: take $x = 2$. (This is the only counterexample.)

    (d) If x is a prime number, then x is even. False: take $x = 3$ (or 5 or 7 or...)

25. (a) If x is a solution of $x + 5 = 9$, then x is odd. False: take $x = 4$.

    (b) If a 3-sided polygon has equal sides, then it is a triangle. True. (We also have non-equilateral triangles.)

    (c) If a person is a calculus student, then that person studies hard. False (unfortunately), but we won't mention names.

    (d) If x is a (real number) solution of $x^2 - 5x + 6 = 0$, then x is even. False: take $x = 3$.

Section 1.0

1. (a) $m = \frac{y-9}{x-3}$. If $x = 2.97$, $m = \frac{-0.1791}{-0.03} = 5.97$. If $x = 3.001$, $m = \frac{0.006001}{0.001} = 6.001$. If $x = 3 + h$,
$$m = \frac{(3+h)^2 - 9}{(3+h) - 3} = \frac{9 + 6h + h^2 - 9}{h} = 6 + h$$
   (b) When h is close to 0, $6 + h$ is close to 6.

3. (a) $m = \frac{y-4}{x-2}$. If $x = 1.99$, $m = \frac{-0.0499}{-0.01} = 4.99$. If $x = 2.004$, $m = \frac{0.020016}{0.004} = 5.004$. If $x = 2 + h$:
$$m = \frac{\left[(2+h)^2 + (2+h) - 2\right] - 4}{(2+h) - 2} = 5 + h$$
   (b) When h is very small, $5 + h$ is very close to 5.

5. All of these answers are **approximate**. Your answers should be close to these numbers.

   (a) average rate of temperature change $\approx$
$$\frac{80° - 64°}{1 \text{ p.m.} - 9 \text{ a.m.}} = \frac{16°}{4 \text{ hours}} = 4 \frac{°}{\text{hour}}$$

   (b) At 10 a.m., temperature was rising about $5°$ per hour; at 7 p.m., its was rising about $-10°/\text{hr}$ (**falling** about $10°/\text{hr}$).

7. All of these answers are **approximate**.

   (a) average velocity $\approx \dfrac{300 \text{ ft} - 0 \text{ ft}}{20 \text{ sec} - 0 \text{ sec}} = 15 \dfrac{\text{ft}}{\text{sec}}$

   (b) average velocity $\approx \dfrac{100 \text{ ft} - 200 \text{ ft}}{30 \text{ sec} - 10 \text{ sec}} = -5 \dfrac{\text{ft}}{\text{sec}}$

   (c) At $t = 10$ seconds, velocity ≈ 30 feet per second (between 20 and 35 ft/sec); at $t = 20$ seconds, velocity ≈ -1 feet per second; at $t = 30$ seconds, velocity ≈ -40 feet per second.

9. (a) $A(0) = 0$, $A(1) = 3$, $A(2) = 6$, $A(2.5) = 7.5$, $A(3) = 9$ (b) The area of the rectangle bounded below by the t-axis, above by the line $y = 3$, on the left by the vertical line $t = 1$ and on the right by the vertical line $t = 4$. (c) Graph of $y = 3x$.

Section 1.1

1. (a) 2 (b) 1 (c) DNE (does not exist) (d) 1

3. (a) 1 (b) -1 (c) -1 (d) 2

5. (a) -7 (b) $\frac{13}{0}$ (DNE)

7. (a) 0.54 (radian mode!) (b) -0.318 (c) -0.54

9. (a) 0 (b) 0 (c) 0

11. (a) 0 (b) -1 (c) DNE

13. The one- and two-sided limits agree at $x = 1$, $x = 4$ and $x = 5$, but not at $x = 2$:

$$\lim_{x \to 1^-} g(x) = 1 \qquad \lim_{x \to 1^+} g(x) = 1 \qquad \lim_{x \to 1} g(x) = 1$$

$$\lim_{x \to 2^-} g(x) = 1 \qquad \lim_{x \to 2^+} g(x) = 4 \qquad \lim_{x \to 2} g(x) \text{ DNE}$$

$$\lim_{x \to 4^-} g(x) = 2 \qquad \lim_{x \to 4^+} g(x) = 2 \qquad \lim_{x \to 4} g(x) = 2$$

$$\lim_{x \to 5^-} g(x) = 1 \qquad \lim_{x \to 5^+} g(x) = 1 \qquad \lim_{x \to 5} g(x) = 1$$

15. (a) 1.0986 (b) 1

17. (a) 0.125 (b) 3.5

19. (a) $A(0) = 0$, $A(1) = 2.25$, $A(2) = 5$, $A(3) = 8.25$ (b) $A(x) = 2x + \frac{1}{4}x^2$ (c) The area of the trapezoid bounded below by the t-axis, above by the line $y = \frac{1}{2}t + 2$, on the left by the vertical line $t = 1$ and on the right by the vertical line $t = 3$.

Section 1.2

1. (a) 2 (b) 0 (c) DNE (d) 1.5

3. (a) 1 (b) 3 (c) 1 (d) ≈ 0.8

5.

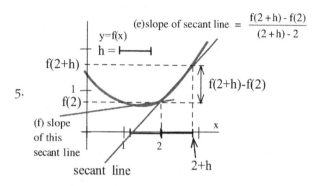

7. (a) 2 (b) -1 (c) DNE (d) 2
   (e) 2 (f) 2 (g) 1 (h) 2 (i) DNE

9. (a) When $v = 0$, $L = A$. (b) 0

11. (a) 4 (b) 1 (c) 2 (d) 0 (e) 1 (f) 1

13. (a) Slope of the line tangent to the graph of $y = \cos(x)$ at the point $(0,1)$. (b) slope $= 0$

15. (a) ≈ 1 (b) ≈ 3.43 (c) ≈ 4

17. At $x = -1$: "connected and smooth"; at $x = 0$: "connected with a corner"; at $x = 1$: "simple hole"; at $x = 2$: "vertical jump"; at $x = 3$: "simple hole"; at $x = 4$: "corner"; at $x = 5$: "smooth"

19. Many lists will work. Here is one example: Put $a_n = 2 + \frac{1}{n}$ so a_n approaches 2 and $\frac{|a_n - 2|}{a_n - 2} = 1$ for all n. Put $b_n = 2 - \frac{1}{n}$ so $b_n \to 2$ and $\frac{|b_n - 2|}{b_n - 2} = -1$ for all n.

21. $-x^2 \le x^2 \cos\left(\frac{1}{x^2}\right) \le x^2$ so limit is 0

23. $-x^2 \le x^2 \sin\left(\frac{1}{x}\right) \le x^2$ so limit is 3

25. $\frac{1}{x^2} - 1 < \left\lfloor \frac{1}{x^2} \right\rfloor \le \frac{1}{x^2}$ so limit is 1

Section 1.3

1. Discontinuous at 1, 3 and 4.

3. (a) Discontinuous at $x = 3$: fails condition (i) there. (b) At $x = 2$: fails (i). (c) Where $\cos(x) < 0$ (for example, at $x = \pi$): fails (i). (d) Where x^2 is

an integer (for example, at $x = 1$ or $x = 2$): fails (ii). (e) Where $\sin(x) = 0$ (for example, at $x = 0$, $x = \pm\pi, \pm 2\pi \ldots$): fails (i). (f) At $x = 0$: fails (i). (g) At $x = 0$: fails (i). (h) At $x = 3$: fails (i). (i) At $x = \frac{\pi}{2}$: fails (i).

5. (a) $f(x) = 0$ for at least 3 values of x in the interval $0 \le x \le 5$. (b) 1 (c) 3 (d) 2 (e) Yes. (It does not have to happen, but it is possible.)

7. (a) $f(0) = 0$, $f(3) = 9$ and $0 \le 2 \le 9$; $c = \sqrt{2} \approx 1.414$ (b) $f(-1) = 1$, $f(2) = 4$ and $1 \le 3 \le 4$; $c = \sqrt{3} \approx 1.732$ (c) $f(0) = 0$, $f(\frac{\pi}{2}) = 1$ and $0 \le \frac{1}{2} \le 1$; $c = \arcsin(\frac{1}{2}) \approx 0.524$ (d) $f(0) = 0$, $f(1) = 1$ and $0 \le \frac{1}{3} \le 1$; $c = \frac{1}{3}$ (e) $f(2) = 2$, $f(5) = 20$ and $2 \le 4 \le 20$; $c = \frac{1 + \sqrt{17}}{2} \approx 2.561$ (f) $f(1) = 0$, $f(10) \approx 2.30$ and $0 \le 2 \le 2.30$; $c = e^2 \approx 7.389$

9. Neither student is correct. The bisection algorithm converges to the root labeled C.

11. (a) D (b) D (c) hits B

13. $[-0.9375, -0.875]$, root ≈ -0.879; $[1.3125, 1.375]$, root ≈ 1.347; $[2.5, 2.5625]$, root ≈ 2.532

15. $[2.3125, 2.375]$, root ≈ 2.32

17. $[-0.375, -0.3125]$, root ≈ -0.32

19.

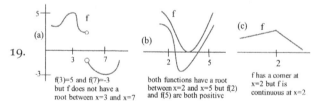

21. (a) $A(2.1) - A(2)$ is the area of the region bounded below by the t-axis, above by the graph of $y = f(t)$, on the left by the vertical line $t = 2$, and on the right by the vertical line $t = 2.1$. $\frac{A(2.1) - A(2)}{0.1} \approx f(2)$ or $f(2.1) \Rightarrow \frac{A(2.1) - A(2)}{0.1} \approx 1$

   (b) $A(4.1) - A(4)$ is the area of the region bounded below by the t-axis, above by the graph of $y = f(t)$, on the left by the vertical line $t = 4$, and on the right by the vertical line $t = 4.1$. $\frac{A(4.1) - A(4)}{0.1} \approx f(4) \approx 2$

23. (a) Yes (you justify). (b) Yes. (c) Try it.

Section 1.4

1. (a) If x is within $\frac{1}{2}$ unit of 3 then $2x + 1$ is within 1 unit of 7. (b) 0.3 (c) 0.02 (d) $\frac{\epsilon}{2}$

3. (a) If x is within $\frac{1}{4}$ unit of 2 then $4x - 3$ is within 1 unit of 5. (b) 0.1 (c) 0.02 (d) $\frac{\epsilon}{4}$

5. In 1, $m = 2$, $\delta = \frac{\epsilon}{2}$; in 2, $m = 3$, $\delta = \frac{\epsilon}{3}$; in 3, $m = 4$, $\delta = \frac{\epsilon}{4}$; in 4, $m = 5$, $\delta = \frac{\epsilon}{5}$. In general: $\delta = \frac{\epsilon}{|m|}$.

7. 0.02 inches

9. (a) Any value of x between $\sqrt[3]{7.5} \approx 1.957$ and $\sqrt[3]{8.5} \approx 2.043$: If x is within 0.043 units of 2 then x^3 will be within 0.5 units of 8. (b) Any x between $\sqrt[3]{7.95} \approx 1.9958$ and $\sqrt[3]{8.05} \approx 2.0042$: If x is within 0.0042 units of 2 then x^3 will be within 0.05 units of 8.

11. (a) Any value of x between 0 and 8: If x is within 3 units of 3 then $\sqrt{1 + x}$ will be within 1 unit of 2. (b) Any x between 2.99920004 and 3.00080004: If x is within 0.00079996 units of 3 then $\sqrt{1 + x}$ will be within 0.0002 units of 2.

13. 0.0059964 inches

15.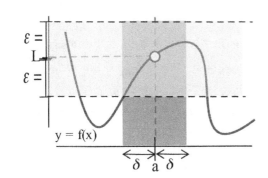

17. On your own.

19. Assume $\lim\limits_{x \to 2} f(x)$ exists. Let $\epsilon = 0.1$, so there must be a δ so that:

$$2 - \delta < x < 2 + \delta \Rightarrow L - 0.1 < f(x) < L + 0.1$$

Now, $f(x) = 4$ for any x with $2 - \delta < x < 2$, so it must be true that:

$$2 - \delta < x < 2 \Rightarrow L - 0.1 < 4 < L + 0.1$$
$$\Rightarrow L > 3.9$$

Similarly, $f(x) = 3$ for any x with $2 < x < 2 + \delta$, so it must be true that:

$$2 < x < 2 + \delta \Rightarrow L - 0.1 < 3 < L + 0.1$$
$$\Rightarrow L < 3.1$$

But no value of L can simultaneously satisfy $L > 3.9$ and $L < 3.1$, so we have reached a contradiction and our assumption must be false: $\lim\limits_{x \to 2} f(x)$ does not exist.

21. Assume $\lim\limits_{x \to 2} f(x)$ exists. Let $\epsilon = 0.1$, so there must be a δ so that:

$$2 - \delta < x < 2 + \delta \Rightarrow L - 0.1 < f(x) < L + 0.1$$

We can assume that $\delta \leq 1$ (if not, we can replace our initial δ with $\delta = 1$ because any smaller value will also work). Now, $f(x) = x$ for any x with $2 - \delta < x < 2$, so it must be true that:

$$2 - \delta < x < 2 \Rightarrow L - 0.1 < x < L + 0.1$$
$$\Rightarrow L < 0.1 + x < 2.1$$

On the other hand, $f(x) = 6 - x$ for any x with $2 < x < 2 + \delta \leq 3$, so it must be true that:

$$2 < x < 2 + \delta \Rightarrow L - 0.1 < 6 - x < L + 0.1$$
$$\Rightarrow L > 5.9 - x \geq 2.9$$

(because $x \leq 3$). But no value of L can simultaneously satisfy $L < 2.1$ and $L > 2.9$, so we have reached a contradiction and our assumption must be false: $\lim\limits_{x \to 2} f(x)$ does not exist.

23. Given any $\epsilon > 0$, we know $\frac{\epsilon}{2} > 0$, so there is a number $\delta_f > 0$ such that $|x - a| < \delta_f \Rightarrow |f(x) - L| < \frac{\epsilon}{2}$. Likewise, there is a number $\delta_g > 0$ such that $|x - a| < \delta_g \Rightarrow |g(x) - L| < \frac{\epsilon}{2}$. Let δ be the smaller of δ_f and δ_g. If $|x - a| < \delta$ then $|f(x) - L| < \frac{\epsilon}{2}$ and $|g(x) - M| < \frac{\epsilon}{2}$ so:

$$|(f(x) - g(x)) - (L - M))|$$
$$= |(f(x) - L) - (g(x) - M)|$$
$$\leq |f(x) - L| + |g(x) - M|$$
$$< \frac{\epsilon}{2} + \frac{\epsilon}{2} = \epsilon$$

so $f(x) - g(x)$ is within ϵ of $L - M$ whenever x is within δ of a.

Section 2.0

1. Values are approximate; your answers may vary.

x	$f(x)$	$m(x)$	x	$f(x)$	$m(x)$
0.0	1.0	1	2.5	0.0	-2
0.5	1.4	$\frac{1}{2}$	3.0	-1.0	-2
1.0	1.6	0	3.5	-1.3	0
1.5	1.4	$-\frac{1}{2}$	4.0	-1.0	1
2.0	1.0	-2			

3. (a) At $x = 1, 3$ and 4. (b) Largest at $x = 4$; smallest at $x = 3$.

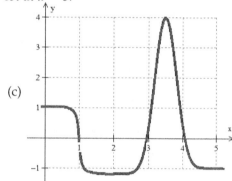

5. (a)

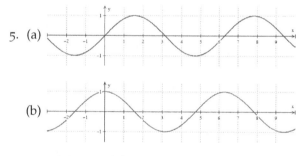

(b)

(c) $m(x) = \cos(x)$

7. Assume we turn off the engine at the point (p, q) on the curve $y = x^2$ and then find values of p and q so the tangent line to $y = x^2$ at the point (p, q) goes through the given point $(5, 16)$: (p, q) is on $y = x^2$ so $q = p^2$ and the tangent line to $y = x^2$ at (p, p^2) is $y = 2px - p^2$ so, substituting $x = 5$ and $y = 16$, we have $16 = 2p(5) - p^2 \Rightarrow p^2 - 10p + 16 = 0 \Rightarrow p = 2$ or $p = 8$. The solution we want (moving left to right along the curve) is $p = 2$, $q = p^2 = 4$ ($p = 8$, $q = 64$ would be the right-to-left solution).

9. Impossible: $(1, 3)$ sits "inside" the parabola.

11. (a) $m_{sec} = \dfrac{[3(x+h) - 7] - [3x - 7]}{(x+h) - x} = \dfrac{3h}{h} = 3$

(b) $m_{tan} = \lim\limits_{h \to 0} m_{sec} = \lim\limits_{h \to 0} 3 = 3$. (c) At $x = 2$, $m_{tan} = 3$. (d) $f(2) = -1$ so the tangent line is $y - (-1) = 3(x - 2)$ or $y = 3x - 7$.

13. (a) $m_{sec} = \dfrac{[a(x+h) + b] - [ax + b]}{(x+h) - x} = \dfrac{ah}{h} = a$

(b) $m_{tan} = \lim\limits_{h \to 0} m_{sec} = \lim\limits_{h \to 0} a = a$. (c) At $x = 2$, $m_{tan} = a$. (d) $f(2) = 2a + b$ so the tangent line is $y - (2a + b) = a(x - 2)$ or $y = ax + b$.

15. (a) $m_{sec} = \dfrac{[8 - 3(x+h)^2] - [8 - 3x^2]}{(x+h) - x} = -6x - 3h$

(b) $m_{tan} = \lim\limits_{h \to 0} m_{sec} = \lim\limits_{h \to 0} [-6x - 3h] = -6x$.

(c) At $x = 2$, $m_{tan} = -6(2) = -12$ (d) $f(2) = -4$ so the tangent line is $y - (-4) = -12(x - 2)$ or $y = -12x + 20$.

17. $a = 1$, $b = 2$, $c = 0 \Rightarrow m_{tan} = (2)(1)(x) + 2 = 2x + 2$ so we need p to satisfy $6 - (p^2 + 2p) = (2p + 2)(3 - p) \Rightarrow p^2 - 6p = 0 \Rightarrow p = 0$ or $p = 6$ so the points are $(0, 0)$ and $(6, 48)$.

Section 2.1

1. derivative (a) of g (b) of h (c) of f

3. (a) $m_{sec} = h - 4 \Rightarrow m_{tan} = \lim\limits_{h \to 0} m_{sec} = -4$

(b) $m_{sec} = h + 1 \Rightarrow m_{tan} = \lim\limits_{h \to 0} m_{sec} = 1$

5. (a) $m_{sec} = 5 - h \Rightarrow m_{tan} = \lim\limits_{h \to 0} m_{sec} = 5$

(b) $m_{sec} = 7 - 2x - h \Rightarrow m_{tan} = 7 - 2x$

7. (a) -1 (b) -1 (c) 0 (d) $+1$ (e) DNE (f) DNE

9. Using the definition:

$$f'(x) = \lim\limits_{h \to 0} \frac{[(x+h)^2 + 8] - [x^2 + 8]}{h}$$

$$= \lim\limits_{h \to 0} \frac{2xh + h^2}{h} = \lim\limits_{h \to 0} [2x + h] = 2x$$

so $f'(3) = 2 \cdot 3 = 6$

11. Using the definition:

$$f'(x) = \lim\limits_{h \to 0} \frac{[2(x+h)^3 - 5(x+h)] - [2x^3 - 5x]}{h}$$

$$= \lim\limits_{h \to 0} \frac{6x^2 h + 6xh^2 + 2h^3 - 5h}{h}$$

$$= \lim\limits_{h \to 0} \left[6x^2 + 6xh + 2h^2 - 5\right] = 6x^2 - 5$$

so $f'(3) = 6 \cdot 3^2 - 5 = 49$

13. For any constant C, if $f(x) = x^2 + C$, then:

$$f'(x) = \lim_{h \to 0} \frac{f(x+h) - f(x)}{h}$$

$$= \lim_{h \to 0} \frac{[(x+h)^2 + C] - [x^2 + C]}{h}$$

$$= \lim_{h \to 0} \frac{2xh + h^2}{h} = \lim_{h \to 0} [2x + h] = 2x$$

The graphs of $f(x) = x^2$, $g(x) = x^2 + 3$ and $h(x) = x^2 - 5$ are "parallel" parabolas: g is f shifted up 3 units; h is f shifted down 5 units.

15. $f'(x) = 2x \Rightarrow f'(1) = 2$ so an equation of the tangent line at $(1,9)$ is $y - 9 = 2(x-1)$ or $y = 2x + 7$; $f'(-2) = -4$ so an equation of the tangent line at $(-2,12)$ is $y - 12 = -4(x+2)$ or $y = -4x + 4$.

17. $f'(x) = \cos(x) \Rightarrow f'(\pi) = \cos(\pi) = -1$ so an equation of the tangent line at $(\pi, 0)$ is $y - 0 = -1(x - \pi)$ or $y = -x + \pi$; $f'\left(\frac{\pi}{2}\right) = \cos\left(\frac{\pi}{2}\right) = 0$ so an equation of the tangent line at $\left(\frac{\pi}{2}, 1\right)$ is $y - 1 = 0\left(x - \frac{\pi}{2}\right)$ or $y = 1$.

19. (a) $y - 5 = 4(x-2)$ or $y = 4x - 3$ (b) $x + 4y = 22$ or $y = -0.25x + 5.5$ (c) $f'(x) = 2x$, so the tangent line is horizontal only if $x = 0$: at the point $(0,1)$. (d) $f'(p) = 2p$ (the slope of the tangent line) so $y - q = 2p(x - p)$ or $y = 2px + (q - 2p^2)$. Since $q = p^2 + 1$, the tangent line is $y = 2px + (p^2 + 1 - 2p^2) = 2px - p^2 + 1$. (e) We need p such that $-7 = 2p(1) - p^2 + 1$, so $p^2 - 2p - 8 = 0 \Rightarrow p = -2$ or $p = 4$. There are two such points: $(-2,5)$ and $(4,17)$.

21. (a) $y'(x) = 2x$, so $y'(1) = 2$ and the angle is $\arctan(2) \approx 1.107$ radians $\approx 63°$ (b) $y'(x) = 3x^2$ so $y'(1) = 3$ and the angle is $\arctan(3) \approx 1.249$ radians $\approx 72°$ (c) $1.249 - 1.107$ radians $= 0.142$ radians (or $71.57° - 63.43° \approx 8.1°$)

23. Units on the horizontal axis are "seconds"; units on the vertical axis are "feet per second."

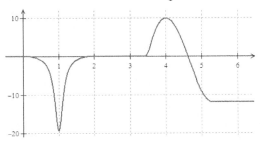

25. (a) $d(4) = 256$ ft; $d(5) = 400$ ft
(b) $d'(x) = 32x \Rightarrow d'(4) = 128$ ft/sec and $d'(5) = 160$ ft/sec.

27. Marginal production cost is $C'(x) = \frac{1}{2\sqrt{x}}$ dollars per golf ball, so $C'(25) = \frac{1}{2\sqrt{25}} = \frac{1}{10} = \$0.10$ per ball and $C'(100) = \frac{1}{2\sqrt{100}} = \frac{1}{20} = \$0.05$ per ball.

29. (a) $A(0) = 0$, $A(1) = \frac{1}{2}$, $A(2) = 2$ and $A(3) = \frac{9}{2}$.
(b) $A(x) = \frac{x^2}{2}$ $(x \geq 0)$ (c) $A'(x) = x$ (d) $A'(x)$ represents the rate at which $A(x)$ is increasing (the rate at which area is accumulating).

31. (a) $9x^8$ (b) $\frac{2}{3x^{\frac{1}{3}}}$ (c) $\frac{-4}{x^5}$ (d) $\pi x^{\pi - 1}$ (e) 1 if $x > -5$ and -1 if $x < -5$

33. $f(x) = x^3 + 4x^2$ (plus any constant)

35. $f(t) = 5 \cdot \sin(t)$ 37. $f(x) = \frac{1}{2}x^2 + \frac{1}{3}x^3$

Section 2.2

1. (a) 0, 1, 2, 3, 5   (b) 0, 3, 5

3. In the table below, **und.** means "undefined":

x	$f(x) \cdot g(x)$	$(f(x) \cdot g(x))'$	$\frac{f(x)}{g(x)}$	$\left(\frac{f(x)}{g(x)}\right)'$
0	2	13	2	-7
1	-15	16	$-\frac{3}{5}$	$\frac{4}{25}$
2	0	-6	0	$-\frac{3}{2}$
3	0	3	**und.**	**und.**

5.

x	$f+g$	$(f+g)'$	$f \cdot g$	$(f \cdot g)'$	$\frac{f}{g}$	$\left(\frac{f}{g}\right)'$
1	5	0	6	2	$\frac{3}{2}$	$-\frac{10}{4}$
2	4	$\frac{1}{2}$	3	$\frac{1}{2}$	$\frac{1}{3}$	$-\frac{1}{18}$
3	4	0	4	0	1	1

7. (a) $\mathbf{D}((x-5)(3x+7)) = (x-5) \cdot 3 + (3x+7) \cdot 1$
(b) $\mathbf{D}(3x^2 - 8x - 35) = 6x - 8$ (the same result)

9. $19 \cdot 3x^2 = 57x^2$ 11. $\cos(x) - \sin(x)$

13. $x^2 \cdot [-\sin(x)] + \cos(x) \cdot 2x$

15. Applying the Product Rule:

$$\mathbf{D}\left(\sin^2(x)\right) = \sin(x) \cdot \cos(x) + \sin(x) \cdot \cos(x)$$

$$= 2\sin(x)\cos(x) = \sin(2x)$$

17. $\dfrac{d}{dx}\left(\dfrac{\cos(x)}{x^2}\right) = \dfrac{x^2(-\sin(x)) - (\cos(x))(2x)}{(x^2)^2} =$

$\dfrac{-x[x\sin(x) + 2\cos(x)]}{x^4} = -\dfrac{x\sin(x) + 2\cos(x)}{x^3}$

19. $\dfrac{(1+x^2)\cdot 0 - 1\cdot 2x}{(1+x^2)^2} = -\dfrac{2x}{(1+x^2)^2}$

21. $\dfrac{\cos(\theta)\cdot 0 - 1\cdot(-\sin(\theta))}{(\cos(\theta))^2} = \dfrac{\sin(\theta)}{\cos^2(\theta)}$

23. $\dfrac{\cos(\theta)\cdot\cos(\theta) - \sin(\theta)\cdot(-\sin(\theta))}{(\cos(\theta))^2} = \dfrac{1}{\cos^2(\theta)}$

25. $40x^4 - 12x^3 + 6x^2 + 14x - 12$

27. $f(x) = ax^2 + bx + c$ so $f(0) = c$, hence $f(0) = 0$ $\Rightarrow c = 0$. $f'(x) = 2ax + b$ so $f'(0) = b$, hence $f'(0) = 0 \Rightarrow b = 0$. Finally, $f'(10) = 20a + b = 20a$ so $f'(10) = 30 \Rightarrow 20a = 30 \Rightarrow a = \frac{3}{2} \Rightarrow$ $f(x) = \frac{3}{2}x^2 + 0x + 0 = \frac{3}{2}x^2$.

29. Their graphs are vertical shifts of each other, and their derivatives are equal.

31. $f(x)\cdot g(x) = k \Rightarrow \mathbf{D}(f(x)\cdot g(x)) = \mathbf{D}(k) = 0$ so $f(x)\cdot g'(x) + g(x)\cdot f'(x) = 0$. If $f(x) \neq 0$ and $g(x) \neq 0$, then $\frac{f'(x)}{f(x)} = -\frac{g'(x)}{g(x)}$.

33. (a) $f'(x) = 2x - 5$, so $f'(1) = -3$
    (b) $f'(x) = 0$ only if $x = \frac{5}{2}$.

35. (a) $f'(x) = 3 + 2\sin(x)$; $f'(1) = 3 + 2\sin(1) \approx 4.7$
    (b) $f'(x) > 0$ because $\sin(x) > -\frac{3}{2}$ for all x.

37. (a) $f'(x) = 3x^2 + 18x = 3x(x+6)$, so $f'(1) = 21$
    (b) $f'(x) = 0$ only if $x = 0$ or $x = -6$.

39. (a) $f'(x) = 3x^2 + 4x + 2$, so $f'(1) = 9$
    (b) $f'(x) > 0$ (the discriminant $4^2 - 4(3)(2) < 0$).

41. (a) $f'(x) = x\cos(x) + \sin(x) \Rightarrow f'(1) = \cos(1) + \sin(1) \approx 1.38$
    (b) The graph of $f'(x)$ crosses the x-axis infinitely often. The root of f' at $x = 0$ is easy to see; approximate others, such as those near $x = 2.03$ and 4.91 and -2.03, using technology.

43. $f'(x) = 3x^2 + 2Ax + B$: the graph of $y = f(x)$ has two distinct "vertices" if $f'(x) = 0$ for two distinct values of x. This occurs if the discriminant of $3x^2 + 2Ax + B > 0$: $(2A)^2 - 4(3)(B) > 0$.

45. Everywhere except at $x = (2k+1)\frac{\pi}{2}$.

47. Everywhere except at $x = 0$ and $x = 3$.

49. Everywhere except at $x = 1$.

51. Everywhere. The only possible difficulty is at $x = 0$: the definition of the derivative gives $f'(0) = 1$ and the derivatives of the "two pieces" of f match (and equal 1) at $x = 0$.

53. Continuity of f at $x = 1$ requires $A + B = 2$. The "left derivative" of f at $x = 1$ is $\mathbf{D}(Ax + B) = A$ and the "right derivative" of f at $x = 1$ is 3 (if $x > 1$ then $\mathbf{D}(x^2 + x) = 2x + 1$), so to achieve differentiability we need $A = 3$ and $B = 2 - A = -1$.

55. (a) $h(x) = 128x - 2.65x^2 \Rightarrow h'(x) = 128 - 5.3x$ so $h'(0) = 128$ ft/sec, $h'(1) = 122.7$ ft/sec and $h'(2) = 117.4$ ft/sec
    (b) $v(x) = h'(x) = 128 - 5.3x$ ft/sec
    (c) $v(x) = 0$ when $x = \frac{128}{5.3} \approx 24.15$ sec
    (d) $h(24.15) \approx 1545.66$ ft
    (e) about 48.3 seconds: 24.15 up and 24.15 down

57. (a) $h(x) = v_0 x - 16x^2 \Rightarrow h'(x) = v_0 - 32x$ ft/sec
    (b) Max height when $x = \frac{v_0}{32}$: max height $=$ $h\left(\frac{v_0}{32}\right) = v_0\left(\frac{v_0}{32}\right) - 16\left(\frac{v_0}{32}\right)^2 = \frac{(v_0)^2}{64}$ ft
    (c) Time aloft $= 2\left(\frac{v_0}{32}\right) = \frac{v_0}{16}$ sec

59. (a) $\frac{(v_0)^2}{64} = 3.75 \Rightarrow v_0 = 8\sqrt{3.75} \approx 15.5$ ft/sec
    (b) $2\left(\frac{v_0}{32}\right) = \frac{8\sqrt{3.75}}{16} \approx 0.97$ sec
    (c) Max lift $= \frac{(v_0)^2}{2g} = \frac{(8\sqrt{3.75})^2}{2(5.3)} = \frac{240}{10.6} \approx 22.64$ ft

61. (a) $y' = -\frac{1}{x^2}$ so $y'(2) = -\frac{1}{4}$ and the tangent line is $y - 1/2 = \left(-\frac{1}{4}\right)(x-2)$ or $y = -\frac{1}{4}x + 1$
    (b) x-intercept at $(4,0)$; y-intercept at $(0,1)$
    (c) $A = \frac{1}{2}(\text{base})(\text{height}) = \frac{1}{2}(4)(1) = 2$

63. Because $(1,4)$ and $(3,14)$ are on the parabola, we need $a + b + c = 4$ and $9a + 3b + c = 14$. Subtracting the first equation from the second, $8a + 2b = 10$; $f'(x) = 2ax + b$ so $f'(3) = 6a + b = 9$, the slope of $y = 9x - 13$. Now solve the system $8a + 2b = 10$ and $6a + b = 9$ to get $a = 2$ and $b = -3$. Then use $a + b + c = 4$ to get $c = 5$: $a = 2, b = -3, c = 5$.

65. (a) $f(x) = x^3$
    (b) $g(x) = x^3 + 1$
    (c) If $h(x) = x^3 + C$ for any constant C, then $\mathbf{D}(h(x)) = 3x^2$.

67. (a) For $0 \le x \le 1$, $f'(x) = 1$ so $f(x) = x + C$; because $f(0) = 0$, we know $C = 0$ and $f(x) = x$. For $1 \le x \le 3$, $f'(x) = 2 - x$ so $f(x) = 2x - \frac{1}{2}x^2 + K$; because $f(1) = 1$, we know $K = -\frac{1}{2}$ and $f(x) = 2x - \frac{1}{2}x^2 - \frac{1}{2}$. For $3 \le x \le 4$, $f'(x) = x - 4$ so $f(x) = \frac{1}{2}x^2 - 4x + L$; because $f(3) = 1$, we know $L = \frac{17}{2}$ and $f(x) = \frac{1}{2}x^2 - 4x + \frac{17}{2}$.

  (b) A vertical shift, up 1 unit, of the graph in (a).

Section 2.3

1. $\mathbf{D}(f^2(x)) = 2 \cdot f^1(x) \cdot f'(x)$; at $x = 1$, $\mathbf{D}(f^2(x)) = 2(2)(3) = 12$. $\mathbf{D}(f^5(x)) = 5 \cdot f^4(x) \cdot f'(x)$; at $x = 1$, $\mathbf{D}(f^5(x)) = 5(2^4)(3) = 240$. $\mathbf{D}(f^{\frac{1}{2}}(x)) = \frac{1}{2} \cdot f^{-\frac{1}{2}}(x) \cdot f'(x)$; at $x = 1$, $\mathbf{D}(f^{\frac{1}{2}}(x)) = (\frac{1}{2})(2^{-\frac{1}{2}})(3) = \frac{3}{2\sqrt{2}} = \frac{3\sqrt{2}}{4}$.

3.

x	$f(x)$	$f'(x)$	$\mathbf{D}(f^2)$	$\mathbf{D}(f^3)$	$\mathbf{D}(f^5)$
1	1	-1	-2	-3	-5
3	2	-3	-12	-36	-240

5. $f'(x) = 5 \cdot (2x - 8)^4 \cdot (2) = 10(2x - 8)^4$

7. $f'(x) = x \cdot 5 \cdot (3x + 7)^4 \cdot 3 + 1 \cdot (3x + 7)^5 = (3x + 7)^4 [15x + (3x + 7)] = (3x + 7)^4 (18x + 7)$

9. $f'(x) = \frac{1}{2}(x^2 + 6x - 1)^{-\frac{1}{2}} \cdot (2x + 6)$, which we can rewrite as $\dfrac{x + 3}{\sqrt{x^2 + 6x - 1}}$

11. (a) Graph of $h(t) = 3 - 2\sin(t)$:

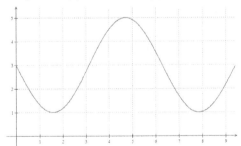

  (b) When $t = 0$, $h(0) = 3$ feet.
  (c) highest: 5 feet above the floor
    lowest: 1 foot above the floor.
  (d) $v(t) = h'(t) = -2\cos(t)$ ft/sec
    $a(t) = v'(t) = 2\sin(t)$ ft/sec^2.
  (e) This spring oscillates forever. The motion of a real spring would "damp out" due to friction.

13. (a) If $h(t) = 5t$, then $v(t) = h'(t) = 5$ and $K(1) = K(2) = \frac{1}{2}m(5^2) = 12.5m$. (b) If $h(t) = t^2$, then $v(t) = h'(t) = 2t$ so $v(1) = 2$ and $v(2) = 4$, hence $K(1) = \frac{1}{2}m(2^2) = 2m$ and $K(2) = \frac{1}{2}m(4^2) = 8m$.

15. $x \cdot (\sin(x))' + \sin(x) \cdot (x)' = x\cos(x) + \sin(x)$

17. $f'(x) = e^x - \sec(x) \cdot \tan(x)$

19. $f'(x) = -e^{-x} + \cos(x)$

21. $f'(x) = 7(x - 5)^6(1)$ so $f'(4) = 7(-1)^6(1) = 7$. Tangent line: $y - (-1) = 7(x - 4)$ or $y = 7x - 29$.

23. $f'(x) = \frac{1}{2}(25 - x^2)^{-\frac{1}{2}}(-2x) = \frac{-x}{\sqrt{25 - x^2}}$ so $f'(3) = \frac{-3}{\sqrt{25 - 9}} = -\frac{3}{4}$. Tangent line: $y - 4 = -\frac{3}{4}(x - 3)$ or $3x + 4y = 25$.

25. $f'(x) = 5(x - a)^4(1)$ so $f'(a) = 5(a - a)^4(1) = 0$. Tangent line: $y - 0 = 0(x - a)$ or $y = 0$.

27. (a) $f'(x) = e^x$ so $f'(3) = e^3$. Tangent line: $y - e^3 = e^3(x - 3)$ or $y = e^3 x - 2e^3$.

  (b) $0 - e^3 = e^3(x - 3) \Rightarrow -1 = x - 3 \Rightarrow x = 2$

  (c) $f'(p) = e^p$ so tangent line at (p, e^p) is $y - e^p = e^p(x - p)$; x-intercept: $0 - e^p = e^p(x - p) \Rightarrow -1 = x - p \Rightarrow x = p - 1$

29. $f'(x) = -\sin(x) \Rightarrow f''(x) = -\cos(x)$

31. $f'(x) = x^2\cos(x) + 2x\sin(x) \Rightarrow f''(x) = -x^2\sin(x) + 2x\cos(x) + 2x\cos(x) + 2\sin(x) = -x^2\sin(x) + 4x\cos(x) + 2\sin(x)$

33. $f'(x) = e^x\cos(x) - e^x\sin(x) \Rightarrow f''(x) = -2e^x\sin(x)$

35. $q' = $ linear, $q'' = $ constant, $q''' = q^{(4)} = \cdots = 0$

37. $p^{(n)} = $ constant $\Rightarrow p^{(n+1)} = 0$

39. $f(x) = 5e^x$

41. $f(x) = (1 + e^x)^5$

43. No. Using the definition of the derivative:

$$\lim_{h \to 0} \frac{f(0 + h) - f(0)}{h} = \lim_{h \to 0} \frac{(0 + h) \cdot \sin\left(\frac{1}{0 + h}\right) - 0}{h}$$

which simplifies to $\lim_{h \to 0} \sin\left(\frac{1}{h}\right)$; to see that this last limit does not exist, graph $\sin\left(\frac{1}{h}\right)$ or evaluate $\sin\left(\frac{1}{h}\right)$ for small values of h.

45. $\left(1 + \dfrac{1}{x}\right)^x \approx 2.718\ldots = e$ when x is large.

47. (a) $s_2 = 2.5$, $s_3 \approx 2.67$, $s_4 \approx 2.708$, $s_5 \approx 2.716$, $s_6 \approx 2.718$, $s_7 \approx 2.71825$, $s8 \approx 2.718178$ (b) They are approaching e.

Section 2.4

1. $f(x) = x^5$, $g(x) = x^3 - 7x \Rightarrow f \circ g(x) = (x^3 - 7x)^5$

3. Setting $f(x) = x^{\frac{5}{2}}$ and $g(x) = 2 + \sin(x)$ yields $f \circ g(x) = (2 + \sin(x))^{\frac{5}{2}}$. (The functions $f(x) = \sqrt{x}$ and $g(x) = (2 + \sin(x))^5$ also work.)

5. $f(x) = |x|$, $g(x) = x^2 - 4 \Rightarrow f \circ g(x) = |x^2 - 4|$

7. $y = u^5$, $u = x^3 - 7x$; $y = u^4$, $u = \sin(3x - 8)$; $y = u^{\frac{5}{2}}$, $u = 2 + \sin(x)$; $y = \frac{1}{\sqrt{u}}$, $u = x^2 + 9$; $y = |u|$, $u = x^2 - 4$; $y = \tan(u)$, $u = \sqrt{x}$

8. & 9.

x	$f \circ g(x)$	$(f \circ g)'(x)$
-2	1	0
-1	1	2
0	0	1
1	2	2
2	-2	-2

11. $g(2) \approx 2$, $g'(2) \approx -1$
$(f \circ g)(2) = f(g(2)) \approx f(2) \approx 1$
$f'(g(2)) \approx f'(2) \approx 0$
$(f \circ g)'(2) = f'(g(2)) \cdot g'(2) \approx 0$

13. $\mathbf{D}\left(\left(1 - \frac{3}{x}\right)^4\right) = 4\left(1 - \frac{3}{x}\right)^3 \cdot \frac{3}{x^2}$

15. $5\left(-\frac{1}{2}\right)(2 + \sin(x))^{-\frac{3}{2}} \cdot \cos(x) = \frac{-5\cos(x)}{2(2+\sin(x))^{\frac{3}{2}}}$

17. $x^2 \cdot \left[\cos(x^2 + 3) \cdot 2x\right] + \sin(x^2 + 3) \cdot 2x = 2x\left[x^2\cos(x^2 + 3) + \sin(x^2 + 3)\right]$

19. $\frac{7}{\cos(x^3 - x)} = 7\sec\left(x^3 - x\right)$ so $\mathbf{D}\left(7\sec\left(x^3 - x\right)\right) = 7(3x^2 - 1) \cdot \sec(x^3 - x) \cdot \tan(x^3 - x)$

21. $\mathbf{D}\left(e^x + e^{-x}\right) = e^x - e^{-x}$

23. (a) $h(0) = 2$ feet above the floor.
(b) $h(t) = 3 - \cos(2t)$ ft, $v(t) = h'(t) = 2\sin(2t)$ ft/sec, $a(t) = v'(t) = 4\cos(2t)$ ft/sec^2
(c) $K = \frac{1}{2}mv^2 = \frac{1}{2}m(2\sin(2t))^2 = 2m\sin^2(2t)$, $\frac{dK}{dt} = 8m \cdot \sin(2t) \cdot \cos(2t)$

25. (a) $P(0) = 14.7$ psi (pounds per square inch), $P(30000) \approx 4.63$ psi
(b) $10 = 14.7e^{-0.0000385h} \Rightarrow \frac{10}{14.7} = e^{-0.0000385h} \Rightarrow h = \frac{1}{-0.0000385}\ln\left(\frac{10}{14.7}\right) \approx 10{,}007$ ft

(c) $\frac{dP}{dh} = 14.7(-0.0000385)e^{-0.0000385h}$ psi/ft. At $h = 2{,}000$ ft, $\frac{dP}{dh} = 14.7(-0.0000385)e^{-0.0000385(2000)}$ psi/ft ≈ -0.000524 psi/ft. Finally, $\frac{dP}{dt} = 500(-0.000524) \approx -0.262$ psi/min.

(d) If temperature is constant, then (from physics!) we know (pressure)(volume) is constant, so decreasing pressure means increasing volume.

27. $\frac{2\cos(z)\left[-\sin(z)\right]}{2\sqrt{1 + \cos^2(z)}} = \frac{-\sin(2z)}{2\sqrt{1 + \cos^2(z)}}$

29. $\frac{d}{dx}\tan(3x + 5) = 3 \cdot \sec^2(3x + 5)$

31. $\mathbf{D}\left(\sin(\sqrt{x + 1})\right) = \cos\left(\sqrt{x + 1}\right) \cdot \frac{1}{2\sqrt{x+1}}$

33. $\frac{d}{dx}\left(e^{\sin(x)}\right) = e^{\sin(x)} \cdot \cos(x)$

35. $f(x) = \sqrt{x} \Rightarrow f'(x) = \frac{1}{2\sqrt{x}}$; $x(t) = 2 + \frac{21}{t} \Rightarrow x'(t) = -\frac{21}{t^2}$. At $t = 3$, $x = 9$ and $x'(t) = -\frac{21}{9} = -\frac{7}{3}$ so $\frac{d}{dt}\left(f(x(t))\right) = \left(\frac{1}{2\sqrt{9}}\right)\left(-\frac{7}{3}\right) = -\frac{7}{18}$.

37. $f(x) = \tan^3(x) \Rightarrow f'(x) = 3 \cdot \tan^2(x) \cdot \sec^2(x)$; $x(t) = 8 \Rightarrow x'(t) = 0$. When $t = 3$, $x = 8$ and $x'(t) = 0$ so $\frac{d}{dt}\left(f(x(t))\right) = 0$.

39. $f(x) = \frac{1}{77}(7x - 13)^{11}$

41. $f(x) = -\frac{1}{2}\cos(2x - 3)$

43. $f(x) = e^{\sin(x)}$

45. $-2\sin(2x) = 2\cos(x)\left[-\sin(x)\right] - 2\sin(x) \cdot \cos(x)$ or $\sin(2x) = 2\sin(x) \cdot \cos(x)$

47. $3\cos(3x) = 3\cos(x) - 12\sin^2(x) \cdot \cos(x)$ so $\cos(3x) = \cos(x)\left[1 - 4\sin^2(x)\right] = \cos(x)\left[1 - 4 + 4\cos^2(x)\right] = 4\cos^3(x) - 3\cos(x)$

49. $y' = 3Ax^2 + 2Bx$

51. $y' = 2Ax \cdot \cos\left(Ax^2 + B\right)$

53. $y' = \frac{Bx}{\sqrt{A + Bx^2}}$

55. $y' = B \cdot \sin(Bx)$

57. $y' = -2Ax \cdot \sin(Ax^2 + B)$

59. $y' = x\left(B \cdot e^{Bx}\right) + e^{Bx} = (Bx + 1) \cdot e^{Bx}$

61. $y' = A \cdot e^{Ax} + A \cdot e^{-Ax}$

63. $y' = \frac{A \cdot \sin(Bx) - Ax \cdot B \cdot \cos(Bx)}{\sin^2(Bx)}$

65. $y' = \frac{(Cx + D)A - (Ax + B)C}{(Cx + D)^2} = \frac{AD - BC}{(Cx + D)^2}$

67. (a) $y' = AB - 2Ax$ (b) $x = \frac{AB}{2A} = \frac{B}{2}$ (c) $y'' = -2A$

69. (a) $y' = 2ABx - 3Ax^2 = Ax(2B - 3x)$

    (b) $x = 0, \frac{2B}{3}$

    (c) $y'' = 2AB - 6Ax$

71. (a) $y' = 3Ax^2 + 2Bx = x(3Ax + 2B)$

    (b) $x = 0, -\frac{2B}{3A}$

    (c) $y'' = 6Ax + 2B$

73. $\dfrac{d}{dx}\left(\arctan(x^2)\right) = \dfrac{2x}{1 + x^4}$

75. $\mathbf{D}\left(\arctan(e^x)\right) = \dfrac{1}{1 + (e^x)^2} \cdot e^x = \dfrac{e^x}{1 + e^{2x}}$

77. $\mathbf{D}(\arcsin(x^3)) = \dfrac{3x^2}{\sqrt{1 - x^6}}$

79. $\dfrac{d}{dt}\left(\arcsin(e^t)\right) = \dfrac{1}{\sqrt{1 - (e^t)^2}} \cdot e^t = \dfrac{e^t}{\sqrt{1 - e^{2t}}}$

81. $\dfrac{d}{dx}\left(\ln(\sin(x))\right) = \dfrac{1}{\sin(x)} \cdot \cos(x) = \cot(x)$

83. $\frac{d}{ds}\left(\ln(e^s)\right) = \frac{1}{e^s} \cdot e^s = 1$ or $\frac{d}{ds}\left(\ln(e^s)\right) = \frac{d}{ds}(s) = 1$

Section 2.5

1. $\mathbf{D}(\ln(5x)) = \dfrac{1}{5x} \cdot 5 = \dfrac{1}{x}$

3. $\mathbf{D}(\ln(x^k)) = \dfrac{1}{x^k} \cdot kx^{k-1} = \dfrac{k}{x}$

5. $\mathbf{D}(\ln(\cos(x))) = \dfrac{1}{\cos(x)} \cdot (-\sin(x)) = -\tan(x)$

7. $\mathbf{D}(\log_2(5x)) = \dfrac{1}{5x} \cdot \dfrac{1}{\ln(2)} \cdot (5) = \dfrac{1}{x\ln(2)}$

9. $\mathbf{D}(\ln(\sin(x))) = \dfrac{1}{\sin(x)} \cdot \cos(x) = \cot(x)$

11. $\mathbf{D}(\log_2(\sin(x))) = \dfrac{1}{\sin(x)} \cdot \dfrac{1}{\ln(2)} \cdot \cos(x) = \dfrac{\cot(x)}{\ln(2)}$

13. $\mathbf{D}(\log_5(5^x)) = \mathbf{D}(x) = 1$

15. $\mathbf{D}(x \cdot \ln(3x)) = x \cdot \dfrac{1}{3x} \cdot 3 + \ln(3x) \cdot 1 = 1 + \ln(3x)$

17. $\mathbf{D}\left(\dfrac{\ln(x)}{x}\right) = \dfrac{x \cdot \frac{1}{x} - \ln(x) \cdot 1}{x^2} = \dfrac{1 - \ln(x)}{x^2}$

19. $\mathbf{D}\left(\ln\left((5x - 3)^{\frac{1}{2}}\right)\right) = \dfrac{1}{(5x - 3)^{\frac{1}{2}}} \cdot \mathbf{D}\left((5x - 3)^{\frac{1}{2}}\right) =$

    $\dfrac{1}{(5x - 3)^{\frac{1}{2}}} \cdot \dfrac{1}{2}(5x - 3)^{-\frac{1}{2}} \cdot \mathbf{D}(5x - 3) = \dfrac{5}{2} \cdot \dfrac{1}{5x - 3}$

21. $\mathbf{D}(\cos(\ln(w))) = -\sin(\ln(w)) \cdot \dfrac{1}{w} = -\dfrac{\sin(\ln(w))}{w}$

23. $\dfrac{d}{dt}\left(\sqrt{\ln(t + 1)}\right) = \dfrac{1}{2(t + 1)}$

25. $\mathbf{D}\left(5^{\sin(x)}\right) = 5^{\sin(x)} \cdot \ln(5) \cdot \cos(x)$

27. $\dfrac{1}{\sec(x) + \tan(x)} \cdot (\sec(x)\tan(x) + \sec^2(x)) = \sec(x)$

29. $f(x) = \ln(x), f'(x) = \frac{1}{x}$. Let $P = (p, \ln(p))$. So we need $y - \ln(p) = \frac{1}{p}(x - p)$ with $x = 0$ and $y = 0$: $-\ln(p) = -1 \Rightarrow p = e$, hence $P = (e, 1)$.

31. $100(-1)(1 + Ae^{-t})^{-2} \cdot (Ae^{-t})(-1) = \dfrac{100Ae^{-t}}{(1 + Ae^{-t})^2}$

33. $f(x) = 8\ln(x) +$ any constant

35. $f(x) = \ln(3 + \sin(x)) +$ any constant

37. $g(x) = \frac{3}{5}e^{5x} +$ any constant

39. $f(x) = e^{x^2} +$ any constant

41. $h(x) = \ln(\sin(x)) +$ any constant

43. (a) When $t = 0$, A is at $(0, 1)$ and B is at $(0, 1)$. When $t = 1$, A is at $(1, 3)$ and B is at $(1, 3)$.

    (b) Both robots traverse the same path (the line segment $y = 2x + 1$ for $0 \le x \le 1$):

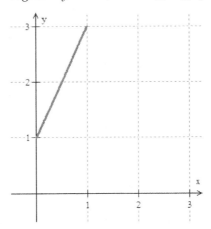

    (c) $\frac{dy}{dx} = 2$ for each, because $y = 2x + 1$.

    (d) A: $\frac{dx}{dt} = 1$, $\frac{dy}{dt} = 2$, speed $= \sqrt{1^2 + 2^2} = \sqrt{5}$
    B: $\frac{dx}{dt} = 2t$, $\frac{dy}{dt} = 4t$, speed $= \sqrt{(2t)^2 + (4t)^2}$
    $= 2\sqrt{5}t$. At $t = 1$, B's speed is $2\sqrt{5}$.

    (e) Moves along the same path $y = 2x + 1$, but to the right and up for about 1.57 minutes, reverses direction and returns to its starting point, then continues left and down for 1.57 minutes, reverses, and continues to oscillate.

45. When $t = 1$: $\frac{dx}{dt} = +$, $\frac{dy}{dt} = -$, $\frac{dy}{dx} = -$.
    When $t = 3$, $\frac{dx}{dt} = -$, $\frac{dy}{dt} = -$, $\frac{dy}{dx} = +$.

47. (a) $x(t) = R(t - \sin(t))$, $y(t) = R(1 - \cos(t))$:

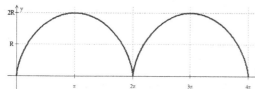

(b) $\frac{dx}{dt} = R(1 - \cos(t))$, $\frac{dy}{dt} = R\sin(t)$, so $\frac{dy}{dx} = \frac{\sin(t)}{1-\cos(t)}$. When $t = \frac{\pi}{2}$, then $\frac{dx}{dt} = R$ and $\frac{dy}{dt} = R$ so $\frac{dy}{dx} = 1$ and speed $= \sqrt{R^2 + R^2} = R\sqrt{2}$. When $t = \pi$, $\frac{dx}{dt} = 2R$ and $\frac{dy}{dt} = 0$ so $\frac{dy}{dx} = 0$ and speed $= \sqrt{(2R)^2 + 0^2} = 2R$.

49. (a) The ellipse $\left(\frac{x}{3}\right)^2 + \left(\frac{y}{5}\right)^2 = 1$.

(b) The ellipse $\left(\frac{x}{A}\right)^2 + \left(\frac{y}{B}\right)^2 = 1$ if $A \neq 0$ and $B \neq 0$. If $A = 0$, the motion is oscillatory along the x-axis; if $B = 0$, the motion is oscillatory along the y-axis.

(c) $(3 \cdot \cos(t), -5 \cdot \sin(t))$ works.

Section 2.6

1. $V = \frac{4}{3}\pi r^3 \Rightarrow \frac{dV}{dt} = 4\pi r^2 \cdot \frac{dr}{dt}$. We know $\frac{dr}{dt}\big|_{r=3} = 2$ in/min, so $\frac{dV}{dt}\big|_{r=3} = 4\pi(3 \text{ in})^2(2 \text{ in/min}) = 72\pi \text{ in}^3/\text{min} \approx 226.19 \text{ in}^3/\text{min}$.

3. (a) $A = \frac{1}{2}bh \Rightarrow \frac{dA}{dt} = \frac{1}{2}\left[b\frac{dh}{dt} + h\frac{db}{dt}\right]$ so with $b = 15$ in, $h = 13$ in, $\frac{db}{dt} = 3\frac{\text{in}}{\text{hr}}$, $\frac{dh}{dt} = -3\frac{\text{in}}{\text{hr}}$:
$$\frac{dA}{dt} = \frac{1}{2}\left[(15 \text{ in})\left(-3\frac{\text{in}}{\text{hr}}\right) + (13 \text{ in})\left(3\frac{\text{in}}{\text{hr}}\right)\right]$$
which is < 0, so A is decreasing.

(b) Hypotenuse $C = \sqrt{b^2 + h^2}$ so:
$$\frac{dC}{dt} = \frac{b\frac{db}{dt} + h\frac{dh}{dt}}{\sqrt{b^2 + h^2}} = \frac{15(3) + 13(-3)}{\sqrt{15^2 + 13^2}}$$
which is > 0 so C is increasing.

(c) Perimeter $P = b + h + C$ so:
$$\frac{dP}{dt} = \frac{db}{dt} + \frac{dh}{dt} + \frac{dC}{dt} = 3 + (-3) + \frac{6}{\sqrt{394}}$$
which is > 0 so P is increasing.

5. (a) Perimeter $P = 2x + 2y \Rightarrow \frac{dP}{dt} = 2\frac{dx}{dt} + 2\frac{dy}{dt} = 2(3 \text{ ft/sec}) + 2(-2 \text{ ft/sec}) = 2 \text{ ft/sec}$.

(b) Area $A = xy \Rightarrow \frac{dA}{dt} = x\frac{dy}{dt} + y\frac{dx}{dt} \Rightarrow \frac{dA}{dt} = (12 \text{ ft})\left(-2\frac{\text{ft}}{\text{sec}}\right) + (8 \text{ ft})\left(3\frac{\text{ft}}{\text{sec}}\right) = 0 \text{ ft}^2/\text{sec}$.

7. Volume $V = \pi r^2 h = \pi r^2\left(\frac{1}{3}\right) \Rightarrow \frac{dV}{dt} = \frac{2\pi}{3}r \cdot \frac{dr}{dt}$. So when $r = 50$ ft and $\frac{dr}{dt} = 6\frac{\text{ft}}{\text{hr}}$, we have $\frac{dV}{dt} = \frac{2\pi}{3}(50 \text{ ft})\left(6\frac{\text{ft}}{\text{hr}}\right) = 200\pi \frac{\text{ft}^3}{\text{hr}} \approx 628.32 \frac{\text{ft}^3}{\text{hr}}$.

9. $w(t) = h(t)$ for all t so $\frac{dw}{dt} = \frac{dh}{dt}$. $V = \frac{1}{3}\pi r^2 h$ and $r = \frac{w}{2} = \frac{h}{2}$ so $V = \frac{1}{3}\pi\left(\frac{h}{2}\right)^2 h = \frac{\pi}{12}h^3 \Rightarrow \frac{dV}{dt} = \frac{\pi}{4}h^2\frac{dh}{dt}$. When $h = 500$ ft and $\frac{dh}{dt} = 2\frac{\text{ft}}{\text{hr}}$, then $\frac{dV}{dt} = \frac{\pi}{4}(500)^2(2) = 125000\pi \frac{\text{ft}^3}{\text{hr}}$.

11. Let x be the distance from the lamp post to the person and L be the length of the shadow (both in feet). By similar triangles, $\frac{L}{6} = \frac{x}{8} \Rightarrow L = \frac{3}{4}x$. We also know that $\frac{dx}{dt} = 3\frac{\text{ft}}{\text{sec}}$.

(a) $\frac{dL}{dt} = \frac{3}{4}\frac{dx}{dt} = \frac{3}{4}\left(3\frac{\text{ft}}{\text{sec}}\right) = 2.25\frac{\text{ft}}{\text{sec}}$.

(b) $\frac{d}{dt}(x + L) = \frac{dx}{dt} + \frac{dL}{dt} = 5.25\frac{\text{ft}}{\text{sec}}$.
(We don't actually need the value of x.)

13. (a) $\sin(35°) = \frac{h}{500} \Rightarrow h = 500\sin(35°) \approx 287$ ft

(b) $L =$ length of the string so $h = L\sin(35°)$ and $\frac{dh}{dt} = \sin(35°)\frac{dL}{dt} \approx (0.57)\left(10\frac{\text{ft}}{\text{sec}}\right) = 5.7\frac{\text{ft}}{\text{sec}}$

15. $V = s^3 - \frac{4}{3}\pi r^3 \Rightarrow \frac{dV}{dt} = 3s^2\frac{ds}{dt} - 4\pi r^2\frac{dr}{dt}$ so when $r = 4$ ft, $\frac{dr}{dt} = 1\frac{\text{ft}}{\text{hr}}$, $s = 12$ ft and $\frac{ds}{dt} = 3\frac{\text{ft}}{\text{hr}}$: $\frac{dV}{dt} = 3(12 \text{ ft})^2\left(3\frac{\text{ft}}{\text{hr}}\right) - 4\pi(4 \text{ ft})^2\left(1\frac{\text{ft}}{\text{hr}}\right) \approx 1094.94\frac{\text{ft}^3}{\text{hr}}$. The volume is increasing at about 1,095 $\frac{\text{ft}^3}{\text{hr}}$.

17. Given: $\frac{dV}{dt} = k \cdot 2\pi r^2$ with k constant. We also have $V = \frac{2}{3}\pi r^3 \Rightarrow \frac{dV}{dt} = 2\pi r^2\frac{dr}{dt}$ so $k \cdot 2\pi r^2 = 2\pi r^2\frac{dr}{dt} \Rightarrow \frac{dr}{dt} = k$. The radius r is changing at a constant rate.

19. (a) $A = 5x$ (b) $\frac{dA}{dx} = 5$ for all $x > 0$. (c) $A = 5t^2$ (d) $\frac{dA}{dt} = 10t$. When $t = 1$, $\frac{dA}{dt} = 10$; when $t = 2$, $\frac{dA}{dt} = 20$; when $t = 3$, $\frac{dA}{dt} = 30$. (e) $A = 10 + 5 \cdot \sin(t) \Rightarrow \frac{dA}{dt} = 5 \cdot \cos(t)$.

21. In (b) and (c) we must use radians because our formulas for derivatives of trigonometric functions assume angles are measured in radians.

(a) $\tan(10°) = \frac{40}{x} \Rightarrow x = \frac{40}{\tan(10°)} \approx 226.9$ ft.

(b) $x = \frac{40}{\tan(\theta)} = 40\cot(\theta) \Rightarrow \frac{dx}{dt} = -40\csc^2(\theta)\frac{d\theta}{dt} \Rightarrow \frac{d\theta}{dt} = -\frac{\sin^2(\theta)}{40}\frac{dx}{dt}$ so when $\theta = 10° \approx 0.1745$ radians and $\frac{dx}{dt} = -25\frac{\text{ft}}{\text{min}}$:
$$\frac{d\theta}{dt} = -\frac{\sin^2(0.1745)}{40}(-25) \approx \frac{(0.1736)^2(25)}{40}$$

$\approx 0.0188 \frac{\text{rad}}{\text{min}} \approx 1.079 \frac{\circ}{\text{min}}.$

(c) $\frac{dx}{dt} = -40 \csc^2(\theta)\frac{d\theta}{dt} = \frac{-40}{\sin^2(\theta)}\frac{d\theta}{dt}$ so when $\theta =$

$10° \approx 0.1745$ rad and $\frac{d\theta}{dt} = 2\frac{\circ}{\text{min}} \approx 0.0349 \frac{\text{rad}}{\text{min}}$:

$$\frac{d\theta}{dt} \approx \frac{-40}{(0.1736)^2}(0.0349) \approx -46.3 \frac{\text{ft}}{\text{min}}$$

(The "-" indicates the distance to the sign is decreasing: you are approaching the sign.) Your speed is $46.3 \frac{\text{ft}}{\text{min}}$.

Section 2.7

1. The locations of x_1 and x_2 appear below:

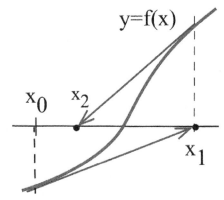

y=f(x)

x_0 x_2

x_1

3. $x_0 = 1$: a; $x_0 = 5$: b

5. $x_0 = 1$: $1, 2, 1, 2, 1, \ldots$
   $x_0 = 5$: x_1 is undefined because $f'(5) = 0$

7. If f is differentiable, then $f'(x_0) = 0$ and $x_1 = x_0 - \frac{f(x_0)}{f'(x_0)}$ is undefined.

9. $f(x) = x^4 - x^3 - 5 \Rightarrow f'(x) = 4x^3 - 3x^2$ so $x_0 = 2 \Rightarrow x_1 = 2 - \frac{3}{20} = \frac{37}{20} = 1.85 \Rightarrow x_2 = 1.85 - \frac{1.85^4 - 1.85^3 - 5}{4(1.85)^3 - 3(1.85)^2} \approx 1.824641$

11. $f(x) = x - \cos(x) \Rightarrow f'(x) = 1 + \sin(x)$ so $x_0 = 0.7 \Rightarrow x_1 = 0.7394364978 \Rightarrow x2 = 0.7390851605$, so root ≈ 0.74.

13. To solve $\frac{x}{x+3} = x^2 - 2$ we search for roots of $f(x) = x^2 - 2 - \frac{x}{x+3}$. If $x_0 = -4$, then the iterates $x_n \to -3.3615$; if $x_0 = -2$, then $x_n \to -1.1674$; if $x_0 = 2$, then the iterates $x_n \to 1.5289$.

15. For $x^5 - 3 = 0$ and $x_0 = 1$, $x_n \to 1.2457$.

17. $f(x) = x^3 - A \Rightarrow f'(x) = 3x^2$ so:

$$x_{n+1} = x_n - \frac{x_n^3 - A}{3x_n^2} = \frac{1}{3}\left[2x_n + \frac{A}{x_n^2}\right]$$

19. (a) $2(0) - \lfloor 2(0) \rfloor = 0 - 0 = 0$

(b) $2(\frac{1}{2}) - \lfloor 2(\frac{1}{2}) \rfloor = 1 - 1 = 0$
$2(\frac{1}{4}) - \lfloor 2(\frac{1}{4}) \rfloor = \frac{1}{2} - 0 = \frac{1}{2} \to 0$
$2(\frac{1}{8}) - \lfloor 2(\frac{1}{8}) \rfloor = \frac{1}{4} - 0 = \frac{1}{4} \to \frac{1}{2} \to 0$
$2(\frac{1}{2^n}) - \lfloor 2(\frac{1}{2^n}) \rfloor = \frac{1}{2^{n-1}} - 0 = \frac{1}{2^{n-1}}$
$\to \frac{1}{2^{n-2}} \to \cdots \to \frac{1}{4} \to \frac{1}{2} \to 0$

21. (a) If $0 \le x \le \frac{1}{2}$, then f stretches x to twice its value, $2x$. If $\frac{1}{2} < x \le 1$, then f stretches x to twice its value $(2x)$ and "folds" the part above the value 1 $(2x - 1)$ to below 1: $1 - (2x - 1) = 2 - 2x$.

(b) $f(\frac{2}{3}) = \frac{2}{3}$
$f(\frac{2}{5}) = \frac{4}{5}$, $f(\frac{4}{5}) = \frac{2}{5}$, and the values continue to cycle.
$f(\frac{2}{7}) = \frac{4}{7}$, $f(\frac{4}{7}) = \frac{6}{7}$, $f(\frac{6}{7}) = \frac{2}{7}$, and the values continue to cycle.
$f(\frac{2}{9}) = \frac{4}{9}$, $f(\frac{4}{9}) = \frac{8}{9}$, $f(\frac{8}{9}) = \frac{2}{9}$, and the values continue to cycle.

(c) $0.1, 0.2, \mathbf{0.4}, 0.8, \mathbf{0.4}, 0.8$, and the pair of values 0.4 and 0.8 continue to cycle.
$0.105, 0.210, 0.42, 0.84, \mathbf{0.32}, 0.64, 0.72, 0.56,$
$0.88, 0.24, 0.48, 0.96, 0.08, 0.16, \mathbf{0.32}, \ldots$
$0.11, 0.22, 0.44, \mathbf{0.88}, 0.24, 0.48, 0.96, 0.08, 0.16,$
$0.32, 0.64, 0.72, 0.56, \mathbf{0.88}, \ldots$

(d) Probably so.

Section 2.8

1. (a) The desired points appear below:

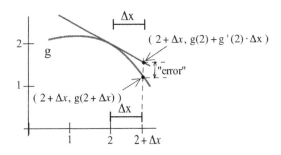

2

g

Δx

$(2 + \Delta x, g(2) + g'(2) \cdot \Delta x)$

"error"

1

$(2 + \Delta x, g(2 + \Delta x))$

Δx

1   2 $2 + \Delta x$

(b) The "error" appears in the figure above.

3. (a) With $f(4) = 2$ and $f'(4) = \frac{1}{2\sqrt{4}} = \frac{1}{4}$, the tangent line is $y - 2 = \frac{1}{4}(x - 4)$ or $y = \frac{1}{4}x + 1$ and $\sqrt{4.2} = f(4.2) \approx \frac{1}{4}(4.2) + 1 = 2.05$

(b) $f(81) = 9$ and $f'(81) = \frac{1}{18}$ so tangent line is $y - 9 = \frac{1}{18}(x - 81)$ or $y = \frac{1}{18}(x - 81) + 9$ and $\sqrt{80} = f(80) \approx \frac{1}{18}(80 - 81) + 9 = \frac{161}{18} \approx 8.944$

(c) $f(0) = 0$ and $f'(0) = 1$ so tangent line is $y = x$ and $\sin(0.3) = f(0.3) \approx 0.3$

4. (a) $f(1) = 0$ and $f'(1) = 1$ so tangent line is $y = x - 1$ and $\ln(1.3) = f(1.3) \approx 1.3 - 1 = 0.3$

(b) $f(0) = 1$ and $f'(0) = 1$ so tangent line is $y = 1 + x$ and $e^{0.1} = f(0.1) \approx 1.1$

(c) $f(1) = 1$ and $f'(1) = 5$ so tangent line is $y - 1 = 5(x - 1)$ or $y = 5x - 4$ and $(1.03)^5 \approx 5(0.03) - 4 = 1.15$

5. $f(x) = (1 + x)^n \Rightarrow f'(x) = n(1 + x)^{n-1}$ so $f(0) = 1$ and $f'(0) = n$, hence tangent line is $y - 1 = n(x - 0)$ or $y = 1 + nx$ and $(1 + x)^n \approx 1 + nx$ (when x is close to 0)

6. (a) $f(x) = (1 - x)^n \Rightarrow f'(x) = -n(1 - x)^{n-1}$ so $f(0) = 1$ and $f'(0) = -n$, hence tangent line is $y - 1 = -n(x - 0)$ or $y = 1 - nx$ and $(1 - x)^n \approx 1 - nx$ (when x is close to 0)

(b) $f(x) = \sin(x) \Rightarrow f'(x) = \cos(x)$ so $f(0) = 0$ and $f'(0) = 1$, hence tangent line is $y = x$ and $\sin(x) \approx x$ (for x close to 0)

(c) $f(x) = e^x \Rightarrow f'(x) = e^x$ so $f(0) = 1$ and $f'(0) = 1$, hence tangent line is $y = x + 1$ and $e^x \approx x + 1$ (for x near 0)

7. (a) $f(x) = \ln(1 + x) \to f'(x) = \frac{1}{1+x}$ so $f(0) = 0$ and $f'(0) = 1$, hence tangent line is $y = x$ and $\ln(1 + x) \approx x$

(b) $f(x) = \cos(x) \Rightarrow f'(x) = -\sin(x)$ so $f(0) = 1$ and $f'(0) = 0$, hence tangent line is $y = 1$ and $\cos(x) \approx 1$

(c) $f(x) = \tan(x) \Rightarrow f'(x) = \sec^2(x)$, so $f(0) = 0$ and $f'(0) = 1$, hence tangent line is $y = x$ and $\tan(x) \approx x$

(d) $f(x) = \sin\left(\frac{\pi}{2} + x\right) \Rightarrow f'(x) = \cos\left(\frac{\pi}{2} + x\right)$ so $f(0) = 1$ and $f'(0) = 0$, hence tangent line is $y = 1$ and $\sin\left(\frac{\pi}{2} + x\right) \approx 1$.

9. (a) Area $A(x) = $ (base)(height) $= x(x^2 + 1) = x^3 + x \Rightarrow A'(x) = 3x^2 + 1 \Rightarrow A'(2) = 13$ so $\Delta A \approx A'(2) \cdot \Delta x = (13)(2.3 - 2) = 3.9$

(b) (base)(height) $= (2.3)((2.3)^2 + 1) = 14.467$ so actual $\Delta A = 14.467 - (2)(2^2 + 1) = 4.467$

11. $V = \pi r^2 h = 2\pi r^2$ and $\Delta V = 2\pi \cdot 2r\Delta r = 4\pi r\Delta r$. Because $V = 2\pi r^2 = 47.3$, we have $r = \sqrt{\frac{47.3}{2\pi}} \approx 2.7437$ cm. We know $|\Delta V| \leq 0.1$ so, using $\Delta V = 4\pi r \Delta r$, we have $0.1 \geq 4\pi(2.7437)\,|\Delta r|$ and $|\Delta r| \leq \frac{0.1}{4\pi(2.7437)} \approx 0.0029$ cm. The required tolerance is ± 0.0029 cm. (Reality check: A coin 2 cm high is quite unusual; 47.3 cm^3 of gold weighs around 2 pounds!)

13. $V = x^3 \Rightarrow \Delta V \approx 3x^2 \Delta x$, $V = 87 \Rightarrow x = \sqrt[3]{87} \approx 4.431$ cm so $\Delta x \approx \frac{\Delta V}{3x^2} \approx \frac{2}{3(4.431)^2} \approx 0.034$ cm

15. With $P = 2\pi\sqrt{\frac{L}{g}}$ and $g = 32\,\frac{\text{ft}}{\text{sec}^2}$:

(a) $P = 2\pi\sqrt{\frac{2}{32}} = \frac{\pi}{2} \approx 1.57$ sec

(b) $1 = 2\pi\sqrt{\frac{L}{32}} \Rightarrow L = \frac{8}{\pi^2} \approx 0.81$ ft

(c) $dP = \frac{2\pi}{\sqrt{32}} \cdot \frac{1}{2\sqrt{2}}\,dL = \frac{2\pi}{16}(0.1) \approx 0.039$ sec

(d) $2\,\frac{\text{in}}{\text{hr}} = \frac{1}{6}\,\frac{\text{ft}}{\text{hr}} = \frac{1}{21600}\,\frac{\text{ft}}{\text{sec}}$ so $\frac{dP}{dt} = \frac{2\pi}{4\sqrt{2}} \cdot \frac{1}{2\sqrt{24}} \cdot \frac{1}{21600} \approx 5.25 \times 10^{-6} > 0$ (increasing)

17. (a) $df = f'(2)\,dx \approx (0)(1) = 0$

(b) $df = f'(4)\,dx \approx (0.3)(-1) = -0.3$

(c) $df = f'(3)\,dx \approx (0.5)(2) = 1$

19. (a) $f'(x) = 2x - 3 \Rightarrow df = (2x - 3)\,dx$

(b) $f'(x) = e^x \Rightarrow df = e^x\,dx$

(c) $f'(x) = 5\cos(5x) \Rightarrow df = 5\cos(5x)\,dx$

(d) $f'(x) = 3x^2 + 2 \Rightarrow df = (3x^2 + 2)\,dx$ so when $x = 1$ and $dx = 0.2$, $df = (3 \cdot 1^2 + 2)(0.2) = 1$

(e) $f'(x) = \frac{1}{x} \Rightarrow df = \frac{1}{x}\,dx$ so when $x = e$ and $dx = -0.1$, $df = \frac{1}{e}(-0.1) = -\frac{1}{10e}$

(f) $f'(x) = \frac{1}{\sqrt{2x+5}} \Rightarrow df = \frac{1}{\sqrt{2x+5}}\,dx$ so when $x = 22$ and $dx = 3$, $df = \frac{1}{\sqrt{49}}(3) = \frac{3}{7}$

Section 2.9

1. (a) $x^2 + y^2 = 100 \Rightarrow 2x + 2y \cdot y' = 0 \Rightarrow y' = -\frac{x}{y}$ so at $(6, 8)$, $y' = -\frac{6}{8} = -\frac{3}{4}$

(b) $y = \sqrt{100 - x^2} \Rightarrow y' = \frac{-x}{\sqrt{100 - x^2}}$ so at $(6, 8)$, $y' = -\frac{6}{\sqrt{100 - 36}} = -\frac{6}{8} = -\frac{3}{4}$

3. (a) $x^2 - 3xy + 7y = 5 \Rightarrow 2x - 3(y + xy') + 7y' = 0 \Rightarrow y' = \dfrac{3y - 2x}{7 - 3x} \Rightarrow y'|_{(2,1)} = \dfrac{3 - 4}{7 - 6} = -1$

(b) $y = \dfrac{5 - x^2}{7 - 3x} \Rightarrow y' = \dfrac{(7 - 3x)(-2x) - (5 - x^2)(-3)}{(7 - 3x)^2}$

so at $(2, 1)$, $y' = \dfrac{(1)(-4) - (1)(-3)}{(1)^2} = -1$

5. (a) $\dfrac{x^2}{9} + \dfrac{y^2}{16} = 1 \Rightarrow \dfrac{2x}{9} + \dfrac{2y}{16}y' = 0 \Rightarrow$

$y' = -\dfrac{16}{9} \cdot \dfrac{x}{y} \Rightarrow y'|_{(0,4)} = 0$

(b) $y = 4\sqrt{1 - \dfrac{x^2}{9}} = \dfrac{4}{3}\sqrt{9 - x^2} \Rightarrow$

$y' = \dfrac{4}{3} \cdot \dfrac{-x}{\sqrt{9 - x^2}} \Rightarrow y'|_{(0,4)} = 0$

7. (a) $\ln(y) + 3x - 7 = 0 \Rightarrow \dfrac{1}{y}y' + 3 = 0 \Rightarrow$

$y' = -3y \Rightarrow y'|_{(2,e)} = -3e$

(b) $y = e^{7-3x} \Rightarrow y' = -3e^{7-3x} \Rightarrow$

$y'|_{(2,e)} = -3e^{7-6} = -3e$

9. (a) $x^2 - y^2 = 16 \Rightarrow 2x - 2yy' = 0 \Rightarrow y' = \dfrac{x}{y} \Rightarrow$

$y'|_{(5,-3)} = -\dfrac{5}{3}$

(b) The point $(5, -3)$ is on the bottom half of the hyperbola so $y = -\sqrt{x^2 - 16} \Rightarrow y' = -\dfrac{x}{\sqrt{x^2 - 16}} \Rightarrow y'|_{(5,-3)} = -\dfrac{5}{\sqrt{25 - 16}} = -\dfrac{5}{3}$

11. $x = 4y - y^2 \Rightarrow 1 = 4y' - 2y \cdot y' \Rightarrow y' = \dfrac{1}{4 - 2y}$.

At $(3, 1)$, $y' = \dfrac{1}{4 - 2(1)} = \dfrac{1}{2}$; at $(3, 3)$, $y' = \dfrac{1}{4 - 2(3)} = -\dfrac{1}{2}$; at $(4, 2)$, $y' = \dfrac{1}{4 - 2(2)}$ is undefined (the tangent line is vertical).

13. $x = y^2 - 6y + 5 \Rightarrow 1 = 2y \cdot y' - 6y' \Rightarrow y' = \dfrac{1}{2y - 6}$. At $(5, 0)$, $y' = \dfrac{1}{2(0) - 6} = -\dfrac{1}{6}$; at $(5, 6)$, $y' = \dfrac{1}{2(6) - 6} = \dfrac{1}{6}$; at $(-4, 3)$, $y' = \dfrac{1}{2(3) - 6}$ is undefined (vertical tangent line).

15. $3y^2 \cdot y' - 5y' = 10x \Rightarrow y' = \dfrac{10x}{3y^2 - 5} \Rightarrow m = y'|_{(1,3)} = \dfrac{10}{22} = \dfrac{5}{11}$

17. $y^2 + \sin(y) = 2x - 6 \Rightarrow 2y \cdot y' + \cos(y) \cdot y' = 2 \Rightarrow y' = \dfrac{2}{2y + \cos(y)} \Rightarrow m = y'|_{(3,0)} = \dfrac{2}{0 + 1} = 2$

19. $e^y + \sin(y) = x^2 - 3 \Rightarrow e^y \cdot y' + \cos(y) \cdot y' = 2x \Rightarrow y' = \dfrac{2x}{e^y + \cos(y)} \Rightarrow m = y'|_{(2,0)} = 2$

21. $x^{\frac{2}{3}} + y^{\frac{2}{3}} = 5 \Rightarrow \dfrac{2}{3}x^{-\frac{1}{3}} + \dfrac{2}{3}y^{-\frac{1}{3}}y' = 0 \Rightarrow y' = -\left(\dfrac{y}{x}\right)^{\frac{1}{3}} \Rightarrow m = y'|_{(8,1)} = -\left(\dfrac{1}{8}\right)^{\frac{1}{3}} = -\dfrac{1}{2}$

23. $2x + x \cdot y' + y + 2y \cdot y' + 3 - 7y' = 0 \Rightarrow y' = \dfrac{-2x - y - 3}{x + 2y - 7} \Rightarrow y'|_{(1,2)} = \dfrac{-2(1) - (2) - 3}{(1) + 2(2) - 7} = \dfrac{7}{2}$

25. Explicitly: $y = Ax^2 + Bx + C \Rightarrow y' = 2Ax + B$

Implicitly: $x = Ay^2 + By + C \Rightarrow 1 = 2Ayy' + By'$

$\Rightarrow y' = \dfrac{1}{2Ay + B}$

27. $Ax^2 + Bxy + Cy^2 + Dx + Ey + F = 0$

$\Rightarrow 2Ax + Bx \cdot y' + By + 2Cy \cdot y' + D + Ey' = 0$

$\Rightarrow y' = \dfrac{-2Ax - By - D}{Bx + 2Cy + E}$

28. $x^2 + y^2 = r^2 \Rightarrow 2x + 2y \cdot y' = 0 \Rightarrow y' = -\dfrac{x}{y}$ so the slope of the tangent line is $-\dfrac{x}{y}$. The slope of the line through the points $(0, 0)$ and (x, y) is $\dfrac{y}{x}$, so the slopes of the lines are negative reciprocals of each other and the lines are perpendicular.

29. From the solution to problem 23, we know that $y' = \dfrac{-2x - y - 3}{x + 2y - 7}$ so $y' = 0$ when $-2x - y - 3 = 0 \Rightarrow y = -2x - 3$. Substituting $y = -2x - 3$ into the original equation, we have:

$$x^2 + x(-2x - 3) + (-2x - 3)^2 + 3x - 7(-2x - 3) + 4 = 0$$

$$\Rightarrow 3x^2 + 26x + 34 = 0 \Rightarrow x = \dfrac{-26 \pm \sqrt{26^2 - 4(3)(34)}}{2(3)}$$

$= \dfrac{-13 \pm \sqrt{67}}{3} \approx -1.605$ or -7.062.

If $x \approx -1.605$ (point A), then $y = -2x - 3 \approx -2(-1.605) - 3 = 0.21$ so coordinates of A are $(-1.605, 0.21)$; if $x \approx -7.062$ (point C), then $y = -2x - 3 \approx -2(-7.062) - 3 = 11.124$ so C is $(-7.062, 11.124)$.

31. From the solution to 29, point C is $(-7.062, 11.124)$. At D, $y' = \dfrac{-2x - y - 3}{x + 2y - 7}$ is undefined so $x + 2y - 7 = 0 \Rightarrow x = 7 - 2y$. Substituting $x = 7 - 2y$ into the original equation:

$$(7 - 2y)^2 + (7 - 2y)y + y^2 + 3(7 - 2y) - 7y + 4 = 0$$

so $3y^2 - 34y + 74 = 0 \Rightarrow y = \dfrac{34 \pm \sqrt{34^2 - 4(3)(74)}}{2(3)} = \dfrac{17 \pm \sqrt{67}}{3} \approx 8.395$ or 2.938.

If $y \approx 2.938$ (point D), then $x = 7 - 2y \approx$ $7 - 2(2.938) = 1.124$ so point D is $(1.124, 2.938)$. Similarly, point B is $(-9.79, 8.395)$.

33. (a) Using the Product Rule:

$$y' = (x^2 + 5)^7 \cdot 4(x^3 - 1)^3 \cdot 3x^2$$
$$+ 7(x^2 + 5)^6 \cdot 2x \cdot (x^3 - 1)^4$$
$$= (x^2 + 5)^6(x^3 - 1)^3(2x)\left[6x^3 + 30x + 7x^3 - 7\right]$$
$$= (x^2 + 5)^6(x^3 - 1)^3(2x)\left[13x^3 + 30x - 7\right]$$

(b) $\ln(y) = 7\ln(x^2 + 5) + 4\ln(x^3 - 1)$ so:

$$\frac{y'}{y} = \frac{14x}{x^2 + 5} + \frac{12x^2}{x^3 - 1}$$

and solving for y' yields:

$$y' = y\left[\frac{14x}{x^2 + 5} + \frac{12x^2}{x^3 - 1}\right]$$
$$= (x^2 + 5)^7(x^3 - 1)^4\left[\frac{14x}{x^2 + 5} + \frac{12x^2}{x^3 - 1}\right]$$

which is the same as part (a). (Really! It is!)

35. (a) $y = x^5(3x + 2)^4$ so:

$$y' = x^5 \mathbf{D}\left((3x + 2)^4\right) + (3x + 2)^4 \mathbf{D}\left(x^5\right)$$
$$= x^5 \cdot 4(3x + 2)^3(3) + (3x + 2)^4 \cdot 5x^4$$

(b) $\ln(y) = 5\ln(x) + 4\ln(3x + 2)$ so:

$$\frac{y'}{y} = \frac{5}{x} + \frac{12}{3x + 2}$$

and solving for y' yields:

$$y' = y\left[\frac{5}{x} + \frac{12}{3x + 2}\right]$$
$$= \left[x^5(3x + 2)^4\right] \cdot \left[\frac{5}{x} + \frac{12}{3x + 2}\right]$$

which is the same as in part (a).

37. (a) $y = e^{\sin(x)} \Rightarrow y' = e^{\sin(x)} \cdot \cos(x)$

(b) $\ln(y) = \sin(x) \Rightarrow \dfrac{y'}{y} = \cos(x)$

$\Rightarrow y' = y \cdot \cos(x) = e^{\sin(x)} \cdot \cos(x)$

39. (a) $y = \sqrt{25 - x^2} \Rightarrow y' = \dfrac{-x}{\sqrt{25 - x^2}}$

(b) $\ln(y) = \frac{1}{2}\ln(25 - x^2)$ so:

$$\frac{y'}{y} = \frac{1}{2} \cdot \frac{-2x}{25 - x^2} = \frac{-x}{25 - x^2}$$

and solving for y' yields:

$$y' = \sqrt{25 - x^2} \cdot \frac{-x}{25 - x^2} = \frac{-x}{\sqrt{25 - x^2}}$$

41. $y = x^{\cos(x)} \Rightarrow \ln(y) = \cos(x) \cdot \ln(x)$
$\Rightarrow \dfrac{y'}{y} = \cos(x) \cdot \frac{1}{x} - \ln(x) \cdot \sin(x)$ so:

$$y' = x^{\cos(x)}\left[\frac{\cos(x)}{x} - \ln(x) \cdot \sin(x)\right]$$

43. $\ln(y) = 4\ln(x) + 7\ln(x - 2) + \ln(\sin(3x))$
$\Rightarrow \dfrac{y'}{y} = \frac{4}{x} + \frac{7}{x-2} + \frac{3\cos(3x)}{\sin(3x)}$ so:

$$y' = x^4(x - 2)^7\sin(3x)\left[\frac{4}{x} + \frac{7}{x - 2} + 3\cot(3x)\right]$$

45. $\ln(y) = x \cdot \ln(3 + \sin(x))$
$\Rightarrow \dfrac{y'}{y} = x \cdot \frac{\cos(x)}{3 + \sin(x)} + \ln(3 + \sin(x))$ so:

$$y' = (3 + \sin(x))^x\left[\frac{x\cos(x)}{3 + \sin(x)} + \ln(3 + \sin(x))\right]$$

47. $f'(1) = 1(1.2) = 1.2$; $f'(2) = 9(1.8) = 16.2$; $f'(3) = 64(2.1) = 134.4$

49. $f'(1) = 5(-1) = -5$; $f'(2) = 2(0) = 0$; $f'(3) = 7(2) = 14$

51. $\ln(f \cdot g) = \ln(f) + \ln(g) \Rightarrow \dfrac{\mathbf{D}(f \cdot g)}{f \cdot g} = \dfrac{f'}{f} + \dfrac{g'}{g}$

so $\mathbf{D}(f \cdot g) = (f \cdot g)\left[\frac{f'}{f} + \frac{g'}{g}\right] = f' \cdot g + g' \cdot f$

52. $\ln\left(\dfrac{f}{g}\right) = \ln(f) - \ln(g) \Rightarrow \dfrac{\mathbf{D}\left(\frac{f}{g}\right)}{\left(\frac{f}{g}\right)} = \dfrac{f'}{f} - \dfrac{g'}{g}$ so:

$$\mathbf{D}\left(\frac{f}{g}\right) = \left(\frac{f}{g}\right)\left[\frac{f'}{f} - \frac{g'}{g}\right] = \frac{f'}{g} - \frac{f \cdot g'}{g^2}$$
$$= \frac{f \cdot g' - g \cdot f'}{g^2}$$

53. $\ln(f \cdot g \cdot h) = \ln(f) + \ln(g) + \ln(h)$
$\Rightarrow \dfrac{\mathbf{D}(f \cdot g \cdot h)}{f \cdot g \cdot h} = \dfrac{f'}{f} + \dfrac{g'}{g} + \dfrac{h'}{h}$ so:

$$\mathbf{D}(f \cdot g \cdot h) = (f \cdot g \cdot h)\left[\frac{f'}{f} + \frac{g'}{g} + \frac{h'}{h}\right]$$
$$= f' \cdot g \cdot h + f \cdot g' \cdot h + f \cdot g \cdot h'$$

54. $\ln(a^x) = x\ln(a) \Rightarrow \dfrac{\mathbf{D}(a^x)}{a^x} = \ln(a) \Rightarrow \mathbf{D}(a^x) = a^x\ln(a)$

Section 3.1

Local maximums at $x = 3$, $x = 5$, $x = 9$ and $x = 13$; global maximum at $x = 13$. Local minimums at $x = 1$, $x = 4.5$, $x = 7$ and $x = 10.5$; global minimum at $x = 7$.

3. $f(x) = x^2 + 8x + 7 \Rightarrow f'(x) = 2x + 8$, which is defined for all values of x; $f'(x) = 0$ when $x = -4$, so $x = -4$ is a critical number. There are no endpoints. The only critical number is $x = -4$, so the only critical point is $(-4, f(-4)) = (-4, -9)$, which is the global (and local) minimum.

5. $f(x) = \sin(x) \Rightarrow f'(x) = \cos(x)$, which is defined for all values of x; $f'(x) = 0$ when $x = \frac{\pi}{2} + k\pi$ so those values are critical numbers. There are no endpoints. $f(x) = \sin(x)$ has local and global maximums at $x = \frac{\pi}{2} + 2k\pi$ and global and local minimums at $x = \frac{3\pi}{2} + 2k\pi$.

7. $f(x) = \sqrt[3]{x} = x^{\frac{1}{3}} \Rightarrow f'(x) = \frac{1}{3}x^{-\frac{2}{3}} = \frac{1}{3\sqrt[3]{x^2}}$, which is defined for all values of x except $x = 0$, and $f'(x) \neq 0$ for all x, so $x = 0$ is the only critical number. There are no endpoints. the function $f(x) = \sqrt[3]{x}$ has no global or local extrema.

9. $f(x) = xe^{5x} \Rightarrow f'(x) = e^{5x}[5x + 1]$, which is defined for all values of x; $f'(x) = 0$ when $5x + 1 = 0 \Rightarrow x = -\frac{1}{5}$ so this is the only critical number. There are no endpoints. The function $f(x) = xe^{5x}$ has a local and global minimum at $x = -\frac{1}{5}$ and no other extrema.

11. $f(x) = (x-1)^2(x-3) \Rightarrow f'(x) = (x-1)^2 + 2(x-1)(x-3) = (x-1)(3x-7)$, which is defined for all values of x; $f'(x) = 0$ when $x = 1$ and $x = \frac{7}{3}$ so those values are critical numbers. There are no endpoints. The only critical points are $(1,0)$, which is a local maximum, and $\left(\frac{7}{3}, -\frac{32}{27}\right)$, which is a local minimum. When the interval is the entire real number line, this function does not have any global extrema.

13. $f(x) = 2x^3 - 96x + 42 \Rightarrow f'(x) = 6x^2 - 96$, which is defined for all values of x; $f'(x) = 6(x+4)(x-4) = 0$ when $x = -4$ and $x = 4$ so those values are critical numbers. There are no endpoints. The only critical points are $(-4, 298)$,

which is a local maximum, and $(4, -214)$, which is a local minimum. When the interval is the entire real number line, this function does not have a global maximum or global minimum.

15. $f(x) = e^{-(x-2)^2} \Rightarrow f'(x) = -2(x-2)e^{-(x-2)^2}$, which is defined for all x; $f'(x) = 0$ only when $x = 2$, so that is a critical number. There are no endpoints. The only critical point is $(2, 1)$, which is a local and global maximum. When the interval is the entire real number line, this function does not have a local or global minimum.

17. $f(x) = \frac{x}{1 + x^2} \Rightarrow f'(x) = \frac{1 - x^2}{(1 + x^2)^2}$, which is defined for all x; $f'(x) = 0$ only when $x = \pm 1$, so these are the only critical numbers. There are no endpoints. The critical point $\left(-1, -\frac{1}{2}\right)$ is a local and global minimum; the critical point $\left(1, \frac{1}{2}\right)$ is a local and global maximum.

19. $f(x) = (x-2)^{\frac{2}{3}} \Rightarrow f'(x) = \frac{2}{3}(x-2)^{-\frac{1}{3}} = \frac{2}{3\sqrt[3]{x-2}}$, which is defined for all values of x except $x = 2$, and $f'(x) \neq 0$ for all x so $x = 2$ is the only critical number. There are no endpoints. The function $f(x) = (x-2)^{\frac{2}{3}}$ has a global and local minimum at $x = 2$ and no other extrema.

21. $f(x) = (x^2 - 4)^{\frac{1}{3}} \Rightarrow f'(x) = \frac{2}{3}(x^2 - 4)^{-\frac{2}{3}} \cdot 2x = \frac{2x}{3\sqrt[3]{(x^2-4)^2}}$, which is defined for all values of x except $x = \pm 2$, and $f'(x) = 0$ only when $x = 0$, so $x = -2$, $x = 0$ and $x = 2$ are the only critical numbers. There are no endpoints. The function $f(x) = \sqrt[3]{x^2 - 4}$ has a global and local minimum at $x = 0$ and no other global or local extrema.

23.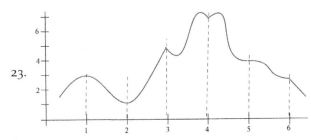

25. $f(x) = x^2 - 6x + 5 \Rightarrow f'(x) = 2x - 6$, which is defined for all values of x; $f'(x) = 0$ only when $x = 3$, so that is a critical number. The endpoints are $x = -2$ and $x = 5$. The critical points are

(3, −4), which is the local and global minimum, (−2, 21), which is a local and global maximum, and (5, 0), which is a local maximum.

27. $f(x) = 2 − x^3 \Rightarrow f'(x) = −3x^2$, which is defined for all values of x; $f'(x) = 0$ only when $x = 0$, so that is a critical number. The endpoints are $x = −2$ and $x = 1$, which are also critical numbers. The critical points are $(−2, 10)$, which is a local and global maximum, $(0, 2)$, which is not a local or global maximum or minimum, and $(1, 1)$, which is a local and global minimum.

29. $f'(x) = 3x^2 − 3 = 3(x − 1)(x + 1)$, which is defined for all x; $f'(x) = 0$ only when $x = −1$ and $x = 1$, so these are critical numbers. The endpoints $x = −2$ and $x = 1$ are also critical numbers. The critical points are $(−2, 3)$, which is a local and global minimum on $[−2, 1]$, the point $(−1, 7)$, which is a local and global maximum on $[−2, 1]$, and the point $(1, 3)$, which is a local and global minimum on $[−2, 1]$.

31. $f'(x) = 5x^4 − 20x^3 + 15x^2 = 5x^2(x^2 − 4x + 3) = 5x^2(x − 3)(x − 1)$, which is defined for all x; $f'(x) = 0$ only when $x = 0$ or $x = 1$ in the interval $[0, 2]$, so each of these values is a critical number. The endpoints $x = 0$ and $x = 2$ are also critical numbers. The critical points are $(0, 7)$, which is a local minimum, $(1, 8)$ which is a local and global maximum, and $(2, −1)$, which is a local and global minimum. (It is also true that $f'(3) = 0$, but $x = 3$ is not in the interval $[0, 2]$.)

33. $f(x) = \frac{1}{x^2+1} \Rightarrow f'(x) = \frac{−2x}{(x^2+1)^2}$, which is defined for all values of x; $f'(x) = 0$ only when $x = 0$, which is not in the interval $[1, 3]$ so $x = 0$ is a not a critical number. The endpoints $x = 1$ and $x = 3$ are critical numbers. The critical points are $\left(1, \frac{1}{2}\right)$, which is a local and global maximum, and $\left(3, \frac{1}{10}\right)$, which is a local and global minimum.

35. $f'(x) = \frac{3}{2}\left(x^2 + 4\right)^{-\frac{1}{2}} \cdot 2x − 1 = \frac{3x}{\sqrt{x^2+4}} − 1$, which exists for all x; $f'(x) = 0$ only when $3x = \sqrt{x^2 + 4} \Rightarrow x = \frac{1}{\sqrt{2}}$. The endpoints are $x = 0$ and $x = 2$, so the critical numbers are $x = 0$, $x = \frac{1}{\sqrt{2}}$ and $x = 2$: $f(0) = 6$, $f\left(\frac{1}{\sqrt{2}}\right) \approx 5.66$ and

$f(2) \approx 6.49$, so $f(x) = 3\sqrt{x^2 + 4} − x$ has a local max at $x = 0$, a global and local min at $x = \frac{1}{\sqrt{2}}$, and a global and local max at $x = 2$.

37. $f(x) = x^3 − \ln(x) \Rightarrow f'(x) = 3x^2 − \frac{1}{x}$, which exists for all x in the interval $\left[\frac{1}{2}, 2\right]$, and $f'(x) = 0 \Rightarrow x = \frac{1}{\sqrt[3]{3}}$. The endpoints are $x = \frac{1}{2}$ and $x = 2$, so the three critical numbers are $x = \frac{1}{2}$, $x = \frac{1}{\sqrt[3]{3}}$ and $x = 2$: $f\left(\frac{1}{2}\right) \approx 0.82$, $f\left(\frac{1}{\sqrt[3]{3}}\right) \approx 0.70$ and $f(2) \approx 7.3$ so $f(x) = x^3 − \ln(x)$ has a local max at $x = \frac{1}{2}$, a global and local min at $x = \frac{1}{\sqrt[3]{3}}$, and a global and local max at $x = 2$.

39. $A(x) = 4x\sqrt{1 − x^2}$ for $0 < x < 1$, so:

$$A'(x) = 4\left[\frac{−x^2}{\sqrt{1 − x^2}} + \sqrt{1 − x^2}\right]$$
$$= 4\left[\frac{1 − 2x^2}{\sqrt{1 − x^2}}\right]$$

hence $A'(x) > 0$ if $0 < x < \frac{1}{\sqrt{2}}$ and $A'(x) < 0$ if $\frac{1}{\sqrt{2}} < x < 1$. Max when $x = \frac{1}{\sqrt{2}}$: $A\left(\frac{1}{\sqrt{2}}\right) = 2$

41. $V = x(8 − 2x)^2$ for $0 \le x \le 4$, so:

$$V' = x(2)(8 − 2x)(−2) + (8 − 2x)^2$$
$$= (8 − 2x)(−4x + 8 − 2x)$$
$$= (8 − 2x)(8 − 6x) = 4(4 − x)(4 − 3x)$$

hence $V' < 0$ if $x > \frac{4}{3}$ and $V' > 0$ if $0 < x < \frac{4}{3}$. $V\left(\frac{4}{3}\right) = \frac{4}{3}\left(8 − \frac{8}{3}\right)^2 = \frac{1024}{27} \approx 37.926$ cubic units is the largest volume. Smallest volume is 0, which occurs when $x = 0$ and $x = 4$.

43. (a) 4: the endpoints and two values of x for which $f'(x) = 0$ (b) 2: the endpoints (c) At most $n + 1$: the two endpoints and the $n − 1$ points x for which $f'(x) = 0$; at least 2: the endpoints.

45. (a) local minimum at $(1, 5)$ (b) no extreme at $(1, 5)$ (c) local maximum at $(1, 5)$ (d) no extreme at $(1, 5)$

47. (a) 0, 2, 6, 8, 11, 12 (b) 0, 6, 11 (c) 2, 8, 12

49. If f does not attain a maximum on $[a, b]$ or f does not attain a minimum on $[a, b]$, then f must have a discontinuity on $[a, b]$.

51. (a) yes, −1 (b) no (c) yes, −1 (d) no (e) yes, $1 − \pi$

53. (a) yes, 0 (b) yes, 0 (c) yes, 0 (d) yes, 0 (e) yes, 0

55. (a) $S(x)$ is minimum when $x \approx 8$

   (b) $S(x)$ is maximum when $x = 2$.

Section 3.2

1. $c \approx 3$, 10 and 13

3. (a) $c = \dfrac{\pi}{2}$ (b) $c = \dfrac{3\pi}{2}, \dfrac{5\pi}{2}, \dfrac{7\pi}{2}, \dfrac{9\pi}{2}$

5. Rolle's Theorem asserts that the velocity $h'(t)$ will equal 0 at some point between the time the ball is tossed and the time you catch it. The ball is not moving as fast when it reaches the balcony after being thrown from below.

7. The function does not violate Rolle's Theorem because the function does not satisfy the hypotheses of the theorem: f is not differentiable at 0, a point in the interval $-1 < x < 1$.

9. No. Velocity is not the same as the rate of change of altitude, which is only one of the components of position. Rolle's Theorem only says there was a time when altitude was not changing.

11. $f'(x) = 3x^2 + 5$, so $f'(x) = 0$ has no real roots. If $f(x) = 0$ for a value of x other than 2, then by the corollary from Problem 10, we would have an immediate contradiction.

13. (a) $f(0) = 0$, $f(2) = 4$, $f'(c) = 2c$, so $\frac{4-0}{2-0} = 2c$ $\Rightarrow c = 1$ (b) $f(1) = 4$, $f(5) = 8$, $f'(c) = 2c - 5$, so $\frac{8-4}{5-1} = 2c - 5 \Rightarrow c = 3$

14. (a) $f(0) = 0$, $f\left(\frac{\pi}{2}\right) = 1$, $f'(c) = \cos(c)$, so $\frac{1-0}{\frac{\pi}{2}-0} = \cos(c) \Rightarrow c = \arccos\left(\frac{2}{\pi}\right) \approx 0.88$

   (b) $f(-1) = -1$, $f(3) = 27$, $f'(c) = 3c^2$, so $\frac{27+1}{3+1} = 3c^2 \Rightarrow c^2 = \frac{7}{3}$, hence $c = \sqrt{\frac{7}{3}}$ (we know $c > -1$, which eliminates $c = -\sqrt{\frac{7}{3}}$)

15. (a) $f(1) = 4$, $f(9) = 2$, $f'(c) = \frac{-1}{2\sqrt{c}}$, so $\frac{2-4}{1-9} = \frac{-1}{2\sqrt{c}} \Rightarrow -\frac{1}{4} = -\frac{1}{2\sqrt{c}} \Rightarrow c = 4$ (b) $f(1) = 3$, $f(7) = 15$, $f'(c) = 2$, so $\frac{15-3}{7-1} = 2$ and any c between 1 and 7 will do.

17. The hypotheses are not all satisfied: $f'(x)$ does not exist at $x = 0$, which is between -1 and 3.

19. $f'(c) = 17$ at some time c does not prove the motorist "could not have been speeding."

21. $f(x) = x^3 + x^2 + 5x + c$, so $f(1) = 7 + c = 10$ when $c = 3$. Therefore $f(x) = x^3 + x^2 + 5x + 3$.

23. (a) $f'(x) = 2Ax$. We need $A \cdot 1^2 + B = 9$ and $2A \cdot 1 = 4$ so $A = 2$ and $B = 7$, hence $f(x) = 2x^2 + 7$.

   (b) $A \cdot 2^2 + B = 3$ and $2A \cdot 2 = -2$ so $A = -\frac{1}{2}$ and $B = 5$, hence $f(x) = -\frac{1}{2}x^2 + 5$.

   (c) $A \cdot 0^2 + B = 2$ and $2A \cdot 0 = 3$, so there is no such A. The point $(0,2)$ is not on the parabola $y = x^2 + 3x - 2$.

25. $f(x) = x^3 + C$, a family of "parallel" curves

27. $v(t) = 300$: Assuming the rocket left the ground at $t = 0$, we have $y(1) = 300$ ft, $y(2) = 600$ ft, $y(5) = 1500$ ft.

29. $f''(x) = 6$, $f'(0) = 4$, $f(0) = -5 \Rightarrow f(x) = 3x^2 + 4x - 5$

31. (a) $A(x) = 3x$ (b) $A'(x) = 3$

33. (a) $A(x) = x^2 + x$ (b) $A'(x) = 2x + 1$

35. $a_1 = 5$, $a_2 = a_1 + 3 = 5 + 3 = 8$, $a_3 = a_2 + 3 = (5+3) + 3 = 11$, $a_4 = a_3 + 3 = (5+3+3) + 3 = 14$: in general, $a_n = 5 + 3(n-1) = 2 + 3n$

Section 3.3

1.

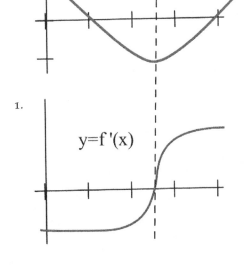

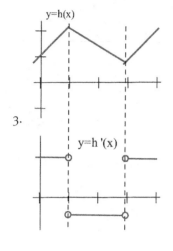

3.

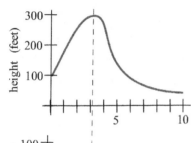

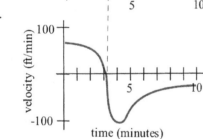

4.

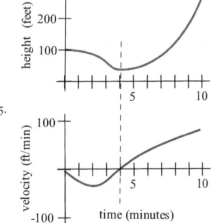

5.

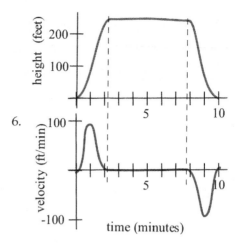

6.

7. A: Q, B: P, C: R

9. $f'(x) = \frac{1}{x} > 0$ for $x > 0$ so $f(x) = \ln(x)$ is increasing on $(0, \infty)$.

11. If f is increasing then $f(1) < f(\pi)$ so $f(1)$ and $f(\pi)$ cannot both equal 2.

13. (a) $x = 3$, $x = 8$ (b) maximum at $x = 8$ (c) none (or only at right endpoint)

14. Relative maximum height at $x = 2$ and $x = 7$; relative minimum height at $x = 4$.

15. Relative maximum height at $x = 6$; relative minimum height at $x = 8$.

17. $f(x) = x^3 - 3x^2 - 9x - 5$ has a relative minimum at $(3, -32)$ and a relative maximum at $(-1, 0)$.

19. $h(x) = x^4 - 8x^2 + 3$ has a relative max at $(0, 3)$ and relative minimums at $(2, -13)$ and $(-2, -13)$.

21. $r(t) = 2(t^2 + 1)^{-1}$ has a relative maximum at $(0, 2)$ and no relative minimums.

23. No positive roots. $f(x) = 2x + \cos(x)$ is continuous and $f(0) = 1 > 0$. Because $f'(x) = 2 - \sin(x) \geq 1 > 0$ for all x, f is increasing and never decreases back to the x-axis (a root).

24. One positive root. $g(x) = 2x - \cos(x)$ is continuous, $g(0) = -1 < 0$ and $g(1) = 2 - \cos(1) \geq 1 > 0$ so by the Intermediate Value Theorem, g has a root between 0 and 1. Because $g'(x) = 2 + \sin(x) \geq 1 > 0$ for all x, g is always increasing and can have only that one root.

25. $h(x) = x^3 + 9x - 10 \Rightarrow h'(x) = 3x^2 + 9 = 3(x^2 + 3) > 0$ for all x so h is always increasing and can cross the x-axis at most at one place. Because the $h(1) = 0$, the graph of h crosses the x-axis at $x = 1$, which is the only root of h.

27. (a) See below left: f is always increasing but $f'(1)$ is undefined.

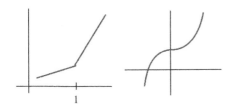

  (b) See above right: $f(x) = x^3 + 1$ is always increasing but $f'(0) = 0$.

  (c) If they travel at the same positive speed in different directions, then the distance between them will not remain constant.

29. (a) $h(x) = x^2$ or $x^2 + 1$ or $x^2 - 7$ or, in general, $x^2 + C$ for any constant C.

  (b) $f(x) = x^2 + C$ and $20 = f(3) = 3^2 + C \Rightarrow C = 20 - 9 = 11$ so $f(x) = x^2 + 11$.

  (c) $g(x) = x^2 + C$ and $7 = g(2) = 2^2 + C \Rightarrow C = 7 - 4 = 3$ so $g(x) = x^2 + 3$.

Section 3.4

1. (a) $f(t) =$ number of workers unemployed at time t; $f'(t) > 0$ and $f''(t) < 0$

  (b) $f(t) =$ profit at time t; $f'(t) < 0$ and $f''(t) > 0$

  (c) $f(t) =$ population at time t; $f'(t) > 0$ and $f''(t) > 0$

3.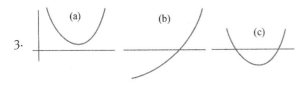

  (d) Not possible.

5. (a) Concave up on $(0,2)$, $(2,3+)$, $(6,9)$. (b) Concave down on $(3+,6)$. (A small technical note: we have defined concavity only at points where the function is differentiable, so we exclude the

endpoints and points where the function is not differentiable from the intervals of concave up and concave down.)

7. $g''(x) = 6x - 6 \Rightarrow g''(-1) < 0$ so $(-1, 12)$ is a local max; $g''(3) > 0$ so $(3, -20)$ is a local min.

9. $f''(x) = 5\left[-\sin^5(x) + 4\sin^3(x) \cdot \cos^2(x)\right] \Rightarrow f''\left(\frac{\pi}{2}\right) < 0$ so $\left(\frac{\pi}{2}, 1\right)$ is a local maximum; $f''\left(\frac{3\pi}{2}\right) > 0$ so $\left(\frac{3\pi}{2}, -1\right)$ is a local minimum; $f''(\pi) = 0$ and f changes concavity at $x = \pi$ so $(\pi, 0)$ is an inflection point.

11. d and e

13. (a) 0 (b) at most 1 (c) at most $n - 2$

15.

x	$f(x)$	$f'(x)$	$f''(x)$
0	$-$	$+$	$+$
1	$+$	0	$-$
2	$-$	$-$	$+$
3	0	$+$	$+$

17.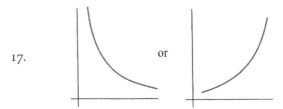

  or

19. $f'(x) = 3x^2 - 42x + 144 = 3(x^2 - 14x + 48) = 3(x - 6)(x - 8)$, which exists everywhere and $f'(x) = 0 \Rightarrow x = 6$ or $x = 8$; meanwhile, $f''(x) = 6x - 42 = 6(x - 7)$, so $f''(6) = -6 < 0$ means that f has a local max at $x = 6$ and $f''(8) = 6 > 0$ means that f has a local min at $x = 8$. The only candidate for an inflection point is where $f''(x) = 0 \Rightarrow x = 7$; $f''(x) < 0$ for $x < 7$ and $f''(x) > 0$ for $x > 7$, so f does have an inflection point at $x = 7$. A window with $2 \le x \le 12$ and $-50 \le y \le 10$ should work.

21. $f'(x) = 7e^{7x} - 5 \Rightarrow f''(x) = 49e^{7x}$ so $f''(x) > 0$ for all x, hence f is always concave up and has no inflection points. Its only critical number is where $7e^{7x} - 5 = 0 \Rightarrow e^{7x} = \frac{5}{7} \Rightarrow x = \frac{1}{7} \cdot \ln\left(\frac{5}{7}\right) \approx -0.05$, which must be a local min. A window with $-1 \le x \le \frac{1}{2}$ and $0 \le y \le 20$ should work.

23. $f'(x) = -3e^{-3x} + 1 \Rightarrow f''(x) = 9e^{-3x} > 0$ for all x, hence f is always concave up and has no inflection points. Its only critical number: $-3e^{-3x} + 1 = 0 \Rightarrow e^{-3x} = \frac{1}{3} \Rightarrow x = \frac{1}{3} \cdot \ln(3) \approx 0.37$, which must be a local min. Try a window with $-1 \le x \le 2$ and $0 \le y \le 3$.

25. $f'(x) = (1 - 3x)e^{-3x} \Rightarrow f''(x) = (9x - 6)e^{-3x}$ so the only critical number is $x = \frac{1}{3}$ and $f''\left(\frac{1}{3}\right) - 3e^{-1} < 0$, so it must be a local max. The only inflection point is where $x = \frac{2}{3}$. Try a window with $-1 \le x \le 2$ and $-2 \le y \le 1$.

27. $f'(x) = \frac{1}{3}x^{-\frac{2}{3}}(4x - 1)$, so the critical numbers are $x = 0$ and $x = \frac{1}{4}$; $f''(x) = \frac{2}{9}x^{-\frac{5}{3}}(2x + 1)$ so the only candidates for an inflection point are $x = 0$ and $x = -\frac{1}{2}$. Because $f''(x) < 0$ when $x < -\frac{1}{2}$ and $x > 0$, the graph of f is concave up on those intervals and concave down for $-\frac{1}{2} < x < 0$; because $f\left(\frac{1}{4}\right) > 0$, f has a local min at $x = \frac{1}{4}$. Try $-1.5 \le x \le 2.5$ and $-1 \le y \le 3$.

29. Critical number at $x = 0$, which is a local min; inflection points at $x = \pm 1$; concave up on $(-1, 1)$, concave down elsewhere; try a window with $-3 \le x \le 3$ and $-1 \le y \le 3$.

31. Critical number at $x = -1$, which is a local min; inflection points at $x = -1 \pm \sqrt{3}$; concave up between those values, concave down elsewhere; try a window with $-6 \le x \le 6$ and $-1 \le y \le 4$.

Section 3.5

1. (a) Using a rectangle with sides of length x and y (see figure below) $2x + 2y = 200 \Rightarrow y = 100 - x$. We want to maximize $A = x \cdot y = x(100 - x) = 100x - x^2$ and $A'(x) = 100 - 2x$ so $A'(x) = 0 \Rightarrow x = 50 \Rightarrow y = 100 - x = 50$; $A''(x) = -2 < 0$ for all x, so $x = 50$ yields the maximum enclosed area. When $x = 50$, $A(50) = 50(100 - 50) = 2500$ square feet.

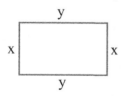

(b) Using a rectangle with sides of length x and y (see figure above) $2x + 2y = P \Rightarrow y = \frac{P}{2} - x$. We want to maximize $A = x \cdot y = x\left(\frac{P}{2} - x\right) = \left(\frac{P}{2}\right)x - x^2$ and $A'(x) = \frac{P}{2} - 2x$ so $A'(x) = 0$ when $x = \frac{P}{4} \Rightarrow y = \frac{P}{2} - \frac{P}{4} = \frac{P}{4}$; $A''(x) = -2 < 0$ for all x, so $x = \frac{P}{4}$ yields the maximum enclosed area. This garden is a $\frac{P}{4}$ by $\frac{P}{4}$ square.

(c) Using a rectangle with sides of length x and y (see figure below) $2x + y = P \Rightarrow y = P - 2x$. We want to maximize $A = xy = x(P - 2x) = Px - 2x^2$ and $A'(x) = P - 4x$ so $A'(x) = 0$ when $x = \frac{P}{4} \Rightarrow y = P - 2\left(\frac{P}{4}\right) = \frac{P}{2}$.

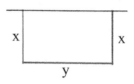

(d) A circle. A semicircle.

3. (a) If x is the horizontal length and y is the vertical length, then $120 = 2x + 5y \Rightarrow y = 24 - \frac{2}{5}x$. We want to maximize $A = xy = x\left(24 - \frac{2}{5}x\right) = 24x - \frac{2}{5}x^2$ and $A'(x) = 24 - \frac{4}{5}x$ so $A'(x) = 0$ when $x = 30$ (and $y = 12$); $A''(x) = -\frac{4}{5} < 0$ for all x so $x = 30$ yields the maximum enclosed area. Area is $(30\,\text{ft})(12\,\text{ft}) = 360$ square feet.

(b) A circular pen divided into four equal stalls by two diameters:

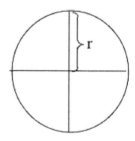

does a better job than a square with area 400 square feet. If the radius is r, then $4r + 2\pi r = 120 \Rightarrow r = \frac{120}{4 + 2\pi} \approx 11.67$; the resulting enclosed area is $A = \pi r^2 \approx \pi(11.67)^2 \approx 427.8$ square feet.

The pen shown here:

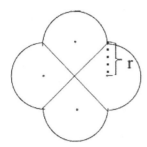

does even better. If each semicircle has radius r, then the figure uses $4\sqrt{2}r + 4\pi r = 120$ feet of fencing so $r = \frac{120}{4\sqrt{2}+4\pi} \approx 6.585$ feet; the resulting enclosed area is

$$A = (\text{square}) + (\text{four semicircles})$$
$$= (2r)^2 + 4\left(\frac{1}{2}\pi r^2\right) \approx 445.90 \text{ ft}^2$$

5. The dimensions of the bottom of the box are $10 - 2x$ and $5 - x$ so we want to maximize $V(x) = x(10 - 2x)(5 - x) = 50x - 20x^2 + 2x^3$; $V'(x) = 50 - 40x + 6x^2 = 2(3x - 5)(x - 5)$ so $V'(x) = 0$ when $x = 5$ or $x = \frac{5}{3}$. When $x = 5$, then $V = 0$ (clearly not a maximum!) so $x = \frac{5}{3}$. The dimensions of the box with the largest volume are $\frac{5}{3}$ inches by $\frac{10}{3}$ inches by $\frac{20}{3}$ inches.

7. (a) Let r and h be the radius and height of the cylindrical can, so the volume is $V = \pi r^2 h = 100 \Rightarrow h = \frac{100}{\pi r^2}$. We want to minimize the cost:

$$C = 2(\text{top area}) + 5(\text{bottom area}) + 3(\text{side area})$$
$$= 2\left(\pi r^2\right) + 5\left(\pi r^2\right) + 3\left(2\pi rh\right)$$
$$= 7\pi r^2 + 6\pi r\left(\frac{100}{\pi r^2}\right) = 7\pi r^2 + \frac{600}{r}$$

So $C'(r) = 14\pi r - \frac{600}{r^2} \Rightarrow C'(r) = 0$ when $r = \sqrt[3]{\frac{600}{14\pi}} \approx 2.39 \Rightarrow h \approx \frac{100}{\pi(2.39)^2} \approx 5.57$.

(b) Let $k = $ top material cost + bottom material cost $= 2\cent + $ bottom cost $> 2\cent + 5\cent = 7\cent$. We want to minimize the cost $C = k\pi r^2 + \frac{600}{r}$; $C'(r) = 2k\pi r - \frac{600}{r^2}$ so $C'(r) = 0$ when $r = \sqrt[3]{\frac{600}{2k\pi}}$. If $k = 8$, then $r \approx 2.29$; if $k = 9$, then $r \approx 2.20$; if $k = 10$, then $r \approx 2.12$. As the cost of the bottom material increases, the radius of the least expensive cylindrical can decreases: the least expensive can becomes narrower and taller.

9. Recall that time $= \frac{\text{distance}}{\text{rate}}$. Let the run distance $= x$ (so $0 \le x \le 60$; why?), hence the run time $= \frac{x}{8}$ and the swim distance $= \sqrt{40^2 + (60 - x)^2}$ so the swim time $= \frac{1}{2}\sqrt{40^2 + (60 - x)^2}$ and the total time is $T(x) = \frac{x}{8} + \frac{1}{2}\sqrt{40^2 + (60 - x)^2}$. Differentiating to compute $T'(x)$ yields:

$$\frac{1}{8} + \frac{1}{2} \cdot \frac{1}{2}\left(40^2 + (60 - x)^2\right)^{-\frac{1}{2}} \cdot 2(60 - x)(-1)$$
$$= \frac{1}{8} - \frac{60 - x}{2\sqrt{40^2 + (60 - x)^2}}$$

so $T'(x) = 0$ when $x = 60 \pm \frac{40}{\sqrt{15}}$. The value $x = 60 + \frac{40}{\sqrt{15}} > 60$ so the least total time occurs when $x = 60 - \frac{40}{\sqrt{15}} \approx 49.7$ meters. In this situation, the lifeguard should run about $\frac{5}{6}$ of the way along the beach before going into the water.

11. (a) Consider a similar problem with a new town D^* located at the "mirror image" of D across the river (below left):

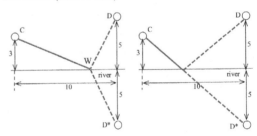

If the water works is built at any location W along the river, then the distances are the same from W to D and from W to D^*: $\text{dist}(W, D) = \text{dist}(W, D^*)$. Then $\text{dist}(C, W) + \text{dist}(W, D) = \text{dist}(C, W) + \text{dist}(W, D^*)$ so the shortest distance from C to D^* is a straight line (see figure above right), and this straight line yields similar triangles with equal side ratios: $\frac{x}{3} = \frac{10 - x}{5} \Rightarrow x = \frac{15}{4} = 3.75$ miles. A consequence of this "mirror image" view of the problem is that "at the best location W the angle of incidence α equals the angle of reflection β."

(b) We want to minimize the cost:

$$C(x) = 3000\,\mathrm{dist}(C, W) + 7000\,\mathrm{dist}(W, D)$$

$$= 3000\sqrt{x^2 + 9} + 7000\sqrt{(10 - x)^2 + 25}$$

so, differentiating gives:

$$C'(x) = \frac{3000x}{\sqrt{x^2 + 9}} + \frac{-7000(10 - x)}{\sqrt{(10 - x)^2 + 25}}$$

and (after much algebra) it turns out that $C'(x) = 0$ when $x \approx 7.82$ miles. As it becomes more expensive to build the pipe from W to D, the cheapest route tends to shorten the distance from W to D.

13. (a) Let x be the length of one edge of the square end. The volume is $V(x) = x^2(108 - 4x) = 108x^2 - 4x^3 \Rightarrow V'(x) = 216x - 12x^2 = 6x(18 - x)$ so $V'(x) = 0$ when $x = 0$ or $x = 18$. Clearly $x = 0$ results in a box of no volume, so the dimensions of the acceptable box with a square end and greatest volume are 18 in by 18 in by 36 in; the volume is 11,664 in^3.

(b) Let x be the length of the shorter edge of the end. The volume $V(x) = 2x^2(108 - 6x) = 216x^2 - 12x^3 \Rightarrow V'(x) = 432x - 36x^2 = 36x(12 - x)$ so $V'(x) = 0$ if $x = 0$ or $x = 12$. Clearly $x = 0$ results in a box of no volume, so the dimensions of the box with an acceptable shape and largest volume are 12 in by 24 in by 36 in; the volume is 10,368 in^3.

(c) Let x be the radius of the circular end. The volume is $V(x) = \pi x^2(108 - 2\pi x) = 108\pi x^2 - 2\pi^2 x^3 \Rightarrow V'(x) = 216\pi x - 6\pi^2 x^2 = 6\pi x(36 - \pi x)$ so $V'(x) = 0$ when $x = 0$ or $x = \frac{36}{\pi} \approx 11.46$ in. The dimensions of the acceptable box with circular end and largest volume are a radius of 11.46 in and a length of 36 in; the volume is 14,851 in^3.

15. **Without calculus:** The area of the triangle is $\frac{1}{2}(\text{base})(\text{height}) = \frac{1}{2}(7)(\text{height})$ and the height is maximum when the angle between the sides is a right angle. **Using calculus:** Let θ be the angle between the sides. Then the area of the triangle is $A = \frac{7}{2}(10\sin(\theta)) = 35\sin(\theta) \Rightarrow A'(\theta) = 35\cos(\theta)$ so $A'(\theta) = 0$ when $\theta = \frac{\pi}{2}$, hence

the triangle is a right triangle with sides 7 and 10. Using either approach, the maximum area of the triangle is $\frac{1}{2}(7)(10) = 35$ square inches, and the third side is the hypotenuse with length $\sqrt{7^2 + 10^2} = \sqrt{149} \approx 12.2$ inches.

17. (a) $A(x) = 2x(16 - x^2) = 32x - 2x^3 \Rightarrow A'(x) = 32 - 6x^2$ so $A'(c) = 0$ when $x = \frac{32}{6} = \frac{16}{3} \approx 2.31$. The dimensions of the rectangle are $2\left(\frac{16}{3}\right) \approx 4.62$ and $16 - \left(\frac{16}{3}\right)^2 = \frac{32}{3} \approx 10.67$.

(b) $A(x) = 2x\sqrt{1 - x^2}$ so, using the Product Rule:

$$A'(x) = 2\left(\sqrt{1 - x^2} - \frac{x^2}{\sqrt{1 - x^2}}\right)$$

and $A'(x) = 0$ when $x = \frac{1}{\sqrt{2}} \approx 0.707$. The dimensions of the rectangle are $2\left(\frac{1}{\sqrt{2}}\right) = \sqrt{2} \approx 1.414$ and $\sqrt{1 - \left(\frac{1}{\sqrt{2}}\right)^2} = \frac{1}{\sqrt{2}} \approx 0.707$.

(c) The graph of $|x| + |y| = 1$ is a "diamond" (a square) with corners at $(1, 0)$, $(0, 1)$, $(-1, 0)$ and $(0, -1)$. For $0 \le x \le 1$:

$$A(x) = 2x \cdot 2(1 - x) = 4x - 4x^2$$

so $A'(x) = 4 - 8x$ and $A'(x) = 0$ when $x = \frac{1}{2}$; $A''(x) = -8 < 0$ for all x, so $x = \frac{1}{2}$ must be a local max. The dimensions of the rectangle are $2\left(\frac{1}{2}\right) = 1$ and $2\left(1 - \frac{1}{2}\right) = 1$.

(d) $A(x) = 2x\cos(x)$, where $0 \le x \le \frac{\pi}{2}$, so $A'(x) = 2\cos(x) - 2x \cdot \sin(x)$ and $A'(x) = 0$ when $x \approx 0.86$. The dimensions of the rectangle are $2(0.86) = 1.72$ and $\cos(0.86) \approx 0.65$.

19. The cross-sectional area is:

$$A = 6\sin\left(\frac{\theta}{2}\right) \cdot 6\cos\left(\frac{\theta}{2}\right)$$

$$= 36 \cdot \frac{1}{2}\sin(\theta) = 18\sin(\theta)$$

which is a maximum when $\theta = \frac{\pi}{2}$. Then the maximum area is $A = 18\sin\left(\frac{\pi}{2}\right) = 18$ in^2. (This problem is similar to Problem 15.)

21. $V = \frac{1}{3}\pi r^2 h$ and $h = \sqrt{9 - r^2}$ so:

$$V(r) = \frac{1}{3}\pi r^2\sqrt{9 - r^2} = \frac{\pi}{3}\sqrt{9r^4 - r^6}$$

Differentiating and simplifying:

$$V'(r) = \frac{\pi}{6} \cdot \frac{36r^3 - 6r^5}{\sqrt{9r^4 - r^6}}$$

so $V'(r) = 0$ when $36r^3 - 6r^5 = 6r^3(6 - r^2) = 0$; $r = 0$ results in zero volume, so $r = \sqrt{6} \approx 2.45$ in $\Rightarrow h = \sqrt{9 - (\sqrt{6})^2} = \sqrt{3} \approx 1.73$ in.

23. Let $n \geq 10$ be the number of passengers. The income is $I(n) = n(30 - (n - 10)) = 40n - n^2$ and the cost is $C(n) = 100 + 6n$, so the profit is $P(n) = I(n) - C(n) = (40n - n^2) - (100 + 6n) = 34n - n^2 - 100 \Rightarrow P'(n) = 34 - 2n$ so $P'(n) = 0$ when $n = 17$: 17 passengers on the flight maximize your profit. (This is an example of treating a naturally discrete variable, the number of passengers, as a continuous variable.)

25. Apply Problem 24 with $R = f$ and $E = g$.

27. (a) Let D be diameter of the can's base and $H =$ the can's height. Then:

$$\theta = \arctan\left(\frac{\text{radius of can}}{\text{height of CG}}\right)$$
$$= \arctan\left(\frac{D/2}{H/2}\right) = \arctan\left(\frac{D}{H}\right)$$

For this can, $D = 5$ cm and $H = 12$ cm so $\theta = \arctan\left(\frac{2.5}{6}\right) \approx 0.395$ radians. The can can be tilted about $22.6°$ before it falls over.

(b) Differentiating $C(x)$ yields:

$$\frac{(60 + 19.2x)(19.2x) - (360 + 9.6x^2)(19.2)}{(60 + 19.2x)^2}$$

so $C'(x) = 0 \Rightarrow (19.2)(9.6x^2 + 60x - 360) = 0$ $\Rightarrow x = 3.75$ or $x = -10$: the height of the cola is 3.75 cm.

(c) $C(3.75) = 3.75$ (The center of gravity is exactly at the top edge of the cola. It turns out that when the CG of a can-and-liquid system is as low as possible then the CG is at the top edge of the liquid.) Then $\theta = \arctan\left(\frac{2.5}{3.75}\right) \approx 0.588$ radians: in this situation, the can can be tilted about $33.7°$ before it falls over.

(d) Less far.

28. See below left for (a), below right for (b):

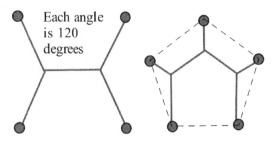

In the solutions to these "shortest path" problems, the roads all meet at $120°$ angles.

29. (a) $A = (\text{base})(\text{height}) = (1 - x)(x^2) = x^2 - x^3$ for $0 \leq x \leq 1$, so $A'(x) = 2x - 3x^2$ and $A'(x) = 0 \Rightarrow x = \frac{2}{3}$ (clearly the endpoints $x = 0$ and $x = 1$ will not yield the largest area). So $A\left(\frac{2}{3}\right) = \frac{1}{3}\left(\frac{2}{3}\right)^2 = \frac{4}{27}$.

(b) $A(x) = (1 - x)(Cx^2) = Cx^2 - Cx^3$ where $0 \leq x \leq 1$, so $A'(x) = 2Cx - 3Cx^2$ and $A'(x) = 0$ if $x = \frac{2}{3}$ (as before, neither endpoint works). So $A\left(\frac{2}{3}\right) = \frac{1}{3} \cdot C\left(\frac{2}{3}\right)^2 = \frac{4C}{27}$.

(c) $A(x) = (B - x)(Cx^2) = BCx^2 - Cx^3$, where $0 \leq x \leq B$, so $A'(x) = 2BCx - 3Cx^2 = Cx(2B - 3x)$ and $A'(x) = 0$ if $x = \frac{2}{3}B$. So $A\left(\frac{2B}{C}\right) = \left(\frac{B}{3}\right)(C)\left(\frac{2B}{3}\right)^2 = \frac{4}{27}B^3C$.

31. (a) Position the lower-left vertex of the triangle at $(0,0)$ so the hypotenuse sits on $y = 20 - \frac{20}{50}x$ and the area of the rectangle is $A = xy = x\left(20 - \frac{2}{5}x\right) = 20x - \frac{2}{5}x^2 \Rightarrow A'(x) = 20 - \frac{4}{5}x$ and $A'(x) = 0$ when $x = 25$. Then $y = 10$ and the area of the rectangle is $25 \cdot 10 = 250$.

(b) Proceeding as in part (a), $y = H - \frac{H}{B}x$ and $A(x) = x\left(H - \frac{H}{B}x\right) = Hx - \frac{H}{B}x^2 \Rightarrow A'(x) = H - \frac{2H}{B}x$ and $A'(x) = 0$ when $x = \frac{B}{2}$. Then $y = \frac{H}{2}$ and area of rectangle is $\frac{B}{2} \cdot \frac{H}{2} = \frac{BH}{4}$.

33. Let r and h be radius and height of can, and $F =$ (top cost) + (bottom cost) + (cost of sides) $= A(\pi r)^2 + B(\pi r)^2 + C(2\pi rh)$. But volume $V = \pi r^2 h \Rightarrow h = \frac{V}{\pi r^2}$ so $F(r) = (A + B)\pi r^2 + \frac{2CV}{r} \Rightarrow F'(r)2(A + B)\pi r - \frac{2CV}{r^2}$, so $F'(r) = 0$ when $r = \sqrt[3]{\frac{CV}{\pi(A+B)}}$. Now you can find h and F.

Section 3.6

1. (a) $h(x)$ has a root at $x = 1$.

 (b) $\lim\limits_{x \to 1^+} h(x) = 0$, $\lim\limits_{x \to 1^-} h(x) = 0$
 $\lim\limits_{x \to 3^+} h(x) = -\infty$, $\lim\limits_{x \to 3^-} h(x) = +\infty$

 (c) $y = h(x)$ has a vertical asymptote at $x = 3$

3. $\lim\limits_{x \to 2^+} h(x) = +\infty$, $\lim\limits_{x \to 2^-} h(x) = -\infty$
 $\lim\limits_{x \to 4^+} h(x) = 0$, $\lim\limits_{x \to 4^-} h(x) = 0$

5. 0 7. -3 9. 0 11. DNE

13. $\frac{2}{3}$ 15. 0 17. -7 19. 3

21. $\cos(0) = 1$ 23. $\ln(1) = 0$

25. (a) $V(t) = 50 + 4t$ gals, $A(t) = 0.8t$ lbs of salt

 (b) $C(t) = \frac{A(t)}{V(t)} = \frac{0.8t}{50+4t}$

 (c) as $t \to \infty$, $C(t) \to \frac{0.8}{4} = 0.2$ lbs per gal

 (d) $V(t) = 200 + 4t$ and $A(t) = 0.8t$, so $C(t) = \frac{0.8t}{200+4t} \to \frac{0.8}{4} = 0.2$ lbs per gal

27. $+\infty$ 29. $-\infty$ 31. $-\infty$ 33. $-\infty$

35. $+\infty$ 37. $-\infty$ 39. 1 41. $-\infty$

43. horizontal: $y = 0$; vertical: $x = 0$

44. horizontal: $y = 0$; vertical: $x = 0$; "hole": $(1,1)$

45. horizontal: $y = 0$; vertical: $x = 3$ and $x = 1$

47. vertical: $x = -1$

49. horizontal: $y = 1$; vertical: $x = 1$

51. slant: $y = 2x + 1$; vertical: $x = 0$

53. other: $y = \sin(x)$; vertical: $x = 2$

55. other: $y = x^2$

57. other: $y = \cos(x)$; vertical: $x = 3$

59. $y = \sqrt{x}$; $x = -3$

Section 3.7

1. $\lim\limits_{x \to 1} \dfrac{x^3 - 1}{x^2 - 1} = \lim\limits_{x \to 1} \dfrac{3x^2}{2x} = \lim\limits_{x \to 1} \dfrac{3}{2}x = \dfrac{3}{2}$

3. $\lim\limits_{x \to 0} \dfrac{\ln(1 + 3x)}{5x} = \lim\limits_{x \to 0} \dfrac{\frac{3}{1+3x}}{5} = \dfrac{3}{5}$

5. $\lim\limits_{x \to 0} \dfrac{x \cdot e^x}{1 - e^x} = \lim\limits_{x \to 0} \dfrac{x \cdot e^x + e^x \cdot 1}{-e^x} = \lim\limits_{x \to 0} \dfrac{x + 1}{-1} = -1$

7. $\lim\limits_{x \to \infty} \dfrac{\ln(x)}{x} = \lim\limits_{x \to \infty} \dfrac{\frac{1}{x}}{1} = 0$

9. $\lim\limits_{x \to \infty} \dfrac{\ln(x)}{x^p} = \lim\limits_{x \to \infty} \dfrac{\frac{1}{x}}{px^{p-1}} = \lim\limits_{x \to \infty} \dfrac{1}{px^p} = 0$

11. $\lim\limits_{x \to 0} \dfrac{1 - \cos(3x)}{x^2} = \lim\limits_{x \to 0} \dfrac{3\sin(3x)}{2x} = \lim\limits_{x \to 0} \dfrac{9\cos(3x)}{2} = \dfrac{9}{2}$

13. For $a \neq 0$: $\dfrac{f'}{g'} = \dfrac{m}{n}x^{m-n} \to \dfrac{m}{n}a^{m-n}$. For $a = 0$:

$$\lim\limits_{x \to a} \dfrac{x^m - a^m}{x^n - a^n} = \begin{cases} 0 & \text{if } m > n \\ 1 & \text{if } m = n \\ +\infty & \text{if } m < n \text{ and } (m - n) \text{ is even} \\ \text{DNE} & \text{if } m < n \text{ and } (m - n) \text{ is odd} \end{cases}$$

15. $\lim\limits_{x \to 0} \dfrac{1 - \cos(x)}{x \cdot \cos(x)} = \lim\limits_{x \to 0} \dfrac{\sin(x)}{-x \cdot \sin(x) + \cos(x)} = 0$

17. $\dfrac{f'}{g'} = \dfrac{pe^{px}}{3} \to \dfrac{p}{3}$ so $p = 3(5) = 15$.

19. (a) All three limits are $+\infty$.

 (b) After applying L'Hôpital's Rule m times:

$$\dfrac{f^{(m)}}{g^{(m)}} = \dfrac{a \cdot b^m \cdot e^{bn}}{c(m)(m - 1)(m - 2) \cdots (2)(1)}$$

$$= \dfrac{\text{constant} \cdot e^{bn}}{\text{another constant}} \to +\infty$$

21. $\lim\limits_{x \to 0^+} \dfrac{\ln(x)}{\csc(x)} = \lim\limits_{x \to 0^+} \dfrac{-\sin(x)}{x} \cdot \tan(x) = -1 \cdot 0 = 0$

23. $\lim\limits_{x \to 0^+} \dfrac{\ln(x)}{x^{-\frac{1}{2}}} = \lim\limits_{x \to 0^+} \dfrac{\frac{1}{x}}{-\frac{1}{2}x^{-\frac{3}{2}}} = \lim\limits_{x \to 0^+} -2\sqrt{x} = 0$

25. Write $\left(1 - \dfrac{3}{x^2}\right)^x = e^{x \cdot \ln\left(1 - \frac{3}{x^2}\right)}$ and compute:

$$\lim\limits_{x \to \infty} \dfrac{\ln\left(1 - \frac{3}{x^2}\right)}{\frac{1}{x}} = \lim\limits_{x \to \infty} \dfrac{-6}{x - \frac{3}{x}} = 0$$

so that the original limit equals $e^0 = 1$.

27. $\lim\limits_{x \to 0} \dfrac{\sin(x) - x}{x \cdot \sin(x)} = \lim\limits_{x \to 0} \dfrac{\cos(x) - 1}{x \cdot \cos(x) + \sin(x)}$

$$= \lim\limits_{x \to 0} \dfrac{-\sin(x)}{-x \cdot \sin(x) + 2\cos(x)} = 0$$

29. Write $\left(\dfrac{x + 5}{x}\right)^{\frac{1}{x}} = e^{\frac{\ln\left(\frac{x+5}{x}\right)}{x}}$ and compute:

$$\lim\limits_{x \to \infty} \dfrac{\ln\left(\frac{x+5}{x}\right)}{x} = \lim\limits_{x \to \infty} \dfrac{-5}{x(x + 5)} = 0$$

so that the original limit equals $e^0 = 1$.

D
Derivative Facts

Basic Patterns

$$\mathbf{D}(k) = 0 \qquad\qquad \mathbf{D}(k \cdot f) = k \cdot \mathbf{D}(f)$$

k represents a constant

$$\mathbf{D}(f + g) = \mathbf{D}(f) + \mathbf{D}(g) \qquad\qquad \mathbf{D}(f - g) = \mathbf{D}(f) - \mathbf{D}(g)$$

$$\mathbf{D}(f \cdot g) = f \cdot \mathbf{D}(g) + g \cdot \mathbf{D}(f) \qquad\qquad \mathbf{D}\left(\frac{f}{g}\right) = \frac{g \cdot \mathbf{D}(f) - f \cdot \mathbf{D}(g)}{g^2}$$

Product Rule and Quotient Rule

Power Rules

$$\mathbf{D}(x^p) = p \cdot x^{p-1} \qquad\qquad \mathbf{D}\left(f^n\right) = n \cdot f^{n-1} \cdot \mathbf{D}(f)$$

Chain Rule

$$\mathbf{D}\left(f(g(x))\right) = f'\left(g(x)\right) \cdot g'(x) \qquad\qquad \frac{dy}{dx} = \frac{dy}{du} \cdot \frac{du}{dx}$$

Exponential and Logarithmic Functions

$$\mathbf{D}\left(e^u\right) = e^u \qquad\qquad \mathbf{D}\left(a^u\right) = a^u \cdot \ln(a)$$

$$\mathbf{D}\left(\ln(|u|)\right) = \frac{1}{u} \qquad\qquad \mathbf{D}\left(\log_a(|u|)\right) = \frac{1}{u \cdot \ln(a)} \qquad\qquad \mathbf{D}\left(\ln(f(x))\right) = \frac{f'(x)}{f(x)}$$

Trigonometric Functions

$$\mathbf{D}\left(\sin(u)\right)) = \cos(u)) \qquad\qquad \mathbf{D}(\tan(u)) = \sec^2(u) \qquad\qquad \mathbf{D}(\sec(u)) = \sec(u) \cdot \tan(u)$$

$$\mathbf{D}(\cos(u)) = -\sin(u) \qquad\qquad \mathbf{D}(\cot(u)) = -\csc^2(u) \qquad\qquad \mathbf{D}(\csc(u)) = -\csc(u) \cdot \cot(u)$$

Inverse Trigonometric Functions

$$\mathbf{D}(\arcsin(u)) = \frac{1}{\sqrt{1 - u^2}}$$

$$\mathbf{D}(\arctan(u)) = \frac{1}{1 + u^2}$$

$$\mathbf{D}(\text{arcsec}(u)) = \frac{1}{|u|\,\sqrt{u^2 - 1}}$$

$$\mathbf{D}(\arccos(u)) = \frac{-1}{\sqrt{1 - u^2}}$$

$$\mathbf{D}(\text{arccot}(u)) = \frac{-1}{1 + u^2}$$

$$\mathbf{D}(\text{arccsc}(u)) = \frac{-1}{|u|\,\sqrt{u^2 - 1}}$$

Hyperbolic Functions

$$\mathbf{D}(\sinh(u))) = \cosh(u))$$

$$\mathbf{D}(\cosh(u)) = \sinh(u)$$

$$\mathbf{D}(\tanh(u)) = \text{sech}^2(u)$$

$$\mathbf{D}(\coth(u)) = -\text{csch}^2(u)$$

$$\mathbf{D}(\text{sech}(u)) = -\text{sech}(u) \cdot \tanh(u)$$

$$\mathbf{D}(\text{csch}(u)) = -\text{csch}(u) \cdot \coth(u)$$

Inverse Hyperbolic Functions

$$\mathbf{D}(\text{argsinh}(u)) = \frac{1}{\sqrt{1 + u^2}}$$

$$\mathbf{D}(\text{argcosh}(u)) = \frac{1}{\sqrt{u^2 - 1}} \quad (\text{for } u > 1)$$

$$\mathbf{D}(\text{argtanh}(u)) = \frac{1}{1 - u^2} \quad (\text{for } |u| < 1)$$

$$\mathbf{D}(\text{argcoth}(u)) = \frac{1}{1 - u^2} \quad (\text{for } |u| > 1)$$

$$\mathbf{D}(\text{argsech}(u)) = \frac{-1}{|u|\,\sqrt{1 - u^2}} \quad (\text{for } 0 < u < 1)$$

$$\mathbf{D}(\text{argcsch}(u)) = \frac{-1}{|u|\,\sqrt{u^2 + 1}} \quad (\text{for } u \neq 0)$$

H
How to Succeed in Calculus

The following comments are based on over 30 years of watching students succeed and fail in calculus courses at universities, colleges and community colleges and of listening to their comments as they went through their study of calculus. This is the best advice we can give to help you succeed.

Calculus takes time. Almost no one fails calculus because they lack sufficient "mental horsepower." Most people who do not succeed are unwilling (or unable) to devote the necessary time to the course. The "necessary time" depends on how smart you are, what grade you want to earn and on how competitive the calculus course is. Most calculus teachers and successful calculus students agree that two (or three) hours every weeknight and six or seven hours each weekend is a good way to begin if you seriously expect to earn an A or B grade. If you are only willing to devote five or 10 hours a week to calculus outside of class, you should consider postponing your study of calculus.

Do NOT fall behind. The brisk pace of the calculus course is based on the idea that "if you are in calculus, then you are relatively smart, you have succeeded in previous mathematics courses, and you are willing to work hard to do well." It is terribly difficult to **catch up** and **keep up** at the same time. A much safer approach is to work very hard for the first month and then evaluate your situation. If you do fall behind, spend a part of your study time catching up, but spend most of it trying to follow and understand what is going on in class.

Go to class, every single class. We hope your calculus teacher makes every idea crystal clear, makes every technique seem obvious and easy, is enthusiastic about calculus, cares about you as a person, and even makes you laugh sometimes. If not, you still need to attend class. You need to hear the vocabulary of calculus spoken and to see how mathematical ideas are strung together to reach conclusions. You need to see how an expert problem-solver approaches problems. You need to hear the announcements about homework and tests. And you need to get to know some of the other students in the class. Unfortunately, when students get a bit behind or confused, they are most likely to miss a class or two (or five). That is absolutely the worst time to miss classes. Attend class anyway. Ask where on campus you can get some free tutoring or counseling. Ask a classmate to help you for an hour after class. If you must miss a class, ask a classmate what material was covered and skim those sections before the next class. Even if you did not read the material, return to class as soon as possible.

Work together. Study with a friend. Work in small groups. It is much more fun and is very effective for doing well in calculus. Recent studies—and our personal observations—indicate that students who regularly work together in small groups are less likely to drop the course and are more likely to get A's or B's. You need lots of time to work on the material alone, but study groups of 3–5 students, working together two or three times a week for a couple hours, seem to help everyone in the group. Study groups offer you a way to get and give help on the material and they can provide an occasional psychological boost ("misery loves company?"); they are a place to use the mathematical language of the course, to trade mathematical tips, and to "cram" for the next day's test. Students in study groups are less likely to miss important points in the course, and they get to know some very nice people—their classmates.

Use the textbook effectively. There are a number of ways to use a mathematics textbook:

- to gain an overview of the concepts and techniques,

- to gain an understanding of the material,

- to master the techniques, and

- to review the material and see how it connects with the rest of the course.

The first time you read a section, just try to see what problems are being discussed. Skip around, look at the pictures, and read some of the problems and the definitions. If something looks complicated, skip it. If an example looks interesting, read it and try to follow the explanation. This is an exploratory phase. Don't highlight or underline at this stage—you don't know what is important yet and what is just a minor detail.

The next time through the section, proceed in a more organized fashion, reading each introduction, example, explanation, theorem and proof. This is the beginning of the "mastery" stage. If you don't understand the explanation of an example, put a question mark (in pencil) in the margin and go on. Read and try to understand each step in the proofs and ask yourself why that step is valid. If you don't see what justified moving from one step to another in the proof, pencil in question marks in the margin. This second phase will go more slowly than the first, but if you don't understand some details, just keep going. Don't get bogged down yet.

Now worry about the details. Go quickly over the parts you already understand, but slow down and try to figure out the parts marked with question marks. Try to solve the example problems before you refer to the explanations. If you now understand parts that were giving you trouble, cross out the question marks. If you still don't understand something, put in another question mark and **write down** your question to ask a teacher, tutor or classmate.

Finally, it is time to try the problems at the end of the section. Many of them are similar to examples in the section, but now *you* need to solve them. Some of the problems are more complicated than the examples, but they still require the same basic techniques. Some of the problems will require that you use concepts and facts from earlier in the course, a combination of old and new concepts and techniques. Working lots of problems is the "secret" of success in calculus.

Work the Problems. Many students read a problem, work it out and check the answer in the back of the book. If their answer is correct, they go on to the next problem. If their answer is wrong, they manipulate (finagle, fudge, massage) their work until their new answer is correct, and then they go on to the next problem. **Do not try the next problem yet!** Before going on, spend a short time, just half a minute, thinking about what you have just done in solving the problem. Ask yourself: "What was the point of this problem?" "What big steps did I need to take to solve this problem?" "What was the **process**?" Do not simply review every single step of the solution process. Instead, look at the outline of the solution, the process, the "big picture." If your first answer was wrong, ask yourself, "What about this problem should have suggested the right process the *first* time?" As much learning and retention can take place in the 30 seconds you spend reviewing the process as took place in the 10 minutes you took to solve the problem. A correct answer is important, but **a correct process — carefully used — will get you many correct answers**.

There is one more step that too many students omit. **Go back and quickly look over the section one more time.** Don't worry about the details, just try to understand the overall logic and layout of the section. Ask yourself, "What was I expected to learn in this section?" Typically, this last step — a review and overview — goes quickly, but it is very valuable. It can help you see and retain the important ideas and connections.

T
Trigonometry Facts

Right Angle Trigonometry

$$\sin(\theta) = \frac{\text{opp}}{\text{hyp}} \qquad \cos(\theta) = \frac{\text{adj}}{\text{hyp}} \qquad \tan(\theta) = \frac{\text{opp}}{\text{adj}}$$

$$\csc(\theta) = \frac{\text{hyp}}{\text{opp}} \qquad \sec(\theta) = \frac{\text{hyp}}{\text{adj}} \qquad \cot(\theta) = \frac{\text{adj}}{\text{opp}}$$

Trigonometric Functions

$$\sin(\theta) = \frac{y}{r} \qquad \cos(\theta) = \frac{x}{r} \qquad \tan(\theta) = \frac{y}{x}$$

$$\cot(\theta) = \frac{x}{y} \qquad \sec(\theta) = \frac{r}{x} \qquad \csc(\theta) = \frac{r}{y}$$

Fundamental Identities

$$\sec(\theta) = \frac{1}{\cos(\theta)} \qquad\qquad \tan(\theta) = \frac{\sin(\theta)}{\cos(\theta)} = \frac{1}{\cot(\theta)}$$

$$\csc(\theta) = \frac{1}{\sin(\theta)} \qquad\qquad \cot(\theta) = \frac{\cos(\theta)}{\sin(\theta)} = \frac{1}{\tan(\theta)}$$

$$\sin^2(\theta) + \cos^2(\theta) = 1 \quad \Rightarrow \quad 1 + \tan^2(\theta) = \sec^2(\theta) \quad \Rightarrow \quad \cot^2(\theta) + 1 = \csc^2(\theta)$$

$$\sin(-\theta) = -\sin(\theta) \qquad \cos(-\theta) = \cos(\theta) \qquad \tan(-\theta) = -\tan(\theta)$$

$$\sin\left(\frac{\pi}{2} - \theta\right) = \cos(\theta) \qquad \cos\left(\frac{\pi}{2} - \theta\right) = \sin(\theta) \qquad \tan\left(\frac{\pi}{2} - \theta\right) = \cot(\theta)$$

Law of Sines: $\dfrac{\sin(A)}{a} = \dfrac{\sin(B)}{b} = \dfrac{\sin(C)}{c}$

Law of Cosines: $c^2 = a^2 + b^2 - 2ab \cdot \cos(C)$

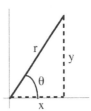

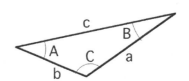

Angle Addition and Subtraction Formulas

$$\sin(x + y) = \sin(x) \cdot \cos(y) + \cos(x) \cdot \sin(y)$$

$$\sin(x - y) = \sin(x) \cdot \cos(y) - \cos(x) \cdot \sin(y)$$

$$\cos(x + y) = \cos(x) \cdot \cos(y) - \sin(x) \cdot \sin(y)$$

$$\cos(x - y) = \cos(x) \cdot \cos(y) + \sin(x) \cdot \sin(y)$$

$$\tan(x + y) = \frac{\tan(x) + \tan(y)}{1 - \tan(x) \cdot \tan(y)} \qquad \tan(x - y) = \frac{\tan(x) + \tan(y)}{1 + \tan(x) \cdot \tan(y)}$$

Product-to-Sum Formulas

$$\sin(x) \cdot \sin(y) = \tfrac{1}{2} \cos(x - y) - \tfrac{1}{2} \cos(x + y)$$

$$\cos(x) \cdot \cos(y) = \tfrac{1}{2} \cos(x - y) + \tfrac{1}{2} \cos(x + y)$$

$$\sin(x) \cdot \cos(y) = \tfrac{1}{2} \sin(x + y) + \tfrac{1}{2} \sin(x - y)$$

Sum-to-Product Formulas

$$\sin(x) + \sin(y) = 2 \sin\left(\frac{x + y}{2}\right) \cdot \cos\left(\frac{x - y}{2}\right)$$

$$\cos(x) + \cos(y) = 2 \cos\left(\frac{x + y}{2}\right) \cdot \cos\left(\frac{x - y}{2}\right)$$

$$\tan(x) + \tan(y) = \frac{\sin(x + y)}{\cos(x) \cdot \cos(y)}$$

Double-Angle Formulas

$$\sin(2x) = 2 \sin(x) \cdot \cos(x)$$

$$\cos(2x) = \cos^2(x) - \sin^2(x) = 2 \cos^2(x) - 1 = 1 - 2 \sin^2(x)$$

$$\tan(2x) = \frac{2 \tan(x)}{1 - \tan^2(x)}$$

Half-Angle Formulas

$$\sin\left(\frac{x}{2}\right) = \pm\sqrt{\frac{1}{2}\left(1 - \cos(x)\right)}$$

The quadrant of $\frac{x}{2}$ determines the $\pm$.

$$\cos\left(\frac{x}{2}\right) = \pm\sqrt{\frac{1}{2}\left(1 + \cos(x)\right)}$$

$$\tan\left(\frac{x}{2}\right) = \frac{1 - \cos(x)}{\sin(x)}$$